CONTROL AND DYNAMIC SYSTEMS

Advances in Theory and Applications

Volume 78

CONTRIBUTORS TO THIS VOLUME

M. AHMADI
SERGIO BITTANTI
BOUALEM BOASHASH
PATRIZIO COLANERI
KAROLOS M. GRIGORIADIS
DALE GROUTAGE
WASSIM M. HADDAD
JOHN TADASHI KANESHIGE
VIKRAM KAPILA
NICHOLAS KOMAROFF
CHRYSOSTOMOS L. NIKIAS
ALAN M. SCHNEIDER
ROBERT E. SKELTON
HAL S. THARP
GEORGE A. TSIHRINTZIS
GUOMING G. ZHU

CONTROL AND DYNAMIC SYSTEMS

ADVANCES IN THEORY AND APPLICATIONS

Edited by

CORNELIUS T. LEONDES

School of Engineering and Applied Science
University of California, Los Angeles
Los Angeles, California

VOLUME 78: DIGITAL CONTROL AND SIGNAL PROCESSING SYSTEMS AND TECHNIQUES

ACADEMIC PRESS

San Diego New York Boston
London Sydney Tokyo Toronto

Find Us on the Web! http://www.apnet.com

This book is printed on acid-free paper. ♾

Academic Press, Inc.
A Division of Harcourt Brace & Company
525 B Street, Suite 1900, San Diego, California 92101-4495

United Kingdom Edition published by
Academic Press Limited
24-28 Oval Road, London NW1 7DX

International Standard Serial Number: 0090-5267

International Standard Book Number: 0-12-012778-4

PRINTED IN THE UNITED STATES OF AMERICA
Transferred to Digital Printing, 2011

CONTENTS

CONTRIBUTORS

Numbers in parentheses indicate the pages on which the authors' contributions begin.

M. Ahmadi (181), *Department of Electrical Engineering, University of Windsor, Ontario, Canada*

Sergio Bittanti (313), *Politecnico di Milano, Dipartimento di Elettronica e Informazione 20133 Milano, Italy*

Boualem Boashash (1), *Signal Processing Research Centre, Queensland University of Technology, Brisbane, Queensland 4000 Australia*

Patrizio Colaneri (313), *Politecnico di Milano, Dipartimento di Elettronica e Informazione 20133 Milano, Italy*

Karolos M. Grigoriadis (229), *Department of Mechanical Engineering, University of Houston, Houston, Texas 77204*

Dale Groutage (71), *David Taylor Research Center, Detachment Puget Sound, Bremerton, Washington 98314*

Wassim M. Haddad (183), *School of Aerospace Engineering, Georgia Institute of Technology, Atlanta, Georgia 30332*

John Tadashi Kaneshige (71), *Mantech NSI Technology Services, Corporation, Sunnyvale, California 94089*

Vikram Kapila (183), *School of Aerospace Engineering, Georgia Institute of Technology, Atlanta, Georgia 30332*

Nicholas Komaroff (275), *Department of Electrical and Computer Engineering, The University of Queensland, Queensland 4072, Australia*

Chrysostomos L. Nikias (341), *Signal and Image Processing Institute, Department of Electrical Engineering-Systems, University of Southern California, Los Angeles, Los Angeles, California 90089*

Alan M. Schneider (71), *Department of Applied Mechanics and Engineering Sciences, University of California, San Diego, La Jolla, California 92093*

Robert E. Skelton (229), *Space Systems Control Laboratory, Purdue University, West Lafayette, Indiana 47907*

Hal S. Tharp (249), *Department of Electrical and Computer Engineering, University of Arizona, Tucson, Arizona 85721*

George A. Tsihrintzis (341), *Communication Systems Laboratory, Department of Electrical Engineering, University of Virginia, Charlottesville, Virginia 22903*

Guoming G. Zhu (229), *Cummins Engine Company, Inc., Columbus, Indiana 47202*

PREFACE

Effective control concepts and applications date back over millennia. One very familiar example of this is the windmill. It was designed to derive maximum benefit from windflow, a simple but highly effective optimization technique. Harold Hazen's 1932 paper in the *Journal of the Franklin Institute* was one of the earlier reference points wherein an analytical framework for modern control theory was established. There were many other notable items along the way, including the MIT Radiation Laboratory Series volume on servomechanisms, the Brown and Campbell book, *Principles of Servomechanisms*, and Bode's book, *Network Analysis and Synthesis Techniques*, all published shortly after mid-1945. However, it remained for Kalman's papers of the late 1950s (which established a foundation for modern state-space techniques) and the tremendous evolution of digital computer technology (which was underpinned by the continuous giant advances in integrated electronics) to establish truly powerful control systems techniques for increasingly complex systems. Today we can look forward to a future that is rich in possibilities in many areas of major significance, including manufacturing systems, electric power systems, robotics, aerospace systems, and many other systems with significant economic, safety, cost, and reliability implications. Separately in the early 1950s, motivated by aerospace systems applications, the field of digital filtering, particularly of telemetry data by mainframe digital computers, started to crystallize. Motivated to a large extent by the aforementioned advances in digital computer technology, this field quickly evolved into what is now referred to as digital signal processing. The field of digital image processing also evolved at this time. These advances began in the 1960s, grew rapidly in the next two decades, and currently are areas of very significant activity, especially regarding their many applications. These fields of digital control and digital signal processing have a number of areas and supplemental advances in common. As a result, this is a particularly appropriate time to devote this volume to the theme of "Digital Control and Signal Processing Systems and Techniques."

Signal analysis is an essential element of both digital control and digital signal processing systems. The first contribution to this volume, "Time Frequency Signal Analysis: Past, Present, and Future Trends," by Boualem

Boashah, one of the leading contributors to this field, provides an in-depth treatment with numerous illustrative examples. Thus it is a most appropriate contribution with which to begin this volume.

Techniques for the conversion of continuous inputs to digital signals for digital control purposes utilize what are referred to as s-to-z mapping functions. Also many digital signal processors are synthesized by the utilization of s-to-z mapping functions as applied directly to analog filters for continuous time signals. The next contribution, "Fundamentals of Higher-Order s-to-z Mapping Functions and Their Application to Digital Signal Processing," by Dale Groutage, Alan M. Schneider, and John Todashi Kaneshige is a comprehensive treatment of these issues. Numerous illustrative examples are also included.

Two-dimensional digital filters cover a wide spectrum of applications including image enhancement, removal of the effects of some degradation mechanisms, separation of features in order to facilitate system identification and measurement by humans, and systems applications. "Design of 2-Dimensional Recursive Digital Filters," by M. Ahmadi, is an in-depth treatment of the various techniques for the design of these filters. In fact, one of these techniques utilizes the mapping functions presented in the previous contribution. Numerous illustrative examples are presented.

Many control systems applications involve continuous-time systems which are subject to digital (discrete-time) control, where the system actuators and senors have differing bandwidths. As a consequence various data rates are utilized, as a practical matter, and this results in multirate control systems design problems. Wassim M. Haddad and Vikram Kapila discuss these problems in "A Periodic Fixed-Architecture Approach to Multirate Digital Control Design." Various examples clearly illustrate the effectiveness of the techniques presented.

Finite digital wordlength problems are common to both digital control and digital signal processing. "Optimal Finite Wordlength Digital Control with Skewed Sampling," by Robert E. Skelton, Guoming G. Zhu, and Karlos M. Grigoriadis, presents techniques for effective system design which take into account the finite wordlengths involved in practical implementations. These techniques are illustrated by several examples.

Pole placement is a problem which occurs in both digital control and digital signal processing. For example, stabilization of digital signal processors is achieved by judiciously moving the poles and zeros on the unit circle to new locations on a circle of radius r where $0<r<1$. By the same token pole relocation or shifting is a major technique utilized for improvement in system performance in control systems. The contribution "Optimal Pole Placement for Discrete-Time Systems" by Hal S. Tharp is an in-depth treatment of techniques available in this important area.

The Discrete Algebraic Ricatti Equation (DARE) plays a fundamental role in a variety of technology fields such as system theory, signal processing, and control theory. "On Bounds for the Solution of the Ricatti Equation for Discrete-Time Control Systems," by Nicholas Komaroff, presents the reasons for the seeking of bounds of DARE, their importance, and their applications. The examples presented illustrate the derivation of bounds and show some implications of various types of bounds.

The long story of periodic systems in signals and control can be traced back to the 1960s. After two decades of study, the 1990s have witnessed an exponential growth of interests, mainly due to the pervasive diffusion of digital techniques in signals (for example, the phenomenon of cyclostationarity in communications and signal processing) and control. The next contribution, "Analysis of Discrete-Time Linear Periodic Systems," by Sergio Bittanti and Patrizio Colaneri, is a comprehensive treatment of the issues in this pervasive area.

In signal processing the choice of good statistical models is crucial to the development of efficient algorithms which will perform the task they are designed for at an acceptable or enhanced level. Traditionally, the signal processing literature has been dominated by the assumption of Gaussian statistics for a number of reasons, and in many cases performance degradation results. Recently, what are referred to as symmetric alpha-stable distributions and random processes have been receiving increasing attention from the signal processing, control system, and communication communities as more accurate models for signals and noises. The result in many applications is significantly improved systems performance which is readily achievable with the computing power that is easily available at low cost today. The contribution "Alpha-Stable Impulsive Interference: Canonical Statistical Models and Design and Analysis of Maximum Likelihood and Moment-Based Signal Detection Algorithms," by George A. Tsihrintzis and Chrysostomos L. Nikias is an in-depth treatment of these techniques and includes an extensive bibliography.

The contributors to this volume are all to be highly commended for comprehensive coverage of digital control and signal processing systems and techniques. They have addressed important subjects which should provide a unique reference source on the international scene for students, research workers, practitioners, and others for years to come.

Time Frequency Signal Analysis:
Past, present and future trends

Boualem Boashash

Signal Processing Research Centre
Queensland University of Technology
2 George street, Brisbane, Qld. 4000, Australia

1 Introduction

This chapter is written to provide both an historical review of past work and an overview of recent advances in time-frequency signal analysis (TFSA). It is aimed at complementing the texts which appeared recently in [1], [2], [3] and [4].

The chapter is organised as follows. Section 1 discusses the need for time-frequency signal analysis, as opposed to either time or frequency analysis. Section 2 traces the early theoretical foundations of TFSA, which were laid prior to 1980. Section 3 covers the many faceted developments which occurred in TFSA in the 1980's and early 1990's. It covers bilinear or *energetic* time-frequency distributions (TFDs). Section 4 deals with a generalisation of bilinear TFDs to multilinear *Polynomial TFDs.* Section 5 provides a coverage of the *Wigner-Ville trispectrum,* which is a particular polynomial TFD, used for analysing Gaussian random amplitude modulated processes. In Section 6, some issues related to multicomponent signals and time-varying polyspectra are addressed. Section 7 is devoted to conclusions.

1.1 An heuristic look at the need for time-frequency signal analysis

The field of time-frequency signal analysis is one of the recent developments in Signal Processing which has come in response to the need to find suitable tools for analysing non-stationary signals. This chapter outlines many of the important concepts underpinning TFSA, and includes an historical perspective of their development. The chapter utilises many concepts and results that were originally reported in [1], [2], [3] and [4] [5], [6], [7].

CONTROL AND DYNAMICS SYSTEMS, VOL. 78

The drawbacks of classical spectral analysis [8], [9], [10], [11], [12], [13] arise largely due to the fact that its principal analysis tool, the Fourier transform, implicitly assumes that the spectral characteristics of the signal are time-invariant, while in reality, signals both natural and man-made, almost always exhibit some degree of non-stationarity. When the important spectral features of the signals are time-varying, the effect of conventional Fourier analysis is to produce an averaged (i.e. smeared or distorted) spectral representation, which leads to a loss in frequency resolution. One way to deal with the spectral smearing is to reduce the effects of the variation in time by taking the spectral estimates over adjacent short time intervals of the signal, centred about particular time instants. Unfortunately, the shortened observation window produces a problem of its own - another smearing caused by the "uncertainty relationship" of time and band-limited signals [14].

Another way to deal with the problem of non-stationarity is to pass the signals through a filter bank composed of adjacent narrow-band bandpass filters, followed by a further analysis of the output of each filter. Again, the same problem described above occurs: the uncertainty principle [14] is encountered this time as a result of the band limitations of the filters. If small bandwidth filters are used, the ability to localise signal features well in time is lost. If large bandwidth filters are used, the fine time domain detail can be obtained, but the frequency resolution becomes poor.

1.2 Problem statement for time-frequency analysis

Classical methods for signal analysis are either based on the analysis of the time signal $s(t)$, or on its Fourier transform defined by

$$S(f) = \int_{-\infty}^{+\infty} s(t)e^{-j2\pi ft}dt \tag{1}$$

The time domain signal reveals information about the presence of a signal, its strengths and temporal evolution. The Fourier transform (FT) indicates which frequencies are present in the signal and their relative magnitudes. For deterministic signals, the representations usually employed for signal analysis are either the instantaneous power (i.e.. the squared modulus of the time signal) or the energy density spectrum (the squared modulus of the Fourier transform of a signal). For random signals, the analysis tools are based on the autocorrelation function (time domain) and its Fourier transform, the power spectrum. These analysis tools have had tremendous success in providing solutions for many problems related to stationary signals. However, they have immediate limitations when applied to non-stationary signals. For example, it is clear that the spectrum gives no indication as to how the frequency content of the signal changes with time, information which is needed when one deals with signals such as frequency modulated

(FM) signals. The chirp signal is an example of such a signal. It is a linear FM signal, used, for example, as a controllable source in seismic processing. It is analogous to a musical note with a steadily rising pitch, and is of the form

$$s(t) = \Pi_T\left(t - \frac{T}{2}\right)\cos\left(f_0 t + \alpha\frac{t^2}{2}\right) \tag{2}$$

where $\Pi_T(t)$ is 1 for $|t| \leq T/2$ and zero elsewhere, f_0 is the centre frequency and α represents the rate of the frequency change.

The fact that the frequency in the signal is steadily rising with time is not revealed by the spectrum; it only reveals a broadband spectrum, (Fig.1, bottom).

It would be desirable to introduce a time variable so as to be able to express the time and frequency-dependence of the signal, as in Fig.1. This figure displays information about the signal in a joint time-frequency domain. The start and stop times are easily identifiable, as is the variation of the spectral behaviour of the signal. This information cannot be retrieved from either the instantaneous power or the spectrum representations. It is lost when the Fourier transform is squared and the phase of the spectrum is thereby discarded. The phase actually contains this information about "the internal organisation" of the signal, as physically displayed in Fig.1. This "internal organisation" includes such details as times at which the signal has energy above or below a particular threshold, and the order of appearance in time of the different frequencies present. The difficulty of interpreting and analysing a phase spectrum makes the concept of a joint time and frequency signal representation attractive. For example, a musician would prefer to interpret a piece of music, which shows the pitch, start time and duration of the notes to be played rather than to be given a magnitude and phase spectrum of that piece of music to decipher [6].

As another illustration of the points raised above, consider the whale signal whose time-frequency (t-f) representation is displayed in Fig.2. By observing this t-f representation, a clear picture of the signal's composition instantly emerges. One can easily distinguish the presence of at least 4 separate components (numbered 1 to 4) that have different start and stop times, and different kinds of energies. One can also notice the presence of harmonics. One could not extract as much information from the time signal (seen at the left hand side in the same figure) or from the spectrum (at the bottom of the same figure).

If such a representation is invertible, the undesirable components of this signal may be filtered out in the time-frequency plane, and the resulting time signal recovered for further use or processing. If only one component of the signal is desired, it can be recognised more easily in such a representation than in either one of the time domain signal or its spectrum. This example illustrates how a time-frequency representation has the potential to be a very powerful tool, due to its ease of interpretation. It is a time-varying extension of the ordinary spectrum which the engineer is comfortable using as an analysis tool.

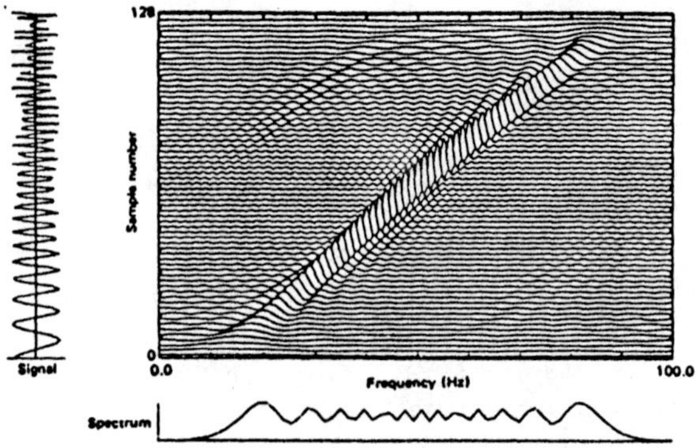

Figure 1. Time-frequency representations of a linear FM signal: the signal appears on the left, and its spectrum on the bottom

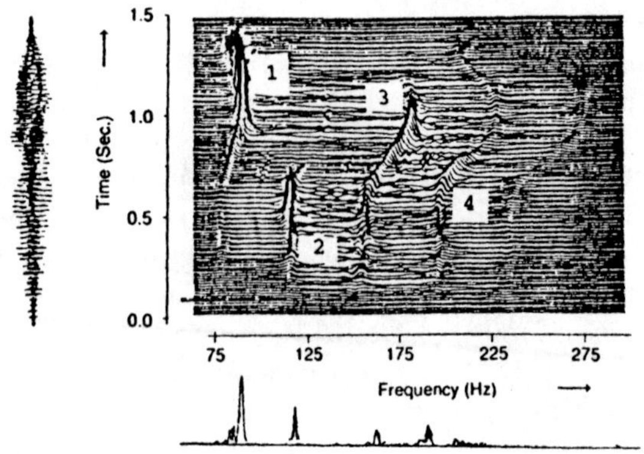

Figure 2. Time-frequency plot of a bowhead whale

2 A review of the early contributions to TFSA

2.1 Gabor's theory of communication

In 1946 Gabor [14] proposed a TFD for the purpose of studying the question of efficient signal transmission. He expressed dissatisfaction with the physical results obtained by using the FT. In particular, the t-f exclusivity of the FT did not fit with his intuitive notions of a time-varying frequency as evident in speech or music. He wanted to be able to represent other signals, not just those limiting cases of a "sudden surge" (delta function) or an infinite duration sinusoidal wave. By looking at the response of a bank of filters which were constrained in time and frequency, Gabor essentially performed a time-frequency analysis. He noted that since there was a resolution limit to the typical resonator, the bank of filters would effectively divide the time-frequency plane into a series of rectangles. He further noted that the dimensions of these rectangles, *tuning width* × *decay time*, must obey Heisenberg's uncertainty principle which translates in Fourier analysis to:

$$\Delta t \cdot \Delta f \geq \frac{1}{4\pi} \tag{3}$$

where Δt and Δf are the equivalent duration and bandwidth of the signal [14]. Gabor believed this relationship to be "at the root of the fundamental principle of communication" [14], since it puts a lower limit on the minimum spread of a signal in time and frequency. The product value of $\Delta t \cdot \Delta f = 1/4\pi$ gives the minimum area unit in this time–frequency information diagram, which is obtained for a complex Gaussian signal.

Gabor's representation divided the time-frequency plane into discrete rectangles of information called logons. Each logon was assigned a complex value, $c_{m,n}$ where m represents the time index and n the frequency index. The $c_{m,n}$ coefficients were weights in the expansion of a signal into a discrete set of shifted and modulated Gaussian windows, which may be expressed as :

$$s(t) = \sum_{m=-\infty}^{\infty} \sum_{n=-\infty}^{\infty} c_{m,n} \psi(t; m, n) \tag{4}$$

where $\psi(t; m, n)$ are Gaussian functions centred about time, m, and frequency, n [14].

Lerner [15] extended Gabor's work by removing the rectangular constraint on the shape of the elementary cells. Helstrom [16] generalised the expansion by replacing the discrete elementary cell weighting with a continuous function, $\xi(\tau, t, f)$. Wavelet theory was later on developed as a further extension of Gabor's work, but with each partition of the time-frequency plane varying so as to yield a constant Q filtering [17].

2.2 The spectrogram

The spectrogram originated from early speech analysis methods and represents the most intuitive approach to spectrum analysis of non-stationary processes. It represents a natural transition from stationary processing towards time-frequency analysis. In this method, a local power spectrum is calculated from slices of the signal centred around the successive time points of interest, as follows:

$$\rho_{spec}(t,f) = |S(t,f)|^2 = \left| \int_{-\infty}^{\infty} s(\tau)h(t-\tau)e^{-j2\pi f\tau}d\tau \right|^2 \tag{5}$$

where $h(t-\tau)$ is the time-limiting analysis window, centred at $t=\tau$, and $S(t,f)$ is referred to as the short-time Fourier transform (STFT). The time-frequency character of the spectrogram is given by its display of the signal as a function of the frequency variable, f, and the window centre time. This is a simple and robust method, and has consequently enjoyed continuing popularity. However, it has some inherent problems. The frequency resolution is dependent on the length of the analysis window and thus degrades significantly as the size of the window is reduced, due to the uncertainty relationships. The spectrogram can also be expressed as a windowed transformation of the signal spectrum as follows:

$$\rho_{spec}(t,f) = |S(t,f)|^2 = \left| \int_{-\infty}^{\infty} S(\nu)H(f-\nu)e^{j2\pi\nu t}d\nu \right|^2 \tag{6}$$

These two representations become identical if $h(t)$ and $H(f)$ are a Fourier transform pair [12]. This indicates that there exists the same compromise for the time resolution; i.e. there is an inherent trade-off between time and frequency resolution. The spectrogram is still one of the most popular tool for TFSA, due to its robustness to noise, linearity property, ease of use and interpretation.

2.3 Page's instantaneous power spectrum

Page [18] was one of the first authors to extend the notion of power spectrum to deal with time-varying signals. He defined the "instantaneous power spectrum" (IPS), $p(t,f)$, which verifies :

$$E_T = \int_{-\infty}^{T} \int_{-\infty}^{\infty} p(t,f)df\,dt \tag{7}$$

where E_T represents the total signal energy contained up to time, T, and where $p(t,f)$ represents the distribution of that energy over time and over the frequency. It is a spectral representation of the signal, which varies as a function of time. In order to obtain an expression for $p(t, f)$, Page first defined a running transform:

$$S_t^-(f) = \int_{-\infty}^{t} s(\tau)e^{-j2\pi f\tau}d\tau \tag{8}$$

which represents the conventional FT of the signal, but calculated only up to time t. This allows the definition of a time-varying FT. He then defined the IPS as the rate of change or gradient in time of $S_t^-(f)$; i.e. the contribution to the overall energy made by each frequency component. This is defined as follows:

$$p(t, f) = \frac{\partial}{\partial t}|S_t^-(f)|^2 \tag{9}$$

It may equivalently be expressed as [18]

$$p(t, f) = 2s(t)\mathcal{R}\{e^{j2\pi ft}S_t^-(f)\} \tag{10}$$

or

$$p(t, f) = 2\int_0^{\infty} s(t)s(t-\tau)\ \cos 2\pi f\tau\ d\tau \tag{11}$$

where $\mathcal{R}$ denotes the real part.

Since $p(t, f)$ is the gradient of a spectrum, it may contain negative values; it redistributes signal energy as time evolves, compensating for previous values which were either too low or too high. The IPS therefore does not localise the information in time and frequency. Turner [19] has shown that the IPS is not unique, since any complementary function which integrates to zero in frequency can be added to it without changing the distribution. He also proved that the IPS is dependent on the initial time of observation. This indicates that the IPS is not a "true" TFD, i.e.. it does not meet some obvious requirements that a signal analyst expects in order to carry out a practical analysis of the signal. Nevertheless, it represented an important step in the development of ideas which led to our current understanding of TFDs.

Levin [20], following Page's work, defined a forward running (or anti-causal) spectrum $S_t^+(f)$, which is based on future time values, by taking a FT from t to $+\infty$. He also defined a time-frequency representation taking an average of the forward and backward IPS to get :

$$p_L(t, f) = \frac{1}{2}\left[\frac{\partial}{\partial t}|S_t^-(f)|^2 + \frac{\partial}{\partial t}|S_t^+(f)|^2\right] \tag{12}$$

$$= 2s(t)\mathcal{R}\{e^{j2\pi ft}S(f)\} \tag{13}$$

By realising that this combination would lead to an overall time-frequency representation which describes better the signal, Levin defined a distribution that is very similar to Rihaczek's [21] which will be discussed next.

It is worthwhile noting here that we will show in section 3.2 that all the TFDs which have been discussed so far can be written using a general framework provided by a formula borrowed from quantum mechanics.

2.4 Rihaczek's complex energy density

Starting from physical considerations, Rihaczek formed a time-frequency energy density function for a complex deterministic signal, $z(t)$, which, he claimed, was a natural extension of the energy density spectrum, $|Z(f)|^2$, and the instantaneous power, $|z(t)|^2$. His reasoning was as follows: the total energy of a complex signal, $z(t)$, is:

$$E = \frac{1}{2}\int_{-\infty}^{\infty} |z(t)|^2 dt \tag{14}$$

Consider a bandlimited portion of the original signal, around f_0, $z_1(t)$ given as

$$z_1(t) = \mathcal{F}^{-1}\{\Pi_{\Delta B}(f - f_0) \cdot Z(f)\} \tag{15}$$

This portion of the signal, $z_1(t)$, contains the energy

$$E_1 = \frac{1}{2}\int_{-\infty}^{\infty} z(t)z_1^*(t)dt \tag{16}$$

If the bandwidth of $z_1(t)$, ΔB is reduced to δB, then $z_1(t)= Z(f_0)\delta B \cdot e^{j2\pi f_0 t}$. Assuming that $Z(f)$ is constant over the spectral band δB, which is reasonable if $\delta B \rightarrow 0$, we then obtain:

$$E_1 = \frac{1}{2}\int_{-\infty}^{\infty} z(t)Z^*(f_0)\delta B e^{-j2\pi f_0 t} dt \tag{17}$$

This quantity in (17) represents the energy in a small spectral band δB, but over all time. To obtain the energy within a small frequency band δB, and a time band ΔT, it suffices to limit the integration in time to ΔT as follows:

$$E_1 = \frac{1}{2}\int_{t_0-\Delta T/2}^{t_0+\Delta T/2} z(t)Z^*(f_0)\delta B e^{-j2\pi f_0 t} dt \tag{18}$$

Taking the limit $\Delta T \rightarrow \delta T$ yields

$$E_1 = \frac{1}{2}\, \delta B\; \delta T\; z(t_0)\; Z^*(f_0)\; e^{-j2\pi f_0 t_0} \tag{19}$$

with the resultant time-frequency distribution function being

$$\rho_R(t, f) = z(t)Z^*(f)e^{-j2\pi f t} \tag{20}$$

which is generally referred to as the Rihaczek Distribution (RD). If $z(t)$ is real, one can see that Levin's TFD (which is based on Page's TFD) is simply twice the real part of Rihaczek's TFD.

It is remarkable to see that different approaches to define a TFD that seem to be all natural and straightforward lead to apparently different definitions of a TFD.

2.5 The Wigner-Ville distribution

Ville's work [22] followed Gabor's contribution; he similarly recognised the insufficiency of time analysis and frequency analysis, using the same analogy of a piece of music. He indicated that since a signal has a spectral structure at any given time, there existed the notion of an "instantaneous spectrum" which had the physical attributes of an energy density. Thus the energy within a small portion of the time-frequency plane, $dt \cdot df$ would be

$$E_\delta = W(t, f)\, dt\, df \tag{21}$$

and its integration over f (respectively over t) should yield the instantaneous power $|s(t)|^2$ (respectively the energy spectral density $|S(f)|^2$) Integration over both t and f would yield the energy E.

$$\int_{-\infty}^{\infty} W(t, f) df = |s(t)|^2 \tag{22}$$

$$\int_{-\infty}^{\infty} W(t, f) dt = |S(f)|^2 \tag{23}$$

$$\int_{-\infty}^{\infty} \int_{-\infty}^{\infty} W(t, f) dt df = E \tag{24}$$

These desirable properties led Ville to draw an analogy with the probability density function (pdf) of quantum mechanics. i.e. consider that:

1. the distribution $\rho(t, f)$ to be found is the joint pdf in time and frequency,
2. the instantaneous power is one marginal probability of $\rho(t, f)$,
3. the spectrum is the other marginal probability of $\rho(t, f)$.

Then, one could form the characteristic function, $F(u, v)$, of this TFD, and equate the marginal results of $|s(t)|^2$ and $|S(f)|^2$ with the moments generated from the characteristic function (using its moment generating properties):

$$W(t, f) = \underset{t \to u}{\mathcal{F}} \quad \underset{f \to v}{\mathcal{F}} \quad F(u, v) \tag{25}$$

Using then the framework of quantum mechanical operator theory [23], Ville established that the proper form for the distribution was:

$$W(t,f) = \int_{-\infty}^{+\infty} z(t+\frac{\tau}{2}) \cdot z^*(t-\frac{\tau}{2}) e^{-j2\pi f\tau} d\tau \tag{26}$$

where $z(t)$ is the analytic complex signal which corresponds to the real signal, $s(t)$ [24]. It is obtained by adding an imaginary part $y(t)$ which is obtained by taking the Hilbert transform of the real signal $s(t)$ [14].

Ville's distribution was derived earlier by Wigner in a quantum mechanical context [25]. For this reason, it is generally referred to as the Wigner-Ville distribution (WVD) and it is the most widely studied of present TFDs. The advantages of the WVD as a signal processing tool are manifold. It is a real joint distribution of the signal in time and frequency. The marginal distributions in time and frequency can be retrieved by integrating the WVD in frequency and time respectively. It achieves optimal energy concentration in the time-frequency plane for linearly frequency modulated signals. It is also time, frequency and scale invariant, and so fits well into the framework of linear filtering theory. The disadvantages of the WVD are chiefly that it is non-positive, that it is "bilinear" and has cross-terms. The non-positivity makes the WVD difficult to interpret as an energy density. The cross-terms cause "ghost" energy to appear mid-way between the true energy components.

A detailed review of the WVD is provided in [2].

3 The second phase of developments in TFSA: 1980's

3.1 Major developments in 1980's

The early research in the 1980's focussed on the WVD as engineers and scientists started to discover that it provided a means to attain good frequency localisation for rapidly time-varying signals. For example, in a seismic context it was shown to be a very effective tool to represent "Vibroseis" chirp signals emitted in seismic processing [26], and hence was used to control the quality of the signal emitted. When the signal emitted was a pure linear FM, the WVD exhibited a sharp peak along the FM law. If the signal was contaminated by harmonic coupling effects and other distortions then this property was lost [27].

The interest in the WVD was fuelled by its good behaviour on chirp signals and by the discovery (and later re-discovery) [25],[22] of its special properties, which made it attractive for the analysis of time-varying signals.

The advance of digital computers also aided its popularity, as the previously prohibitive task of computing a two-dimensional distribution came within practical reach [28][1].

The WVD of a signal, $z(t)$, is constructed conceptually as the Fourier transform of a "bilinear" kernel[2], as

$$W_z(t,f) = \underset{\tau \rightarrow f}{\mathcal{F}} \; [K_z(t,\tau)] \tag{27}$$

where $\underset{\tau \rightarrow f}{\mathcal{F}}$ represents a Fourier transformation with respect to the τ variable, and where $K_z(t,\tau)$ is the "bilinear" kernel defined by

$$K_z(t,\tau) \triangleq z(t+\frac{\tau}{2})z^*(t-\frac{\tau}{2}) \tag{28}$$

Most of the early research in the WVD concentrated on the case of deterministic signals, for which the WVD is interpreted as a distribution of the signal in the time-frequency domain. For random signals, it was shown [29] that the expected value of the WVD equals the FT of the time-varying auto-correlation function (when these quantities exist). This gave the WVD an important interpretation as a time-varying Power Spectral Density (PSD) and sparked significant research efforts along this direction.

Filtering and Signal synthesis. It was also realised early that the WVD could be used as a time-varying filter [30]. A simple algorithm was devised which masked (i.e.. filtered) the WVD of the input signal and then performed a least-squares inversion of the WVD to recover the filtered signal [2] [30]. It was also shown that the input-output convolution relationships of filters were preserved when one used the WVD to represent the signals.

Implementation The computational properties of the WVD were further studied and this led to an efficient real-time implementation which exploits the symmetry properties of the Wigner-Ville kernel sequence [31].

Signal Detection, Estimation and Classification. The areas of detection and estimation saw significant theoretical developments based on the WVD [32], [33], [34], motivated by the belief that signal characterisation should be more accurate in a joint t-f domain. A key property helped motivate this interest: the WVD is a unitary (energy preserving) transform.

[1]To the author's best knowledge, the first WVD programme was written by him in APL language in September 1978, for the processing of Vibroseis chirp data [28].

[2]The kernel is actually bilinear only for the cross WVD, introduced in Sec.3.1.

Therefore, many of the classical detection and estimation problem solutions had alternate implementations based on the WVD. The two-dimensional t-f nature of the implementation, however, allowed greater flexibility than did the classical one [35], [36].

The theory and important properties of the WVD which prompted so much of the investigations outlined above were reviewed in detail in [2], and will be briefly summarised in Section 3.3.

A mistake that was made by many of the early researchers was to "sell" uninhibitedly the method as a universal tool, whereas its field of application is really quite specialised. As the WVD became increasingly exposed to the signal processing community, users started to discover the limitations of the method, which are presented below.

Non-linearities One of the main limitations is that the WVD is a method which is non-linear. The WVD performs a "bilinear" transformation of the signal, which is equivalent to a "dechirping" operation. There are serious consequences for multicomponent signals, that is, composite signals such as a sum of FM signals [2]. For such signals, the bilinear nature of the WVD causes it to create cross-terms (or artifacts) which occur in between individual components. This can often render the WVD very difficult to interpret, such as in cases where there are many components or where components are not well separated. In addition, the bilinearity exaggerates the effect of additive noise by creating cross-terms between the signal component and the noise component[3]. At low signal-to-noise ratio (SNR), where the noise dominates, this leads to a very rapid degradation of performance.

Limited duration. Another drawback sometimes attributed to the WVD is that it performs well only for infinite duration signals. Since it is the FT of a bilinear kernel, it is tuned to the presence of infinite duration complex exponentials in the kernel, and hence to linear FM components in the signal. Real life signals are usually time limited, therefore a simple FT of the bilinear kernel does not provide a very effective analysis of the data. There is a need to take a windowed FT [31] of the bilinear kernel or to replace the FT by a high-resolution model-based spectral analysis such as Burg's algorithm [37] [38].

Cohen's bilinear class of smoothed WVDs. A lot of effort went into trying to overcome the drawbacks of the WVD [39], [40], [41], [42], [43], [2]. Researchers realised that although the WVD seemed to be theoretically the best of the TFSA tools available, for practical analysis other TFDs were better because they reduced the cross-terms. They borrowed Cohen's formula from quantum mechanics so as to link all the bilinear TFDs together

[3]This has led to some jokes nick-naming the WVD as a noise generator!

by a 2-D smoothing of the WVD. They then tried to find an optimum smoothing, optimum in the sense of facilitating the job of the analyst in the use of TFSA methods. Since most of the known TFSA methods were obtainable by a 2-D smoothing of the WVD, there was perhaps an ideal TFSA method not yet known which could be discovered by a proper choice of the smoothing window (the spectrogram, Page's running Spectrum,[18], Rihaczek's complex energy distribution [21] could all be achieved via this type of smoothing [26], [11]).

Filtering out cross-terms in the Ambiguity domain. Many researchers turned to 2-D Gaussian smoothing functions to reduce the artifacts [39], [40], because of the Gaussian window property of minimising the bandwidth-time product.

A key development in a more effective effort at trying to reduce artifacts was to correlate this problem with a result from radar theory using the fact that in the ambiguity domain (Doppler-lag), the cross-terms tended to be distant from the origin, while the auto-terms were concentrated around the origin [44], [30]. Understanding this link was very helpful since the WVD was known to be related to the ambiguity function via a 2-D Fourier Transform [2]. The natural way of reducing the cross-terms of the WVD was then simply to filter them out in the ambiguity domain, followed by a 2-D FT inversion.

This led to greater refinements and thought in the design of TFDs. Using this approach Choi-Williams designed a TFD with a variable level smoothing function, so that the artifacts could be reduced more or less depending on the application [43]. Zhao, Atlas and Marks designed smoothing functions in which the artifacts folded back onto the auto-terms [42]. Amin [41] came to a similar result in the context of random signals, with this providing inspiration for the work reported in [45]. There it was shown how one could vary the shape of the cross-terms by appropriate design of the smoothing function.

By making the smoothing function data dependent, Baraniuk and Jones produced a signal dependent TFD which achieved high energy concentration in the t-f plane. This method was further refined by Jones and Boashash [46] who produced a signal dependent TFD with a criterion of local adaptation.

Cross WVD (XWVD). Another approach to reduce or nullify the presence of cross-terms was based on replacing the WVD by the XWVD in order to obtain a distribution which is linear in the signal. The XWVD could be interpreted as an extension of the cross-correlation function for non-stationary signals. The XWVD is defined as:

$$W_{12}(t,f) = \underset{\tau \rightarrow f}{\mathcal{F}} [K_{12}(t,\tau)] \tag{29}$$

where

$$K_{12}(t,\tau) = z_1(t+\frac{\tau}{2})z_2^*(t-\frac{\tau}{2}) \tag{30}$$

where $z_1(t)$ is a reference signal and $z_2(t)$ is the signal under analysis. There were then systematic efforts in trying to substitute the use of the XWVD in all areas of application of the WVD. In many cases, this was straightforward, because a reference signal was available. Thus, the XWVD was proposed for optimal detection schemes [32], for sonar and radar applications [47], and for seismic exploration [48]. These schemes were seen to be equivalent to traditional matched filter and ambiguity function based schemes, but their representation in another domain allowed for some flexibility and variation. In other cases, where reference signals were not available, the XWVD could not in general be applied, a fact which prevented the further spread of the XWVD as a replacement for the WVD.

In some applications, however, it is possible to define reference signals from filtered estimates of the original signal, and then use these as if they were the true signal. The filtering procedure uses the IF as a critical feature of the signal. Jones and Parks [49] implicitly used a similar philosophy to estimate their data dependent distributions. They estimated their reference signal as that signal component which maximised the energy concentration in the distribution.

Wideband TFDs. The problems relating to the WVD's poor performance with short duration or wideband signals were addressed in several ways. One way was as mentioned earlier, to use autoregressive modelling techniques. Much more attention, though, was given to designing wideband or *affine* time-frequency representations. The first to be considered was the wavelet transform, which is linear. It was like the Gabor transform in that it obtained its coefficients by projecting the signal onto basis functions corresponding to different positions in time-frequency. The wavelet transform differed from the Gabor transform in that its basis functions all had the same shape. They were simply dilated (or scaled) and time shifted versions of a *mother* wavelet. This feature causes the representation to exhibit a *constant Q* filtering characteristic. That is, at high frequencies the resolution in time is good, while the resolution in frequency is poor. At low frequencies, the converse is the case. Consequently, abrupt or step changes in time may be detected or analysed very well.

Subsequent efforts aimed at incorporating these wideband analysis techniques into bilinear TFDs. One of the early attempts was in [39], and used the Mellin transform (rather than the Fourier transform) to analyse the bilinear kernel. The Mellin transform is a scale-invariant transform, and as a consequence, is suited to constant Q analysis. A significant contribution was also made by the Bertrands, who used a rigourous application of Group Theory to find the general bilinear class of scale-invariant TFDs [50]. Others showed that this class of TFDs could be considered to be smoothed (in the affine sense) WVDs, and that many properties might be found which were analogous to those of Cohen's class [51].

These techniques were extended for use in wideband sonar detection applications [52], and in speech recognition [53].

Instantaneous Frequency. The development of time-frequency analysis was parallelled by a better understanding of the notion of instantaneous frequency (IF), since in most cases TFSA methods were aiming at a better IF estimation of the FM laws comprising the signal of interest.

For analysts who were used to time-invariant systems and signal theory, the simultaneous use of the terms *instantaneous* and *frequency* appeared paradoxical and contradictory. A comprehensive review of the conceptual and practical aspects of progressing from frequency to instantaneous frequency was given in [3], [4].

The IF is generally defined as the derivative of the phase. For discrete-time signals, the IF is estimated by such estimators as the "central finite difference" (CFD) of the phase [3]. Recently, this was extended by defining a general phase difference estimate of the IF [4]. This allowed us to understand why the WVD performed well for linear FM signals only - it has an in-built CFD IF estimator which is unbiased for linear FM signals only. It is therefore optimally suited for linear FM signals, but only for this class of signals. The remainder of this section will deal with the class of bilinear TFDs which is suited to the analysis of such signals. Section 4 will present a new class of TFDs referred to as polynomial TFDs, which are suited to the analysis of non-linear polynomial FM signals.

3.2 Bilinear class of TFDs

From an engineering point of view, Ville's contribution laid the foundations for a class of TFSA methods which developed in the 1980s. A key step was to realise that if one wanted to implement the WVD, then in practice one can only use a windowed version of the signal and operate within a finite bandwidth. Another key step was to realise that the WVD may be expressed as the FT of the bilinear kernel:

$$W_z(t,f) = \underset{\tau \to f}{\mathcal{F}} \{z(t+\frac{\tau}{2}) \cdot z^*((t-\frac{\tau}{2})\} \tag{31}$$

If one replaces $z(t)$ by a time windowed and a band-limited version of $z(t)$, then, after taking the FT, one would obtain a new TFD.

$$\rho(t,f) = W_z(t,f) ** \gamma(t,f) \tag{32}$$

where the smoothing function $\gamma(t,f)$ describes the time limitation and frequency limitation of the signal, and $**$ denotes convolution in time and frequency.

If one then decides to vary $\gamma(t,f)$ according to some criteria so as to refine some measurement, one obtains a general time-frequency distribution which could adapt to the signal limitations. These limitations may be inherent to the signal or may be caused by the observation process. If we write in full the double convolution, this then leads to the following formulation:

$$\rho(t,f) = \int_{-\infty}^{\infty}\int_{-\infty}^{\infty}\int_{-\infty}^{\infty} e^{j2\pi\nu(u-t)} g(\nu,\tau) z(u+\frac{\tau}{2})\cdot z^*(u-\frac{\tau}{2}) e^{-j2\pi f\tau}\, d\nu\, du\, d\tau \tag{33}$$

which is generally referred to as Cohen's formula, since it was defined by Cohen in 1966 [54] in a quantum mechanics environment[4]. The function $g(\nu,\tau)$ in (33) is the double FT of the smoothing function $\gamma(t,f)$.

This formula may also be expressed in terms of the signal FT, $Z(f)$, as:

$$\rho_z(t,f) = \int_{-\infty}^{\infty}\int_{-\infty}^{\infty}\int_{-\infty}^{\infty} e^{j2\pi\tau(\eta-f)} g(\nu,\tau) Z(\eta+\frac{\nu}{2})\cdot Z^*(\eta-\frac{\nu}{2}) e^{-j2\pi t\nu}\, d\tau\, d\eta\, d\nu \tag{34}$$

or in the time-lag domain as

$$\rho_z(t,f) = \int_{-\infty}^{\infty}\int_{-\infty}^{\infty} G(t-u,\tau) z(u+\frac{\tau}{2})\cdot z^*(u-\frac{\tau}{2}) e^{-j2\pi f\tau}\, du\, d\tau \tag{35}$$

where $G(t,\tau)$ is the inverse Fourier transform of $g(\nu,\tau)$ in the ν variable. Finally, in the frequency-Doppler domain it is given by:

$$\rho_z(t,f) = \int_{-\infty}^{\infty}\int_{-\infty}^{\infty} \Gamma(f-\eta,\nu) Z(\eta+\frac{\nu}{2})\cdot Z^*(\eta-\frac{\nu}{2}) e^{j2\pi\nu t}\, d\eta\, d\nu \tag{36}$$

The smoothing functions are related by the FT as shown in the following diagram:

[4] It is only recently (about 1984) that Leon Cohen became aware that the formula he devised for quantum mechanics in 1966 [54], was being used by engineers. Since then, he has taken a great and active interest in the use of his formula in signal processing, and has brought the fresh perspective of a physicist to the field of time-frequency signal analysis.

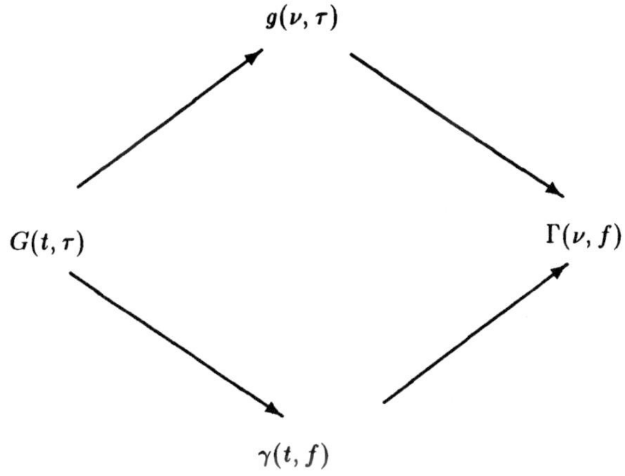

It was shown in the 1980s that nearly all the then known bilinear TFDs were obtainable from Cohen's formula by appropriate choice of the smoothing function, $g(\nu, \tau)$. In this chapter we will refer to this class as the bilinear class of TFDs. Most of the TFDs proposed since then are also members of the bilinear class. Some of those TFDs were discussed earlier. Others have been studied in detail and compared together in [2], [1]. Table 1 lists some of the most common TFDs and their determining functions in the discrete-form, $G(n, m)$. Knowing $G(n, m)$, each TFD can be calculated as:

$$\rho_z(n, k) = \sum_m \sum_p G(p, m)\, z(n + p + m)\, z^*(n + p - m)\, e^{-j4\pi mk/N} \quad (37)$$

3.3 Properties and limitations of the bilinear class of TFDs

The properties of the bilinear class of TFDs will be listed and compared with the "desirable" properties expected by signal analysts for a TFD to be used as a practical tool. A more complete treatment is provided in [1] and [2].

Realness. To represent the variation in signal energy, the TFD should be real.

Marginal conditions. It is usually desired that integrating the TFD over time (respectively frequency) yields the power spectrum (respectively the instantaneous power). Naturally, the integration over both time and frequency should yield the signal energy.

Time–Frequency Representation	$G(n,m)$
Windowed Discrete WVD	$\delta(n) \quad m \in [\frac{-(M-1)}{2}, \frac{(M-1)}{2}]$ $0 \quad$ otherwise
Pseudo WVD using a rectangular window of odd length P	$\frac{1}{P} \quad n \in [\frac{-(P-1)}{2}, \frac{(P-1)}{2}]$ $0 \quad$ otherwise
Rihaczek–Margenau	$\frac{1}{2}[\delta(n+m) + \delta(n-m)]$
STFT using a Rectangular Window of odd length P.	$\frac{1}{P} \quad \lvert m+n\rvert \leq \frac{(P-1)}{2}$ $0 \quad$ otherwise
Born–Jordan–Cohen	$\frac{1}{\lvert m\rvert+1} \quad \lvert m\rvert \leq \lvert n\rvert$ $0 \quad$ otherwise
Choi–Williams (parameter σ)	$\frac{\sqrt{\sigma/\pi}}{2m} e^{-\sigma n^2/4m^2}$

Table 1: Some TFDs and their determining functions $G(n,m)$. Knowing $G(n,m)$, each TFD can be calculated using eq.(37).

Positivity. It is normally expected that a TFD would be positive. However, we have seen that by construction, Page's TFD was not. It was also shown [12], that:

1) for a TFD to be positive, the smoothing function had to be an ambiguity function, or a linear combination of ambiguity functions, and

2) the property of positivity was incompatible with verifying the marginal condition.

An interpretation of TFDs which is compatible with their non- positivity aspect, is to consider that they provide a measure of the energy flow through the spectral band $[f - \Delta f, f + \Delta f]$ during the time $[t - \Delta t, t + \Delta t]$, [2].

Time shifting and frequency shifting. Time and frequency shifts of amounts, (t_0, f_0), of a signal, $s(t)$, should be reflected in its TFD; i.e. if $\rho_s(t, f)$ is the TFD of $s(t)$, then signal $s(t - t_0)e^{j2\pi f_0 t}$ has a TFD which is $\rho_s(t - t_0, f - f_0)$. All TFDs of the bilinear class satisfy this property.

Input-output relationship of a linear filter. Bilinear TFDs are consistent with linear system theory; i.e. if

$$y(t) = s(t) * h(t), \quad \text{that is if} \quad Y(f) = S(f) \cdot H(f)$$

then

$$\rho_y(t, f) = \rho_s(t, f) \underset{t}{*} \rho_h(t, f)$$

Equivalently, if

$$Y(f) = S(f) * H(f), \quad \text{that is if} \quad y(t) = s(t) \cdot h(t)$$

then

$$\rho_y(t, f) = \rho_s(t, f) \underset{f}{*} \rho_h(t, f)$$

Finite support. If a TFD is to represent the instantaneous energy in time and frequency, then it should be zero when the signal is zero; i.e. if $s(t) = 0$ for $t_a < t < t_b$, and $S(f) = 0$ for $f_a < f < f_b$, then, $\rho_s(t, f) = 0$

for $t_a < t < t_b$ or $f_a < f < f_b$. The finite support in time holds provided that the function $g(\nu, \tau)$ introduced in (33) satisfies [11]:

$$\mathop{\mathcal{F}^{-1}}_{\nu \to t} \{g(\nu, \tau)\} = 0 \qquad \text{for } |t| > |\tau|/2 \tag{38}$$

and in frequency holds when

$$\mathop{\mathcal{F}}_{\tau \to f} \{g(\nu, \tau)\} = 0 \qquad \text{for } |f| > |\nu|/2 \tag{39}$$

Instantaneous frequency and group delay. For FM signals, the non-stationary or time-varying aspect is directly quantified by the variation of the IF, $f_i(t)$, or the group delay (GD), $\tau_g(f)$, expressed as follows:

$$f_i(t) = \frac{1}{2\pi} \frac{d}{dt} arg\{z(t)\} \tag{40}$$

$$\tau_g(f) = -\frac{1}{2\pi} \frac{d}{df} arg\{Z(f)\}. \tag{41}$$

One of the most desirable properties of TFDs is that their first moments in time (respectively frequency) give the GD (respectively the IF), as follows:

$$f_i(t) = \frac{\int_{-\infty}^{\infty} f\rho(t, f)df}{\int_{-\infty}^{\infty} \rho(t, f)df} \tag{42}$$

$$\tau_g(f) = \frac{\int_{-\infty}^{\infty} t\rho(t, f)dt}{\int_{-\infty}^{\infty} \rho(t, f)dt}. \tag{43}$$

These conditions are respected if the partial derivatives of the smoothing function are zero at the origin, $\nu = \tau = 0$. However, they are incompatible with the positivity requirement [2]. The IF and its relationship with TFDs has been the subject of a two-part review paper [3], [4].

Cross-terms, interference terms, artifacts. Since bilinear TFDs are a smoothed version of the FT of the bilinear kernel, $K_z(t, \tau)$ given by (28), i.e. as follows:

$$K_z(t, \tau) = z(t + \frac{\tau}{2}) \cdot z^*(t - \frac{\tau}{2}) \tag{44}$$

there always exist cross-terms which are created in the TFD as a consequence of the interaction of different components of the signal. Consider a signal composed of two complex linear FM signal components

$$z_3(t) = z_1(t) + z_2(t) \tag{45}$$

The bilinear kernel of the signal in the TFD is:

$$K_{z_3}(t,\tau) = K_{z_1}(t,\tau) + K_{z_2}(t,\tau) + K_{z_1 z_2}(t,\tau) + K_{z_2 z_1}(t,\tau) \tag{46}$$

where the cross-kernels, $K_{z_1 z_2}(t,\tau)$, and $K_{z_2 z_1}(t,\tau)$ are defined respectively by:

$$K_{z_1 z_2}(t,\tau) = z_1(t+\frac{\tau}{2}) \cdot z_2^*(t-\frac{\tau}{2}) \tag{47}$$

$$K_{z_2 z_1}(t,\tau) = z_2(t+\frac{\tau}{2}) \cdot z_1^*(t-\frac{\tau}{2}) \tag{48}$$

$$\tag{49}$$

The third and fourth kernel terms comprise the cross-terms, which often manifest in the t-f representation in a very inconvenient and confusing way. Consider the WVD of the signal consisting of two linear FM components, given in Fig.3. It shows three components, when one expects to find only two. The component in the middle exhibits large (positive and negative) amplitude terms in between the linear FM's signal energies where it is expected that there should be no energy at all. These are the cross-terms resulting from the bilinear nature of the TFD, which are often considered to be the fundamental limitations which have prevented more widespread use of time-frequency signal analysis. TFDs have been developed which reduce the magnitude of the cross-terms, but they inevitably compromise some of the other properties of TFDs, such as resolution. The cross-term phenomenon is discussed in more details in [2] and [1].

The analytic signal. We have seen above that the bilinearity causes the appearance of cross-terms for composite signals, i.e. signals with multiple components. A real signal can be decomposed into two complex signals with symmetric spectra. The TFD of a real signal would therefore exhibit cross-terms at the zero frequency position. The analytic signal is normally used in the formation of TFDs to avoid this problem; it is constructed by adding an imaginary part, $y(t)$, to the real signal, $s(t)$, such that $y(t)$ is the Hilbert transform of $s(t)$. This ensures that the resulting complex signal has no energy content for negative frequencies.

Apart from this property, the analytic signal is useful for interpretive reasons and for efficient implementation of TFDs. A full treatment of this question is given in [2], [1], [3].

Multilinearity. Section 4 will show that the "bilinearity" of TFDs makes them optimal only for linear FM signals. For non-linear FM signals, a new class of multilinear TFDs is defined, and presented in the next section.

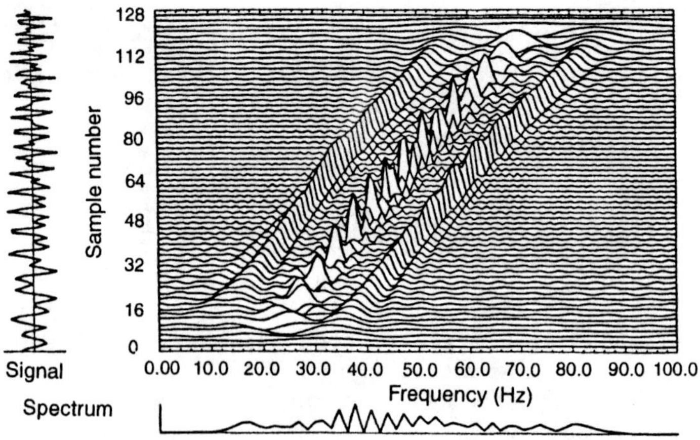

Figure 3.The WVD of a two component signal formed by summing two linear FM signals

Algorithms for bilinear TFDs. In considering the general usefulness of TFDs, it is important to consider the properties of their discrete-time equivalents, and their ease of implementation.

The discrete-time equivalent of the time-lag definition given in (35) leads to the simplest way of implementing TFDs [2, p.444]:

$$\rho_z(n,k) = \underset{m \to k}{\mathcal{F}} \left\{ G(n,m) \underset{n}{*} K_z(n,m) \right\} \tag{50}$$

The implementation of this discrete-time bilinear TFDs requires three steps:

1. Formation of the bilinear kernel

$$K_z(n,m) = z(n+m) \cdot z^*(n-m)$$

2. Discrete convolution in discrete time n with the smoothing function $G(n,m)$.

3. Discrete FT with respect to discrete delay m.

The implementation of steps 1, 2 and 3 may be simplified by taking advantage of the many symmetries which exist, as explained in [31], [55].

The WVD does not require step 2 because its smoothing function $G(n,m)$ equals $\delta(n)$. Further details of the implementation are contained in [1, chapter 7].

The list of properties of discrete TFDs is given in [2, p.445-451]. In summary the characteristics of $G(n,m)$ determine the properties of a bilinear TFD, such as maximising energy concentration or reducing artifacts. The source code for the implementation of TFDs is listed in [1, chapter 7].

4 Polynomial TFDs.

It was indicated at the end of Section 3.1 that the investigation of the notion of the IF led us to realise that the bilinearity of the WVD makes it suitable for the analysis of linear FM signals only. This observation motivated some work which led to the definition of Polynomial WVDs (PWVDs) which were constructed around an in-built IF estimator which is unbiased for non-linear FM signal [56], [57]. This development led to the definition of a new class of TFDs which behave well for non-linear FM signals, and that are able to solve problems that Cohen's bilinear class of TFDs cannot [58] [59]. Another property of the PWVDs is that they are related to the notion of higher order spectra [57].

The reason why one might be interested in polynomial TFDs and/or time-varying higher order spectra is that 1) many practical signals exhibit some form of non-stationarity *and* some form of non-linearity, and that 2) higher order spectra have in theory the ability to reject Gaussian noise. Since TFDs have been widely used for the analysis of non-stationary signals, and higher order spectra are likewise used for the study of non-linearities and non-Gaussianities, it seems natural to seek the use of time-varying higher order spectra to deal with both phenomena simultaneously. There are many potential applications. For example, in underwater surveillance, ship noise usually manifests itself as a set of harmonically related narrowband signals; if the ship is changing its speed, then the spectral content of these signals is varying over time. Another example is an acoustic wave produced by whales (See the time-frequency representation of a typical whale signal shown in Fig.2). Similar situations occur in vibration analysis, radar surveillance, seismic data processing and other types of engineering applications. For the type of signals described above, the following questions often need to be addressed: what are the best means of analysis, detection and classification; how can one obtain optimal estimates for relevant signal parameters such as the instantaneous frequency of the fundamental component and its harmonics, and the number of non-harmonically related signals ; how can one optimally detect such signals in the presence of noise?

In this section we present first some specific problems and show how one may obtain solutions using the concepts described above. In the remainder of the section, the Polynomial WVDs are introduced as a tool for the analysis of polynomial (non-linear) FM signals of arbitrary order. The next section presents a particular member of this class, referred to as the Wigner-Ville trispectrum. This representation is useful for the analysis of FM signals affected by multiplicative noise.

4.1 Polynomial Wigner-Ville distributions

The problem: In this section we address the problem of representation and analysis of deterministic signals, possibly corrupted by additive noise with high SNR, and which can be modelled as signals having constant amplitude and polynomial phase and frequency modulation.
The model: We consider the signal:

$$z(t) = \Pi_T(t)\; A\; e^{j\phi(t)} + w(t) \tag{51}$$

where

$$\phi(t) = \sum_{i=0}^{p} a_i t^i \tag{52}$$

and $w(t)$ is complex noise. Since the polynomial form of the phase, $\phi(t)$, uniformly approximates any continuous function on a closed interval (Weierstrass approximation theorem [60]), the above model can be applied to any continuous phase signal. We are primarily interested in the IF, defined in (40), and rewritten here as:

$$f_i(t) = \frac{1}{2\pi} \frac{d\phi(t)}{dt} \tag{53}$$

The solution: The case where the phase polynomial order, p is equal to 1, belongs to the field of *stationary* spectrum and frequency estimation [61].

The case $p = 2$ corresponds to the class of signals with linear FM and can be handled by the WVD.

The case $p > 2$, corresponds to the case of *non-linear* FM signals, for which the Wigner-Ville transform becomes inappropriate (see Section 4.2) and the ML approach becomes computationally very complicated. However, signals having a non-linear FM law do occur in both nature and engineering applications. For example, the sonar system of some bats often uses *hyperbolic* and *quadratic* FM signals for echo-location [62]. In radar, some of the pulse compression signals are *quadratic* FM signals [63]. In geophysics, in some modes of long-propagation of seismic signals, non-linear signals may occur from earthquakes or underground nuclear tests [64]. In passive acoustics, the estimation of the altitude and speed of a propeller driven aircraft is based on the instantaneous frequency which is a non-linear function of time [65]. Non-linear FM signals also appear in communications, astronomy, telemetry and other engineering and scientific disciplines. It is therefore important to find appropriate analysis tools for such signals.

The problem of representing signals with polynomial phase was also studied by Peleg and Porat in [66], [67]. They defined the polynomial phase transform (PPT) for estimating the coefficients of the phase, $\phi(n)$, and calculated the Cramer-Rao (CR) lower bounds for the coefficients. A more recent work on the same subject can be found in [68].

A different, yet related approach was taken by this author and his co-workers, extending the conventional WVD to be able to process polynomial FM signals effectively [56], [57]. This approach is developed below.

4.2 The key link: the Wigner-Ville distribution and its inbuilt IF estimator

The WVD has become popular because it has been found to have optimal concentration in the time-frequency plane for linear FM signals [2]. That is, it yields a continuum of delta functions along the signal's instantaneous frequency law as $T \to \infty$ [2], [3]. For non-linear FM signals this optimal concentration is no longer assured and the WVD based spectral representations become smeared.

We describe in this section the key link between the WVD and the IF that makes it possible to design *Polynomial Wigner-Ville distributions (PWVDs)*[5] which will exhibit a continuum of delta functions along the IF law for polynomial FM signals.

To explain how this is achieved, one needs to look closely at the mechanism by which the WVD attains optimal concentration for a linear FM signal. Consider a unit amplitude analytic signal, $z(t) = e^{j\phi(t)}$. Given that the WVD of this signal is defined by (27) and (28), substitution of $z(t) = e^{j\phi(t)}$, and equation (28) into (27) yields

$$W_z(t,f) = \underset{\tau \to f}{\mathcal{F}} \left[e^{j(\phi(t+\tau/2)-\phi(t-\tau/2))} \right] \tag{54}$$

Note that the term $\phi(t+\tau/2) - \phi(t-\tau/2)$ in (54) can be re-expressed as

$$\phi(t+\tau/2) - \phi(t-\tau/2) = 2\pi\tau \hat{f}_i(t,\tau) \tag{55}$$

where $\hat{f}_i(t,\tau)$ can be considered to be an instantaneous frequency estimate. This estimate is the difference between two phase values divided by $2\pi\tau$, where τ is the separation in time of the phase values. This estimator is simply a scaled finite difference of phases centrally located about time

[5] In earlier publications, polynomial WVDs were referred to as generalised WVDs.

instant t, and is known as the *central finite difference* estimator [69], [3]. The estimator follows directly from eq.(53):

$$f_i(t) = \frac{1}{2\pi} \lim_{\tau \to 0} \left[\frac{\phi(t+\tau/2) - \phi(t-\tau/2)}{\tau} \right] \tag{56}$$

Eq. (54) can therefore be rewritten as

$$W_z(t,f) = \underset{\tau \to f}{\mathcal{F}} \left[e^{j2\pi\tau \hat{f}_i(t,\tau)} \right] \tag{57}$$

Thus the WVD's bilinear kernel is seen to be a function which is reconstructed from the central finite difference derived IF estimate. It now becomes apparent why the WVD yields good energy concentration for linear FM signals. Namely, the central finite difference estimator is known to be unbiased for such signals [3], and in the absence of noise, $\hat{f}_i(t,\tau) = f_i(t)$. Thus linear FM signals are transformed into sinusoids in the WVD kernel with the frequency of the sinusoid being equal to the instantaneous frequency of the signal, $z(t)$, at that value of time. Fourier transformation of the bilinear kernel then becomes

$$W_z(t,f) = \delta(f - f_i(t)) \tag{58}$$

that is, a row of delta functions along the true IF of the signal. The above equation is valid only for unit amplitude linear FM signals of infinite duration in the absence of noise. For non-linear FM signals a different formulation of the WVD has to be introduced in order to satisfy (58) under the same conditions.

4.3 The design of Polynomial Wigner-Ville distributions

The design of Polynomial WVDs which yield (58) for a non-linear FM signal, is based on replacing the central finite difference estimator, which is inherent in the definition of the WVD, cf.(57), by an estimator which would be unbiased for polynomial FM signals. The general theory of polynomial phase difference coefficients [69], [70] [71] describes the procedure for deriving unbiased IF estimators for arbitrary polynomial phase laws. It is presented below.

4.3.1 Phase difference estimators for polynomial phase laws of arbitrary order

Analysis of the problem. We consider now the discrete-time case where:

$$z(n) = Ae^{j\phi(n)} + w(n), \qquad n = 0, \ldots, N-1 \tag{59}$$

Practical requirements generally necessitate that the IF be determined from discrete time observations, and this means that a discrete approximation to the differentiation operation must be used. This is done by using an FIR differentiating filter. Thus for the discrete time signal whose phase is defined by

$$\phi(n) = \sum_{m=0}^{p} a_m n^m, \tag{60}$$

the IF is computed using the relation:

$$f_i(n) = \frac{1}{2\pi}\phi(n) * d(n) \tag{61}$$

where $d(n)$ is the differentiating filter, which leads to the following estimator:

$$\hat{f}_i(n) = \frac{1}{2\pi}\phi(n) * \hat{d}(n)$$

This section addresses the design of the differentiating filter $d(n)$. For phase laws which are linear or quadratic (i.e. for complex sinusoids or linear FM signals), the differentiating filter needs only to be a two tap filter. It is, in fact, a simple scaled phase difference, known as the *central finite difference*. As the order of the phase polynomial increases, so does the number of taps required in the filter. The filter then becomes a weighted sum of phase differences. The following derivation determines the exact form of these higher order phase difference based IF estimators.

Derivation of the estimates. For the discrete polynomial phase sequence given in (60) the IF is:

$$f_i(n) = \frac{1}{2\pi}\sum_{m=1}^{p} m a_m n^{m-1} \tag{62}$$

For a signal with polynomial phase of order p, a more generalised form of the phase difference estimator is required to perform the desired differentiation. It is defined as [72]:

$$\hat{f}_i^{(q)}(n) = \sum_{l=-q/2}^{q/2} \frac{1}{2\pi} d_l \phi(n+l) \tag{63}$$

where q is the order of the estimator. The d_l coefficients are to be found so that in the absence of noise, $\hat{f}_i^{(q)}(n) = f_i(n)$ for any n, that is:

$$\sum_{l=-q/2}^{q/2} d_l \phi(n+l) = \sum_{i=1}^{p} i a_i n^{i-1} \tag{64}$$

q	d_{-3}	d_{-2}	d_{-1}	d_0	d_1	d_2	d_3
2			-1/2	0	1/2		
4		1/12	-2/3	0	2/3	-1/12	
6	-1/60	3/20	-3/4	0	3/4	-3/20	1/60

Table 2: The values of differentiating filter coefficients for $q = 2, 4, 6$.

$$\sum_{l=-q/2}^{q/2} d_l \sum_{i=0}^{p} a_i \cdot (n+l)^i = \sum_{i=1}^{p} i\, a_i n^{i-1} \tag{65}$$

Because a polynomial's order is invariant to the choice of its origin, without any loss of generality we can set $n = 0$. Then (65) becomes

$$\sum_{l=-q/2}^{q/2} d_l \sum_{i=0}^{p} a_i l^i = a_1 \tag{66}$$

Then by equating coefficients for each of the a_i, we obtain $p + 1$ equations. In matrix form this is given by:

$$Q\underline{d} = \underline{\xi} \tag{67}$$

where

$$Q = \begin{bmatrix} 1 & 1 & \dots & 1 \\ (-q/2) & (-q/2+1) & \dots & (q/2) \\ \dots & \dots & \dots & \dots \\ (-q/2)^p & (-q/2+1)^p & \dots & (q/2)^p \end{bmatrix} \tag{68}$$

$$\underline{d} = [d_{-q/2} \ \dots \ d_0 \ \dots \ d_{q/2}]^T \tag{69}$$

$$\underline{\xi} = [0\ 1\ 0\ \dots\ 0]^T \tag{70}$$

The matrix equation, (67), has a unique solution for $q = p$. The coefficients of the differentiating filter are given in Table 2, and are derived by solving the matrix equation (67) for $p = q = 2, 4, 6$. It is obviously most convenient if q is an even number. This will ensure that for estimating the IF at a given value of n, one does not need to evaluate the phase at non-integer sample points. That is, one does not need to interpolate the signal to obtain the required phase values. In practice, the use of estimators with odd valued q brings unnecessary implementational problems without any benefit in return. Therefore only even valued q estimators are used.

For the case where $p > q$, the matrix equation (67) can be approximately solved assuming that Q has a full rank. This represents an overdetermined problem, and the least-squares solution is given by:

$$\underline{d} = (Q^T Q)^{-1} Q^T \underline{\xi} \tag{71}$$

Choice of p and q. In analysing practical signals, the first task is to estimate the true order of the signal's polynomial phase p. This may involve a priori information, some form of training scheme, or even an educated guess. Once p has been estimated, the order q of the estimator $\hat{f}_i^{(q)}(n)$, has to be chosen. For an exact solution of (67), the rule is to chose q to be the least even number which is greater than p. In some situations, however, it may be preferable to use a lower value of q (and hance only to approximate the solution of eq.(67)) because the differentiating filter will be less susceptible to noise. This is due to the fact that as the polynomial order increases, there is increased likelihood that the noise will be modelled as well as the signal.

The next section uses these generalised (or polynomial) phase difference IF estimators, to replace the central finite difference based IF estimator which is built in to the WVD. The result of this replacement is a class of *polynomial WVDs* which ideally concentrate energy for polynomial phase signals.

4.3.2 Non-integer powers form for Polynomial WVDs (form I)

The q-th order unbiased IF estimator for polynomial phase signals can be expressed by [73]:

$$\hat{f}_i^{(q)}(t) = \frac{1}{2\pi\tau} \sum_{l=-q/2}^{q/2} d_l \, \phi(t + l\tau/q) \tag{72}$$

where $q \geq p$. Now it is straightforward to define Polynomial Wigner-Ville distributions with fractional powers as a generalisation of eq.(57):

$$W_z^{(q)}(t,f) = \underset{\tau \to f}{\mathcal{F}} \left\{ \exp\{j2\pi\tau \hat{f}_i^{(q)}(t,\tau)\} \right\} = \underset{\tau \to f}{\mathcal{F}} \{K_z^{(q)}(t,\tau)\} \tag{73}$$

where $\hat{f}_i^{(q)}(t,\tau)$ is the estimator given by eq.(72), centrally located about time instant, t. For a unit amplitude signal, i.e. $A = 1$ in (51), it follows from (73) and (72) that:

$$\begin{aligned} K_z^{(q)}(t,\tau) &= \exp\left\{ j \sum_{l=-q/2}^{q/2} d_l \, \phi(t + l\tau/q) \right\} = \prod_{l=-q/2}^{q/2} \exp\{j d_l \phi(t + l\tau/q)\} \\ &= \prod_{l=-q/2}^{q/2} [z(t + l\tau/q)]^{d_l} \end{aligned} \tag{74}$$

We refer to this form of Polynomial WVDs as the "fractional powers form" since the coefficients d_l are in general, rational numbers[6].

Example 1: Suppose the order of polynomial in (52) is $p = 2$ (linear FM signal). Then for $q = 2$ and $A = 1$, we get from (74):

$$K_z^{(2)}(t,\tau) = z(t+\tau/2)z^*(t-\tau/2) \tag{75}$$

Thus, the PWVD of order $q = 2$ is identical to the conventional WVD.

Example 2: Suppose $p = 3$ (quadratic FM) or $p = 4$ (cubic FM). Then if we set $q = 4$ (such that $q \geq p$) and for $A = 1$ we obtain from (74):

$$K_z^{(4)}(t,\tau) = \left[z^*(t+\frac{\tau}{2})\right]^{1/12} \left[z(t-\frac{\tau}{2})\right]^{1/12} \left[z(t+\frac{\tau}{4})\right]^{2/3} \left[z^*(t-\frac{\tau}{4})\right]^{2/3} \tag{76}$$

It is easy to verify that for cubic FM signals when $T \to \infty$,

$$W_z^{(4)}(t,f) = \delta(f - \frac{1}{2\pi}(a_1 + 2a_2 t + 3a_3 t^2 + 4a_4 t^3)) = \delta(f - f_i(t)) \tag{77}$$

Although derived for a polynomial phase signal model, the PWVD with a fixed value of q, (PWVD_q) can be used as a non-parametric analysis tool, in much the same way as the short-time Fourier transform or the conventional WVD is used.

Discrete implementation For computer implementation, it is necessary that the discrete form for the signal kernel and resulting polynomial WVD be used. The discrete form for the kernel and distribution are therefore presented here. The discrete form for the multilinear kernel is

$$K_z^{(q)}(n,m) = \prod_{l=-q/2}^{q/2} [z(n+lm)]^{d_l} \tag{78}$$

where $n = tf_s$, $m = \tau f_s$ and f_s is the sampling frequency assumed to be equal to 1 for simplification. The resulting time-frequency distribution is

[6]Note also that $K_z^{(q)}(t,\tau)$ is a *multi-linear kernel* if the coefficients d_l are integers. While the WVD's (bilinear) kernel transforms linear FM signals into sinusoids, the PWVD (multi-linear) kernel can be designed to transform higher order polynomial FM signals into sinusoids. These sinusoids manifest as delta functions about the IF when Fourier transformed. Thus the WVD may be interpreted as a method based on just the first order approximation in a polynomial expansion of phase differences.

given by:

$$W_z^{(q)}(n,k) = \underset{m \to k}{\mathcal{F}} \left\{K_z^{(q)}(n,m)\right\} = \underset{m \to k}{\mathcal{F}} \left\{ \prod_{l=-q/2}^{q/2} [z(n+lm)]^{d_l} \right\} \tag{79}$$

where k is the discrete frequency variable.

Implementation difficulties. The implementation of the Polynomial WVD with signal, $z(n)$, raised to fractional powers, requires the use of phase unwrapping procedure (calculation of the phase sequence from the discrete-time signal $z(n)$). However, phase unwrapping performs well only for high SNRs and mono-component signals.

Since the implementation of the "non-integer powers" form of the PWVD is problematic and since its expected value cannot be interpreted as conventional time-varying polyspectra, we present an alternative form of the PWVD, where the signal, $z(n)$, is raised to integer powers.

4.3.3 Integer powers form for polynomial WVDs (form II)

The alternative way of implementing "unbiased" IF estimators for arbitrary polynomial phase laws requires that we weight the phases at unequally spaced samples and then take their sum. This allows the weights, (b_l), to be prespecified to integer values. The IF estimator of this type is defined as [56]:

$$\hat{f}_i^{(q)}(t,\tau) = \frac{1}{2\pi\tau} \sum_{l=-q/2}^{q/2} b_l \, \phi(t + c_l \tau) \tag{80}$$

Here c_l are coefficients which control the separation of the different phase values used to construct the IF estimator. Coefficients b_l and c_l may be varied to yield unbiased IF estimates for signals with an arbitrary polynomial FM law. The procedure for determining the b_l and c_l coefficients for the case $q = 4$ is illustrated in the Example 3, given below. While the b_l may theoretically take any values, they are practically constrained to be integers, since the use of integer b_l enables the expected values of the PWVD to be interpreted as time-varying higher order spectra. This important fact will make the form II of the PWVD preferable, and further on in the text, form II will be assumed unless otherwise stated.

The Polynomial Wigner-Ville distributions which result from incorporating the estimator in (80), are defined analogously to (73), again assuming constant amplitude A. The multilinear kernel of the PWVD is given by

$$K_z^{(q)}(t,\tau) = \prod_{l=-q/2}^{q/2} [z(t + c_l\tau)]^{b_l} \tag{81}$$

The above expression for the kernel may be rewritten in a symmetric type form according to:

$$K_z^{(q)}(t,\tau) = \prod_{l=0}^{q/2} [z(t + c_l\tau)]^{b_l} \, [z^*(t + c_{-l}\tau)]^{-b_{-l}} \tag{82}$$

The discrete time version of the PWVD is given by the Discrete FT of:

$$K_z^{(q)}(n,m) = \prod_{l=0}^{q/2} [z(n + c_l m)]^{b_l} \, [z^*(n + c_{-l} m)]^{-b_{-l}} \tag{83}$$

where $n = tf_s$, $m = \tau f_s$ and f_s is the sampling frequency.

We have already mentioned earlier that the conventional WVD is a special case of the Polynomial WVD and may be recovered by setting $q = 2$, $b_{-1} = -1$, $b_0 = 0$, $b_1 = 1$, $c_{-1} = -1/2$, $c_0 = 0$,$c_1 = 1/2$.

Example 3: Design of the PWVD form II, for quadratic and cubic FM signals

Since $p = 3$ for quadratic FM, or $p = 4$ for cubic FM, we set $q = 4$ to account for both cases. The set of coefficients b_l and c_l must be found to completely specify the new kernel. In deciding on integer values to be assigned to the b_l it is also desired that the sum of all the $|b_l|$ be as small as possible. This criteria is used because the greater the sum of the $|b_l|$ the greater will be the deviation of the kernel from linearity, since the b_l coefficients which multiply the phases translate into powers of $z(t + c_l\tau)$. The extent of the multilinearity in the kernel should be limited as much as possible to prevent excessively poor performance of the PWVD in noise.

To be able to transform second order phase law signals into sinusoids (via the conventional WVD's kernel), it is known that the b_l must take on the values, $b_{-1} = -1$, $b_0 = 0$ and $b_1 = +1$. To transform third and fourth order phase law signals into sinusoids, it is necessary to incorporate two extra b_l terms (i.e. the phase differentiating filter must have two extra taps [69]). An attempt to adopt ± 1 for these two extra b_l terms values would fail since the procedure for determining a_l coefficients, eq. (89) would yield an inconsistent set of equations. As a consequence, the IF estimator would be biased. The simplest values that these terms can assume are $+2$ and -2, and therefore the simplest possible kernel satisfying the criteria specified above is characterised by:

$$b_2 = -b_{-2} = 1, \;\; b_1 = -b_{-1} = 2, \;\; b_0 = 0 \tag{84}$$

The c_l coefficients must then be found such that the PWVD kernel transforms unit amplitude cubic, quadratic or linear frequency modulated signals into sine waves. The design procedure necessitates setting up a system of equations which relate the polynomial IF of the signal to the IF estimates obtained from the polynomial phase differences, and solving for the c_l. It is described below.

In setting up the design equations it is assumed that the signal phase in discrete-time form is a p-th order polynomial, given by:

$$\phi(n) = \sum_{i=0}^{p} a_i\, n^i \tag{85}$$

where the a_i are the polynomial coefficients. The corresponding IF is then given by [69]:

$$f_i(n) = \frac{1}{2\pi} \sum_{i=1}^{p} i\, a_i\, n^{i-1} \tag{86}$$

A q-th order phase difference estimator ($q \geq p$) is applied to the signal and it is required that, at any discrete-time index, n, the output of this estimator gives the true IF. The required system of equations to ensure this is:

$$\frac{1}{2\pi m} \sum_{l=-q/2}^{q/2} b_l\, \phi(n + c_l m) = f_i(n) \tag{87}$$

that is:

$$\frac{1}{2\pi m} \sum_{l=-q/2}^{q/2} b_l \sum_{i=0}^{p} a_i\, (n + c_l m)^i = \frac{1}{2\pi} \sum_{i=1}^{p} i\, a_i\, n^{i-1} \tag{88}$$

Note that because of the invariance of a polynomial's order to its origin, n may be set equal to zero without loss of generality. Setting n equal to zero in (88), then, yields

$$\frac{1}{m} \sum_{i=0}^{p} a_i\, m^i \sum_{l=-q/2}^{q/2} b_l\, c_l^i = a_1 \tag{89}$$

All of the a_i coefficients on the left and right hand side of (89) may be equated to yield a set of $p+1$ equations. Doing this for the values of b_l

specified in (84) and for $p = q = 4$ yields:

$$a_0 \left[1 - 1 + 2 - 2\right] = 0 \times a_0 \tag{90}$$

$$a_1 \left[c_2 - c_{-2} + 2c_1 - 2c_{-1}\right] = 1 \times a_1 \tag{91}$$

$$a_2 \left[c_2^2 - c_{-2}^2 + 2c_1^2 - 2c_{-1}^2\right] = 0 \times a_2 \tag{92}$$

$$a_3 \left[c_2^3 - c_{-2}^3 + 2c_1^3 - 2c_{-1}^3\right] = 0 \times a_3 \tag{93}$$

$$a_4 \left[c_2^4 - c_{-2}^4 + 2c_1^4 - 2c_{-1}^4\right] = 0 \times a_4 \tag{94}$$

It is obvious that (90) is always true, and if $c_1 = -c_{-1}$ and $c_2 = -c_{-2}$, eqns. (92) and (94) are satisfied too. This condition amounts to verifying the symmetry property of the FIR filter. Solving for c_1, c_{-1}, c_2 and c_{-2} then becomes straightforward from (91) and (93) subject to the condition that $c_1 = -c_{-1}$ and $c_2 = -c_{-2}$. The solution is:

$$c_1 = -c_{-1} = \frac{1}{2(2 - 2^{1/3})} \approx 0.675 \tag{95}$$

$$c_2 = -c_{-2} = -2^{1/3} c_1 \approx -0.85 \tag{96}$$

The resulting discrete-time kernel is then given by:

$$K_z^{(4)}(n, m) = \left[z(n + 0.675m)\, z^*(n - 0.675m)\right]^2 z^*(n + 0.85m)\, z(n - 0.85m) \tag{97}$$

Note. It was recently shown that for $p = q = 4$, the solution given by (95) and (96) is just one of an infinite number of possible solutions. The full details appeared in [74].

Fig.4(a) and 4(b) illustrate the conventional WVD and the PWVD$_{q=4}$ (form I or form II) of the same quadratic FM signal (noiseless case) respectively. The superior behaviour of the latter is indicated by the sharpness of the peaks in Fig.4(b). From the peaks of the PWVD the quadratic IF law can be recovered easily. The conventional WVD, on the other hand, shows many oscillations that tend to degrade its performance.

Implementation. Several important points need to be made concerning the practical implementation of the kernel in (97). Firstly, to form the discrete kernel one must have signal values at non-integer time positions. The signal must therefore be sampled or interpolated reasonably densely. The interpolation can be performed by use of an FFT based interpolation filter. Secondly, it is crucial to use analytic signals, so that the multiple artifacts between positive and negative frequencies are suppressed [24]. Thirdly, the PWVD is best implemented by actually calculating a frequency scaled version of the kernel in (97) and then accounting for the scaling in the Fourier transform operation on the kernel. That is, the PWVD is best implemented as

$$W_z^{(4)}(n, k) = \underset{m \to \frac{1}{0.85}k}{DFT} \left\{\left[z(n + 0.794m) z^*(n - 0.794m)\right]^2 z^*(n + m) z(n - m)\right\} \tag{98}$$

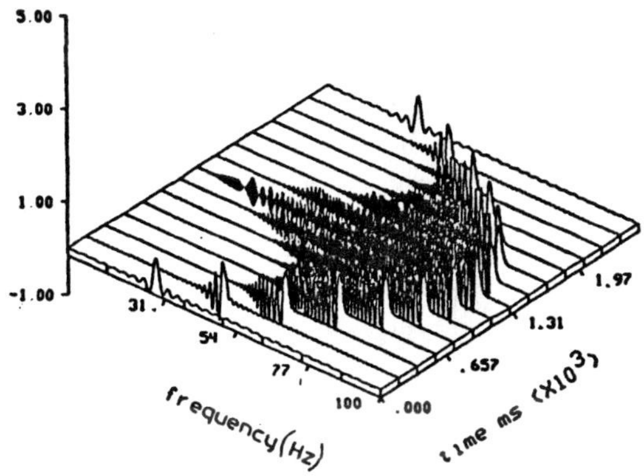

(a)

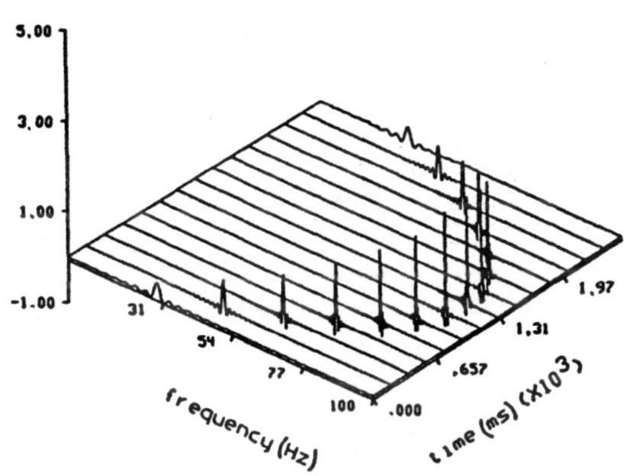

(b)

Figure 4.Time-frequency representations of a quadratic FM signal: (a) WVD; (b) $PWVD_{q=4}$ (form I or form II)

This formulation, because it causes some of the terms within the kernel to occur at integer lags, reduces errors which arise from inaccuracies in the interpolation process.

4.4 Some properties of a class of PWVDs

The Polynomial WVD preserves and/or generalises most of the properties that characterise the WVD. In the derivation of the following properties we have assumed that the PWVD kernel is given by

$$K_z^{(q)}(t,\tau) = \prod_{l=1}^{q/2} [z(t+c_l\tau)]^{b_l} \, [z^*(t+c_{-l}\tau)]^{-b_{-l}} \tag{99}$$

and that the following conditions apply:

$$b_i = -b_{-i} \qquad i = 1, \ldots, q/2 \tag{100}$$

$$c_i = -c_{-i} \qquad i = 1, \ldots, q/2 \tag{101}$$

$$\sum_{i=1}^{q/2} b_i c_i = 1/2 \tag{102}$$

These limitations define the class of PWVDs we are considering. For consistency of notations used in higher-order spectra, it is important to introduce a parameter, which is alternative to the order, q. The parameter used here is defined as:

$$k = 2 \cdot \sum_{i=1}^{q/2} b_i \tag{103}$$

and corresponds to the order of multi-linearity of the PWVD kernel, or in the case of random signals, the order of polyspectra. Note that this represents a slight change of notation.

The following properties, are valid $\forall k \in \mathrm{N}$, and $\forall (t,f) \in \mathrm{R}^2$ (see Appendix C for proofs):

P-1. The PWVD is real for any signal $x(t)$:

$$\left[W^{(k)}_{\{x(t)\}}(t,f)\right]^* = W^{(k)}_{\{x(t)\}}(t,f) \tag{104}$$

P-2. The PWVD is an even function of frequency if $x(t)$ is real:

$$W^{(k)}_{\{x^*(t)\}}(t,-f) = W^{(k)}_{\{x(t)\}}(t,f) \tag{105}$$

P-3. A shift in time by t_0 and in frequency by f_0 (i.e. modulation by $e^{j2\pi f_0 t}$) of a signal $x(t)$ results in the same shift in time and frequency of the PWVD ($\forall t_0, f_0 \in R$):

$$W^{(k)}_{\{x(t-t_0)e^{j2\pi f_0(t-t_0)}\}}(t,f) = W^{(k)}_{\{x(t)\}}(t-t_0, f-f_0) \tag{106}$$

P-4. If $y(t) = w(t)z(t)$ then:

$$W^{(k)}_{\{y(t)\}}(t,f) = W^{(k)}_{\{w(t)\}}(t,f) *_f W^{(k)}_{\{z(t)\}}(t,f) \tag{107}$$

where $*_f$ denotes the convolution in frequency.

P-5. Projection of the $W^{(k)}_{\{x(t)\}}(t,f)$ to the time axis (time marginal):

$$\int_{-\infty}^{\infty} W^{(k)}_{\{x(t)\}}(t,f)df = |x(t)|^k \tag{108}$$

P-6. The local moment of the PWVD in frequency gives the instantaneous frequency of the signal $x(t)$:

$$\frac{\int_{-\infty}^{\infty} fW^{(k)}_{\{x(t)\}}(t,f)df}{\int_{-\infty}^{\infty} W^{(k)}_{\{x(t)\}}(t,f)df} = \frac{1}{2\pi}\frac{d\phi(t)}{dt} \tag{109}$$

P-7. Time-frequency scaling: for $y(t) = \sqrt[k]{|a|} \cdot x(at)$

$$W^{(k)}_{\{y(t)\}}(t,f) = W^{(k)}_{\{x(t)\}}(at, \frac{f}{a}) \tag{110}$$

P-8. Finite time support: $W^{(k)}_{\{x(t)\}}(t,f) = 0$ for t outside $[t_1, t_2]$ if $x(t) = 0$ outside $[t_1, t_2]$.

4.5 Instantaneous frequency estimation at high SNR

Consider signal $z(n)$ as given by (59) and (60), where additive noise $w(n)$ is complex white Gaussian.

Since the WVD is very effective for estimating the IF of signals with linear ($p = 2$) FM laws [3], a natural question which arises is whether the PWVD can be used for accurate estimation of the IF of non-linear polynomial FM signals in noise, as expected by construction. The peaks of the PWVD can in fact be used for computationally efficient IF estimation. This is shown in Appendix A for polynomial phase laws up to order $p = 4$. For higher order polynomial phase laws the SNR operational thresholds can become quite high and methods based on unwrapping the phase are often simpler [69].

Fig.5 summarises results for a quadratic ($p = 3$) FM signal in complex additive white Gaussian noise ($N = 64$ points), as specified in eqn.(59). The curves showing the reciprocal value of the mean square IF estimate error for $PWVD_{q=4}$ (solid line) and the WVD (dashed line) were obtained by Monte Carlo simulations and plotted against the Cramer-Rao (CR) lower variance bound for the IF [7]. One can observe that the $PWVD_{q=4}$ peak based IF estimates meet the CR bound at high SNRs and thus it shows that PWVD peak based IF estimates provide a very accurate means for IF estimation. On the other hand, the WVD peak based IF estimate is biased and that is why its mean square error (MSE) is always greater than the CR lower bound.

For polynomial phase laws of order higher than $p = 4$, the SNR threshold for PWVD based IF estimation becomes comparatively high. As mentioned earlier, in these circumstances, alternative computationally simpler methods based on unwrapping the phase (or a smoothed version of it) tend to be just as effective [3], [69].

The question remains as to how much we loose by choosing the order q of the PWVD higher than necessary. In Fig.6 we summarise the results for a linear FM signal ($p = 2$) in additive Gaussian noise ($N = 64$ points). The dashed curve shows the performance of the conventional WVD (or $PWVD_{q=2}$), while the solid line shows the inverse of the MSE curve for $PWVD_{q=4}$. Both curves were obtained by Monte Carlo simulations[8] and plotted against the CR bound (which is in this case about 8 dB higher than for $p = 3$, Fig.5). One can observe from Fig.6 that if the value of q is chosen higher than required ($q = 4$), the variance of the $PWVD_{q=4}$ based estimate never reaches the CR bound (it is about 8 dB below) and its SNR threshold appears at a higher SNR. This observation is not surprising since going to higher-order non-linearities always causes higher variances of estimates.

4.6 Higher order TFDs

In the same way as the WVD can be interpreted as the core of the class of bilinear time-frequency distributions, the PWVD can be used to define a class of multilinear or higher-order TFDs [56]. Alternative forms of higher-order TFDs, as extensions of Cohen's class, were proposed by several authors [76], [77], [78], [79]. Note that the general class of higher order TFDs can be defined in the multitime-multifrequency space, as a method for time-varying higher order spectral analysis. However, in our approach, we choose to project the full multidimensional space onto the t-f subspace, in order to obtain specific properties (such as t-f representation of polynomial FM signals). The implication of the projection operation is currently

---

[7] Expressions for CR bounds can be found in [69] and [67].

[8] The performance of the WVD peak IF estimator is also confirmed analytically in [75]

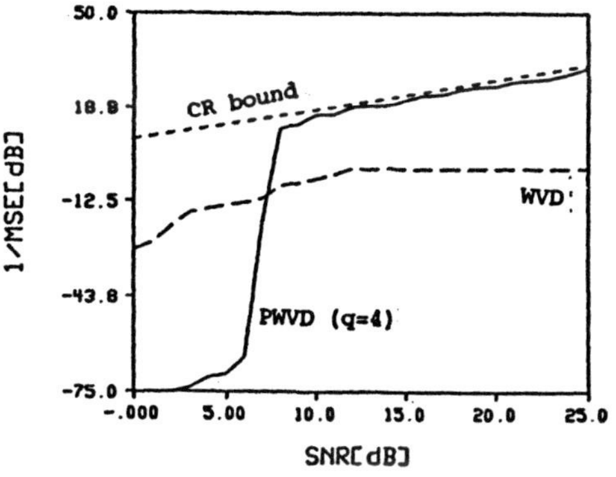

Figure 5. Statistical performance of the $\mathrm{PWVD}_{q=4}$ (solid line) and the WVD (dashed line) IF estimators vs CR bound. The signal is a *quadratic* FM in additive white Gaussian noise (64 points).

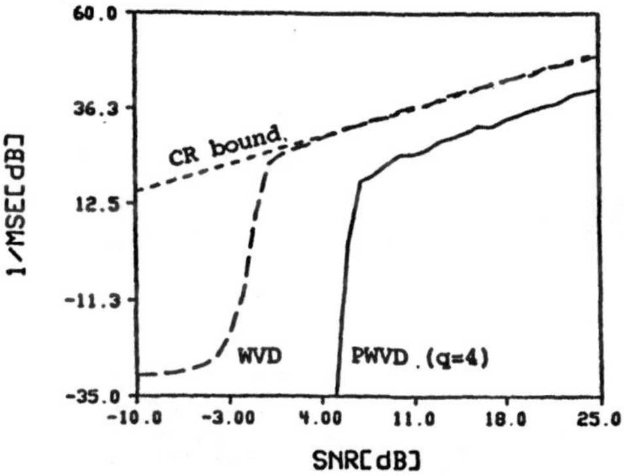

Figure 6. Statistical performance of the $\mathrm{PWVD}_{q=4}$ (solid line) and the WVD (dashed line) IF estimators vs CR bound. The signal is a *linear* FM in additive white Gaussian noise (64 points).

under investigation and further results are expected to appear in [74]. Section 6 will briefly discuss the case of multicomponent (or composite) signals. Before that, the next section will present one particular PWVD.

5 The Wigner-Ville trispectrum

This section presents a particular member of the class of polynomial TFDs which can solve a problem for which bilinear TFDs are ineffective: namely the analysis and representation of FM signals affected by multiplicative noise.

5.1 Definition

Many signals in nature and in engineering applications can be modelled as amplitude modulated FM signals. For example, in radar and sonar applications, in addition to the non-stationarity, the signal is subjected to an amplitude modulation which results in Doppler spreading. In communications, the change of the reflective characteristics of channel during the signal interval, causes amplitude modulation referred to as time-selective fading. Recently Dwyer [80] showed that a reduced form (or slice) of the trispectrum clearly reveals the presence of Gaussian amplitude modulated (GAM) tones. This was shown to be the case even if the noise was white. The conventional power spectrum is unable to perform this discrimination, because the white noise smears the spectrum. The Wigner-Ville distribution (WVD) would have the same limitation being a second-order quantity. A fourth order TFD, however, is able to detect not only GAM tones, but also GAM linear FM signals. Ideally one would like to detect GAM signals of arbitrarily high order polynomial phase signals. This, however, is beyond the scope of this chapter.

This extension of Dwyer's fourth order to a higher order TFD which could reveal GAM linear FM signals has since been called the *Wigner-Ville trispectrum* (WVT) [58], [7].

The WVT thus defined is a member of the class of Wigner-Ville polyspectra based on the PWVDs. The values of the parameters q, b_i and c_i can be derived by requiring that the WVT is an "optimal" t-f representation for linear FM signals and at the same time a fourth-order spectrum (as it name suggests). Furher discussion of these two requirements follows.

(i) $k = 4$

The fourth-order spectrum or the trispectrum, was shown [80] to be very effective for dealing with amplitude modulated sinusoids. The lowest value of q that can be chosen in (99) is $q = 2$. Then we have: $|b_1| + |b_{-1}| = k = 4$. In order to obtain a real WVT, condition (100) should be satisfied and thus we get: $b_1 = -b_{-1} = 2$.

(ii) Optimality for linear FM signals.

The WVT of a noiseless deterministic linear FM signal, $y(t)$, with a unit amplitude and infinite duration should give a row of delta impulses along the signal's instantaneous frequency:

$$W_y^{(4)}(t,f) = \delta(f - f_i(t)) \tag{111}$$

Suppose a signal, $y(t)$, is given by:

$$y(t) = e^{j2\pi(f_0+\alpha t)t} \tag{112}$$

Then:

$$K_y^{(4)}(t,\tau) = y^2(t + c_1\tau)[y^*(t + c_{-1}\tau)]^2 \tag{113}$$

that is:

$$\arg\{K_y^{(4)}(t,\tau)\} = 2\pi[f_0\tau(2c_1 - 2c_{-1}) + 2\alpha t\tau(2c_1 - 2c_{-1}) + \alpha\tau^2(2c_1^2 - 2c_{-1}^2)] \tag{114}$$

In order to satisfy (111) notice that the following must hold:

$$\arg\{K_y^{(4)}(t,\tau)\} = 2\pi\tau(f_0 + 2\alpha t) \tag{115}$$

From (114) and (115) we obtain a set of two equations:

$$2c_1 - 2c_{-1} = 1 \tag{116}$$

$$2c_1^2 - 2c_{-1}^2 = 0 \tag{117}$$

Solving for c_1 and c_{-1} we get $c_1 = -c_{-1} = 1/4$. Thus the remaining two conditions for the properties of the PWVDs to be valid, namely (101) and (102), are thus satisfied.

Definition. The Wigner-Ville trispectrum of a random signal, $z(t)$, is defined as:

$$\mathcal{W}_z^{(4)}(t,f) = \mathcal{E}\left\{\int_{-\infty}^{\infty} z^2(t + \frac{\tau}{4})\,[z^*(t - \frac{\tau}{4})]^2 e^{-j2\pi f\tau} d\tau\right\} \tag{118}$$

where $\mathcal{E}$ is the expected value.

Note. The WVT is actually a reduced form of the full Wigner-Ville trispectrum that was defined in [81] and [79] as follows:

$$\begin{aligned}\mathcal{W}_z^{(4)}(t,f_1,f_2,f_3) &= \mathcal{E}\int_{\tau_1}\int_{\tau_2}\int_{\tau_3} z^*(t-\alpha_3)z(t-\alpha_3+\tau_1) \\ &\quad z(t-\alpha_3+\tau_2)z^*(t-\alpha_3+\tau_3)\prod_{i=1}^{3} e^{-j2\pi f_i\tau_i} d\tau_i \end{aligned} \tag{119}$$

where $\alpha_3 = (\tau_1 + \tau_2 + \tau_3)/4$. Equation (118) is obtained by selecting: $\tau_1 = \tau_2 = \tau/2$; $\tau_3 = 0$, and $f_1 = f_2 = f_3 = f$ in (119). For simplification, we use the term WVT to refer to the reduced form. The WVT satisfies all the properties listed in Sec.4.4. Its relationship with the signal group delay is given in Appendix B.

Cumulant based 4th order spectra. There are a number of ways of forming a cumulant based WVT. One definition has been provided in [76]. A second way to define the cumulant based WVT, assuming a zero-mean random signal, $z(t)$, was given in [74], where the time-varying fourth order cumulant function is given by:

$$\begin{aligned} \mathcal{C}_z^{(4)}(t,\tau) &= \mathcal{E}\{z^2(t+\frac{\tau}{4})\,[z^*(t-\frac{\tau}{4})]^2\} - \\ & \quad 2\left[\mathcal{E}\{z(t+\frac{\tau}{4})z^*(t-\frac{\tau}{4})\}\right]^2 - \mathcal{E}\{z^2(t)\}\cdot\mathcal{E}\{z^*(t)^2\} \end{aligned}$$

The corresponding cumulant WVT then is defined as:

$$c\mathcal{W}_z^{(4)}(t,f) = \underset{\tau \to f}{\mathcal{F}} \left\{\mathcal{C}_z^{(4)}(t,\tau)\right\} \tag{120}$$

This definition has the advantage that it is a natural extension of Dwyer's fourth order spectrum, and hence can detect GAM linear FM signals.

5.2 Analysis of FM signals affected by Gaussian multiplicative noise

Let us assume a signal model:

$$z(t) = a(t)e^{j\phi(t)} \tag{121}$$

where $a(t)$ is a real zero-mean Gaussian multiplicative noise process with covariance function, $R_a(\tau) = v\lambda e^{-2\lambda|\tau|}$ and $\phi(t)$ is the phase of the signal, given by: $\phi(t) = 2\pi(\theta + f_0 t + \alpha t^2)$. The covariance function was chosen in order to describe a white noise process as $\lambda \longrightarrow \infty$. The initial phase θ is a random variable uniformly distributed on $(-\pi, \pi]$. In radar the real Gaussian modulating process, $a(t)$, represents a model for a time-fluctuating target where the pulse length of transmitted signal is longer than the reciprocal of the reflection process [82].

The problem that we investigate is that of recovering the IF of the signal, $z(t)$. For this class of signals we show that for an asymptotically white process, $a(t)$, describing the envelope of $z(t)$, we have:

(A.) The expected value of the WVD is:

$$\mathcal{W}_z^{(2)}(t,f) = v \tag{122}$$

(B.) The Wigner-Ville trispectrum is:

$$\mathcal{W}_z^{(4)}(t,f) = v^2\lambda^2\delta(f - f_i(t)) + 2v^2\lambda \tag{123}$$

Proof: The power spectral density of $a(t)$ is $S_a(f) = v\,\lambda^2/(\lambda^2 + \pi^2 f^2)$. For $\lambda \to \infty$ (asymptotically white $a(t)$), $R_a(\tau) = v\delta(\tau)$ and $S_a(f) = v$.
(A.)

$$\begin{aligned}\mathcal{K}_z^{(2)}(t,\tau) &= \mathcal{E}\{z(t+\tau/2)z^*(t-\tau/2)\} && (124)\\ &= R_a(\tau)e^{j2\pi(f_0+2\alpha t)\tau} && (125)\\ &= v\delta(\tau) \qquad (\lambda \to \infty) && (126)\end{aligned}$$

The FT of $\mathcal{K}_z^{(2)}(t,\tau)$ with respect to τ gives (122).
(B.)

$$\begin{aligned}\mathcal{K}_z^{(4)}(t,\tau) &= \mathcal{E}\{z^2(t+\tau/4)[z^*(t-\tau/4)]^2\} && (127)\\ &= [R_a^2(0) + 2R_a^2(\tau/2)]\,e^{j2\pi(f_0+2\alpha t)\tau} && (128)\\ &= [v^2\lambda^2 + 2v^2\lambda^2 e^{-2\lambda|\tau|}]\,e^{j2\pi(f_0+2\alpha t)\tau} && (129)\\ &= [v^2\lambda^2 + 2v^2\lambda\delta(\tau)]\,e^{j2\pi(f_0+2\alpha t)\tau} \quad (\lambda \to \infty) && (130)\end{aligned}$$

Since the IF of the signal, $z(t)$, is $f_0 + 2\alpha t$, the FT of the above expression leads to (123).

In summary, the WVD of a signal, $z(t)$, affected by white multiplicative noise cannot describe the instantaneous frequency law of the signal in the time-frequency plane, while the WVT can. In order to confirm statements (A) and (B), the following experiment was performed:

Experiment. A complex signal, $e^{j2\pi(\theta + f_0 t + \alpha t^2)}$, is simulated with parameters, $f_0 = 50$ Hz, $\alpha = 47.25$ Hz/s, and where θ is a random variable uniformly distributed over $(-\pi, \pi]$. The sampling frequency and the number of samples are 400 Hz and 256 respectively. The signal is modulated by white Gaussian noise, and the real part of the resulting AM signal, $z(t)$, is shown in Fig.7. A gray-scale plot of the WVD (single realization) of the signal, $z(t)$, is shown in Fig.8(a). Notice that no clear feature can be extracted from this time-frequency representation. The WVT (single realization) of the same signal is presented in Fig.8(b). The linear time-varying frequency component appears clearly. Figs.9(a) and (b) show the WVD and the WVT (respectively) averaged over 10 realizations.

Additional results relevant to this section can be found in [74], where the cumulant WVT was studied and compared to the (moment) WVT defined in (118).

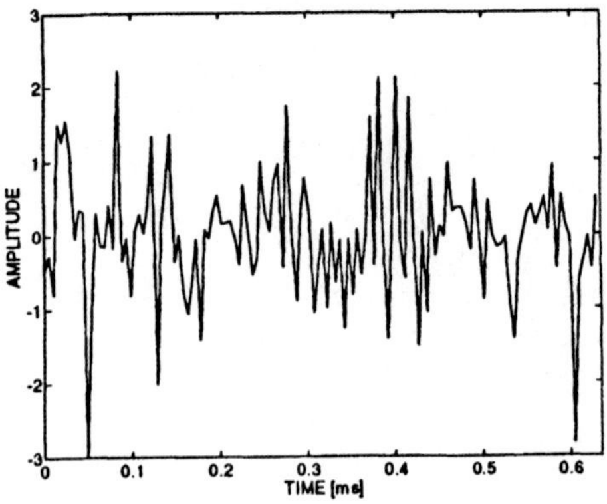

Figure 7. A linear FM signal modulated by white Gaussian noise

5.3 Instantaneous frequency estimation in the presence of multiplicative and additive Gaussian noise

This section discusses the problem of estimating the instantaneous frequency law of a discrete-time complex signal embedded in noise, as follows:

$$z(n) = a(n)e^{j\phi(n)} + w(n), \qquad n = 0, \ldots, N-1 \tag{131}$$

where $w(n)$ is complex white Gaussian noise with zero-mean and variance σ_w^2; $\phi(n)$ is the signal phase with a quadratic polynomial law:

$$\phi(n) = 2\pi(\theta + f_0 n + \alpha n^2) \tag{132}$$

and $a(n)$ is real white Gaussian noise with mean, μ_a, and variance, σ_a^2, independent of $w(n)$. A further assumption is that only a single set of observations, $z(n)$, is available. The instantaneous frequency is estimated from the peak of the discrete WVT. Three separate cases are considered:

1. $a(n) = \mu_a = A = const$, that is $\sigma_a^2 = 0$;
2. $\mu_a = 0$; $\sigma_a^2 \neq 0$.
3. $\mu_a \neq 0$; $\sigma_a^2 \neq 0$.

Case 3 describes a general case. In the first case, multiplicative noise is absent. In the second case the multiplicative noise plays a dominant role.

5.3.1 Performance of the estimator for case 1

Expressions for the variance of the estimate of the instantaneous frequency (IF), for signals given by (131) and (132) and for the case $a(n) = A$, are

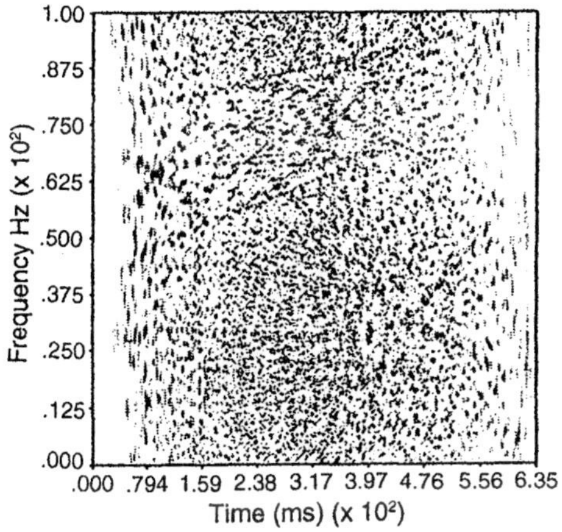

(a)

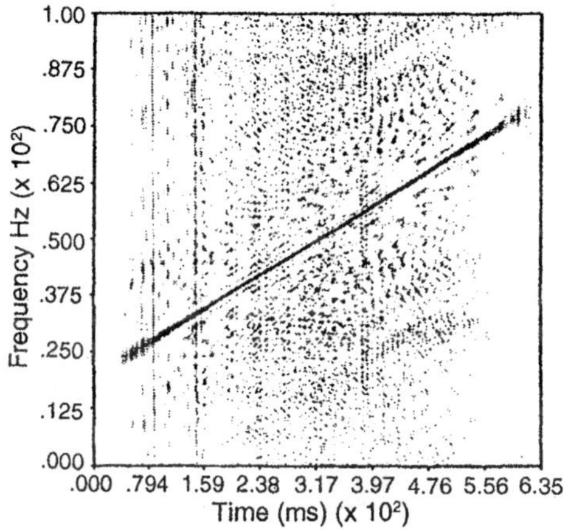

(b)

Figure 8. The WVD (a) and the WVT (b) of a linear FM modulated by white Gaussian noise (one realization)

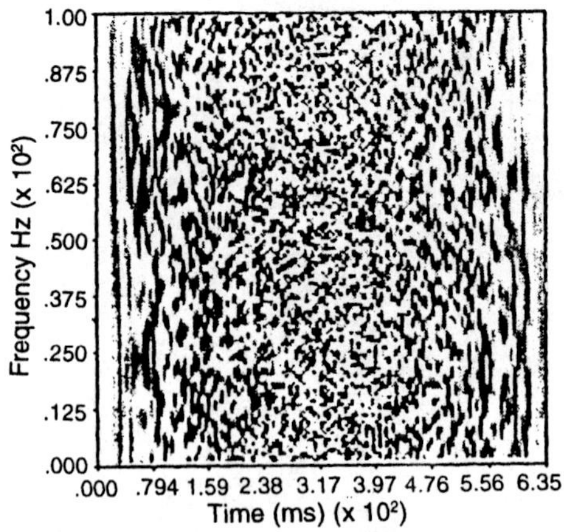

(a)

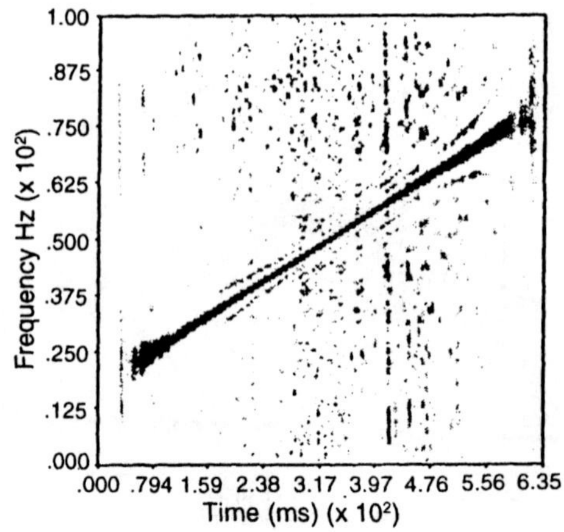

(b)

Figure 9. The WVD (a) and the WVT (b) of a linear FM modulated by white Gaussian noise (ten realizations)

derived in [75] using the peak of the discrete WVD. These results are also confirmed in [83]. Following a similar approach, we derived expressions for the algorithm based on the peak of the WVT [59]. The variance of the WVT peak estimator of the IF (for a linear FM signal in additive white Gaussian noise) is shown to be [59]:

$$\sigma^2_{\hat{f}_i} = \frac{6\sigma^2_w}{A^2N(N^2-1)(2\pi)^2} \tag{133}$$

This expression is identical to the CR lower bound of the variance of an estimate of the frequency for a stationary sinusoid in white Gaussian noise as given in [84]. The same result is obtained for a discrete WVD peak estimate [75]. As SNR decreases, a threshold effect occurs due to the non-linear nature of the estimation algorithm. As reported in [84], this threshold appears when $SNR_{DFT} \approx 15dB$, for a frequency estimate of a stationary sinusoid in noise. The threshold SNR for the discrete WVT peak estimate is shown to be [59]:

$$SNR = (27 - 10\log N)dB \tag{134}$$

This equation can be used to determine the minimum input SNR (for a given sample length, N) which is required for an accurate IF estimation.

Computer simulations were performed in order to verify the results given by eqs. (133) and (134). The results are shown in Fig.10 for $N = 128$ points. The x-axis shows the input SNR, defined as $10\log A^2/\sigma^2_w$. The y-axis represents the reciprocal of the mean-square error in dB. The curves for the WVD and WVT based estimates are obtained by averaging over 100 trials. Notice that the variance of the WVT based estimate meets the CR bound as predicted by (133) and that the threshold SNR for the WVT appears at $6dB$, as predicted by (134). This threshold is about $6dB$ higher than the threshold SNR for the WVD peak estimate, the higher threshold being due to the increased non-linearity of the kernel.

5.3.2 Performance of the estimator for case 2

Suppose that $a(n)$ is a real white Gaussian process with zero-mean and variance, σ^2_a, such that $a(n) \neq 0$, $(n = 0, \ldots, N-1)$. It is shown in [59] that the expression for the variance of the IF estimate is:

$$\sigma^2_{\hat{f}_i} = \frac{18\sigma^2_w}{(2\pi)^2\sigma^2_a N(N^2-1)} \tag{135}$$

Computer simulations have confirmed expression (135). The results are shown in Fig.11 for $N = 128$ points. The axes are the same as in Fig.10, except that the input SNR in dB is assumed to be $10\log\sigma^2_a/\sigma^2_w$.The curves for the WVT and WVD were obtained over 100 trials. We observe that the variance of the IF estimate given by (135) is three times ($4.7dB$) greater

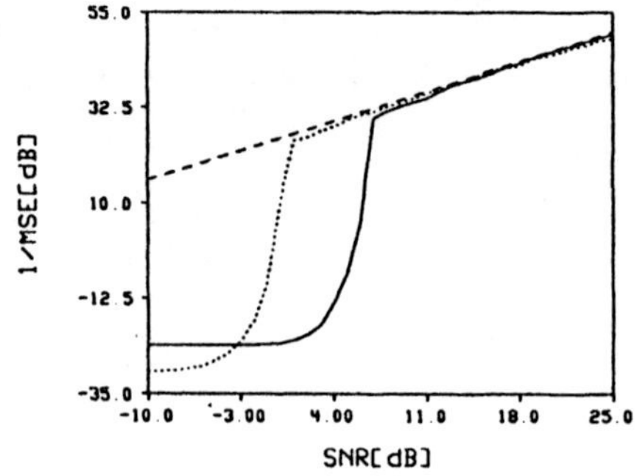

Figure 10. Statistical performance of the WVD (dotted line) and the WVT (solid line) vs CR bound (dashed line) for a constant amplitude linear FM signal in additive white Gaussian noise

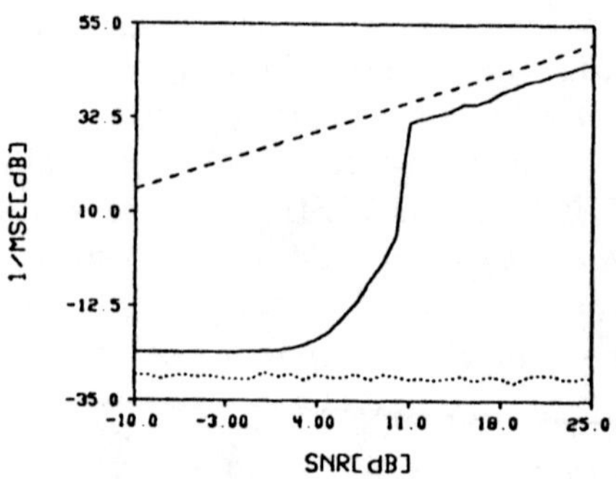

Figure 11. Statistical performance of the WVD (dotted line) and the WVT (solid line) vs CR bound (dashed line) for a linear FM signal modulated by real white zero-mean Gaussian noise and affected by additive white Gaussian noise

than the one expressed by (133). The SNR threshold for the WVT is at $10dB$.

5.3.3 Performance of the estimator for case 3

In the general case, where $a(n) \sim \mathcal{N}(\mu_a, \sigma_a)$, the expression for the variance of the IF estimate is given in [5] as:

$$\sigma_{\hat{f}_i}^2 = \frac{6\,(3\sigma_a^4 + 6\mu_a^2\sigma_a^2 + \mu_a^4) \cdot \sigma_w^2}{(2\pi)^2(\mu_a^2 + \sigma_a^2)^3\, N\,(N^2 - 1)} \tag{136}$$

This expression was confirmed by simulations, with the results being shown in Fig.12(a). There the reciprocal of the MSE is plotted as a function of the input SNR defined as: $10\log(\mu_a^2 + \sigma_a^2)/\sigma_w^2$ and the quantity R defined as: $R = \sigma_a/(\sigma_a + \mu_a)$. Note that $R = 0$ and $R = 1$ correspond to the case 1 and 2 respectively. Fig.12(b) shows the behaviour of the WVD peak IF estimator for this case. One can observe that for $R > 0.25$ (i.e. $\mu_a < 3\sigma_a$) the WVT outperforms the WVD based IF estimator.

In summary, random amplitude modulation (here modelled as a Gaussian process) of an FM signal, behaves as multiplicative *noise* for second-order statistics, while in the special case of the fourth-order statistics it contributes to the signal power. In practical situations, the choice of the method (second- or fourth-order) depends on the input SNR and the ratio between the mean (μ_a) and the standard deviation (σ_a) of Gaussian AM.

6 Multicomponent signals and Polynomial TFDs

Until now we have considered only single component FM signals; that is, signals limited to only one time-varying feature in the frequency domain. In both natural and man-made signals, it is much more common to encounter *multicomponent* signals. It is therefore important to see if and how Polynomial TFDs can be used for multicomponent signal analysis. This is the problem which is investigated in this section[9].

6.1 Analysis of the cross-terms

Let us consider a composite FM signal which may be modelled as follows:

$$z_M(t) = \sum_{i=1}^{M} a_i(t) e^{j[\Theta_i + 2\pi \int_0^t f_i(u)du]} = \sum_{i=1}^{M} y_i(t) \tag{137}$$

[9]The results presented in this section were obtained while finishing this manuscript

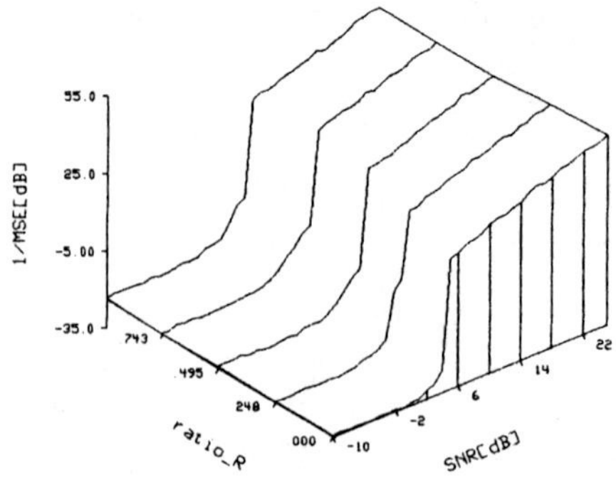

(a)

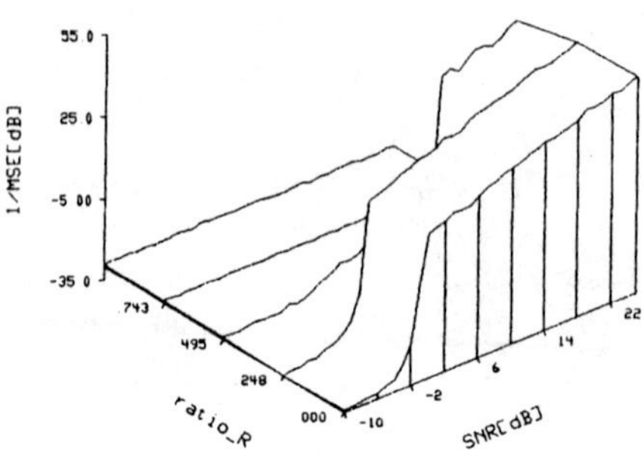

(b)

Figure 12. Statistical performance of the peak based IF estimator for a Gaussian multiplicative process with mean μ_a and variance σ_a^2, where $R = \sigma_a/(\sigma_a + \mu_a)$: (a) WVT; (b) WVD

where each $y_i(t)$ is an FM signal with random amplitude $a_i(t)$; Θ_i are random variables, such that $\Theta_i \sim \mathcal{U}[-\pi, \pi)$; and $a_i(t)$ and Θ_i are all mutually independent for any i and t. The (moment) WVT given by eq.(118), of $z_M(t)$ can be expressed as:

$$\begin{aligned} W_{z_M}^{(4)}(t,f) &= \sum_{i=1}^{M} W_{y_i}^{(4)}(t,f) + \sum_{i=1}^{M} \sum_{j=1, j\neq i}^{M} W_{y_i,y_i,y_j,y_j}^{(4)}(t,f) + \\ &\quad +4 \sum_{i=1}^{M} \sum_{j=i+1}^{M} W_{y_i y_j, y_i, y_j}^{(4)}(t,f) \end{aligned} \tag{138}$$

while the cumulant WVT defined in (120) is given by:

$$cW_{z_M}^{(4)}(t,f) = \sum_{i=1}^{M} cW_{y_i}^{(4)}(t,f) \tag{139}$$

since all components $y_i(t)$ are zero-mean and mutually independent. Hence, only the moment WVT is affected by the cross-terms which are represented by the second and the third summand in (138). The cross-terms are artificially created by the method and they have no physical correspondent in the signal. Hence, the cross-terms are generally treated as undesirable in time-frequency analysis as discussed in section 3.3 [2]. The cross-terms of the (moment) WVT (138) can be divided into two groups:

- those given by:

$$\sum_{i=1}^{M} \sum_{j=1, j\neq i}^{M} W_{y_i,y_i,y_j,y_j}^{(4)}(t,f)$$

  which have *oscillatory* amplitude in the time-frequency plane.

These cross-terms can be suppressed by time-frequency smoothing of the WVT using methods equivalent to that of Choi-Williams or Zhao-Atlas-Marks [85]. These cross-terms correspond to well-studied cross-terms generated by quadratic t-f methods.

- those with *constant* amplitude in the t-f plane.

This class of cross-terms can be expressed in the form:

$$4 \sum_{i=1}^{M} \sum_{j=i+1}^{M} W_{y_i y_j, y_i, y_j}^{(4)}(t,f) \tag{140}$$

and as (140) suggests, they have 4 time greater amplitude than the auto-terms, and frequency contents along:

$$f_{ij}(t) = \frac{f_i(t) + f_j(t)}{2} \qquad (i = 1, \ldots, M; \;\; j = i+1, \ldots, M)$$

Note that if $a_i(t)$ $(i = 1, \ldots, M)$ are white processes, then each cross-term:

$$W^{(4)}_{y_i y_j, y_i, y_j}(t, f) = \text{const}$$

Then these cross-terms will be spread in the entire t-f plane, rather than concentrated along $f_{ij}(t)$.

The most serious problem in the application of the moment WVT to composite FM signals is that of distinguishing these constant or "non-oscillating" cross-terms from the auto-terms.

In the next subsection, we consider methods for elimination of the "non-oscillating" cross-terms based on alternative forms of reducing the tri-frequency space to the frequency subspace.

6.2 Non-oscillating cross-terms and slices of the moment WVT

Consider the WVT, defined in the time–tri-frequency space (t, f_1, f_2, f_3) [81], of a deterministic signal $z_M(t)$. One definition is given in (119), and repeated here:

$$\begin{aligned} W^{(4)}_{z_M}(t, f_1, f_2, f_3) &= \int_{\tau_1}\int_{\tau_2}\int_{\tau_3} z^*_M(t - \alpha_3) z_M(t + \tau_1 - \alpha_3) z_M(t + \tau_2 - \alpha_3) \\ & \quad z^*_M(t + \tau_3 - \alpha_3) \prod_{i=1}^{3} e^{-j2\pi f_i \tau_i} d\tau_i \end{aligned} \tag{141}$$

where $\alpha_3 = (\tau_1 + \tau_2 + \tau_3)/4$. The WVT can be equivalently expressed in terms of $Z_M(f)$, FT of $z_M(t)$ as [81]:

$$\begin{aligned} W^{(4)}_{z_M}(t, f_1, f_2, f_3) &= \int_{\nu} Z^*_M(f_1 + f_2 + f_3 - \frac{\nu}{4})\, Z_M(f_1 + \frac{\nu}{4})\, Z_M(f_2 + \frac{\nu}{4}) \\ & \quad Z^*_M(-f_3 - \frac{\nu}{4})\, e^{j2\pi\nu t} d\nu \end{aligned} \tag{142}$$

We postulate that:

Postulate 1 *If the signal $z_M(t)$ has no overlap between its components in the* time *domain, then all slices of the WVT expressed by (141) such that:*

$$\tau_1 + \tau_2 - \tau_3 = \tau \tag{143}$$

are free from "non-oscillating cross-terms".

In addition, one can show that if

$$\tau_1 - \tau_2 \pm \tau_3 = 0 \tag{144}$$

then for any deterministic signal $z_1(t) = e^{j2\pi(f_0 t+\alpha t^2/2)}$, the WVT sliced as above yields: $W_{z_1}^{(4)} = \delta[f - (f_0 + \alpha t)]$. Obviously, the WVT defined by (118) can be derived from (141) satisfying both (143) and (144) by selecting $\tau_1 = \tau_2 = \tau/2$ and $\tau_3 = 0$. We refer to this form of the WVT (given below):

$$W_{z_M}^{(4)}(t,f) = \int_\tau [z_M(t+\frac{\tau}{4})]^2\, [z_M(t-\frac{\tau}{4})]^2 e^{-j2\pi f\tau} d\tau \tag{145}$$

as the *lag-reduced form*.

Postulate 2 *If the signal $z_M(t)$ has no overlap between its components in the* frequency *domain, then a (single) slice of the WVT expressed by (142) along $f_1 = f_2 = -f_3 = \Omega$ and given by:*

$$W_{z_M}^{(4)}(t,\Omega) = \int_\nu [Z_M(\Omega+\frac{\nu}{4})]^2\, [Z_M^*(\Omega-\frac{\nu}{4})]^2 e^{j2\pi\nu t} d\nu \tag{146}$$

is free from "non-oscillating cross-terms".

We refer to this form of the WVT as to the *frequency-reduced form*.

Example 1. Consider the particular deterministic composite signal given by

$$z_2(t) = e^{j2\pi F_1 t} + e^{j2\pi F_2 t} \tag{147}$$

[$M = 2$, $a_i = 1$, $\Theta_i = 0$, $f_i(t) = F_i$ in (137)]. This is an example of a signal with non-overlapping frequency content. It is straightforward to show that:

$$\int_t W_{z_2}^{(4)}(t,f)dt = \delta(f-F_1) + \delta(f-F_2) + 4\delta[f-(F_1+F_2)/2]$$

where the third summand above is the non-oscillating cross-term. Smoothing the WVT $W_{z_M}^{(4)}(t,f)$ (in the t-f plane) cannot eliminate this cross-term. On the contrary, the frequency reduced form of the WVT yields:

$$\int_t W_{z_M}^{(4)}(t,\Omega)dt = \delta(\Omega-F_1) + \delta(\Omega-F_2)$$

which is the correct result. Smoothing $W_{z_M}^{(4)}(t,\Omega)$ is necessary to suppress the oscillating cross-terms. Fig.13 illustrates a similar example of two *linear FM* signals with non-overlapping components in frequency. Smoothing of the WVT is performed using an adaptive kernel based on the Radon transform [86]. As postulate 1 claims, only the t-f representation in Fig.13(b) allows an accurate description and analysis of the signal.

Example 2. Consider a composite signal given by

$$z_2'(t) = \delta(t-T_1) + \delta(t-T_2)$$

which is dual to (147). This is an example of a signal with non-overlapping content in the time domain. One can show that:

$$\int_f W_{z_2'}^{(4)}(t,f)df = \delta(t-T_1) + \delta(t-T_2)$$

The frequency reduced form of the WVT yields:

$$\int_\Omega W_{z_2'}^{(4)}(t,\Omega)d\Omega = \delta(t-T_1) + \delta(t-T_2) + 4\delta[t-(T_1+T_2)/2]$$

Fig.14 illustrates a similar example of two *linear FM* signals with non-overlapping components in the time domain. Smoothing of the WVT is performed by the same method as in Fig.13. As postulate 2 claims, only the t-f representation in Fig.14(a) allows an accurate description and analysis of the signal.

For general composite FM signals with possible time *and* frequency overlap it is necessary to initially perform an automatic segmentation of data[2] so that the problem is either reduced to the monocomponent case or to one of the two cases covered by the postulates stated above. The general case will appear elsewhere. Additional material related to multicomponent signals and PWVDs can be found in [5].

7 Conclusions

This chapter has presented a review of the important issues of time-frequency analysis, and an overview of recent advances based on multilinear representations.

It was shown in this chapter that the bilinear class of TFDs is suited only to the analysis of linear FM signals, i.e. for signals with a first order degree of non-stationarity. These bilinear TFDs, however, are not appropriate for the analysis of non-linear FM signals. For these signals, *Polynomial TFDs* have been proposed which are suitable for the analysis of such signals. In this chapter we have considered in particular, a sub-class of Polynomial TFDs, namely the Wigner-Ville trispectrum, which revealed to be a very efficient tool for the analysis of FM signals affected by multiplicative noise. The issue of multicomponent signal analysis using time-varying polyspectra has been briefly addressed.

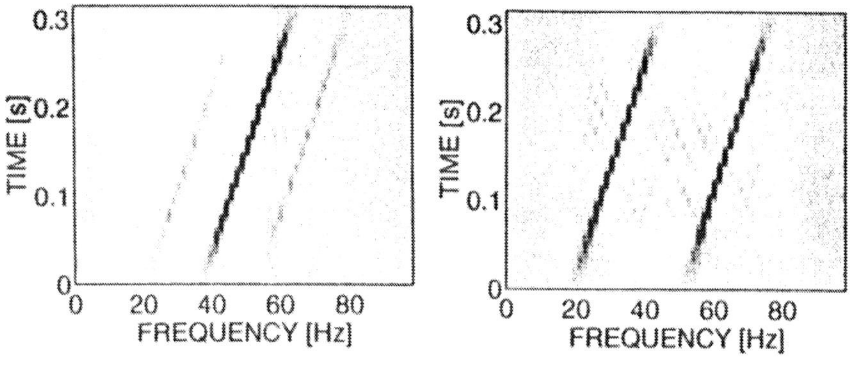

(a) (b)

Figure 13. Smoothed moment WVT in (a) the lag-reduced form; (b) the frequency-reduced form of a signal with two linear FMs with frequency non-overlapping content

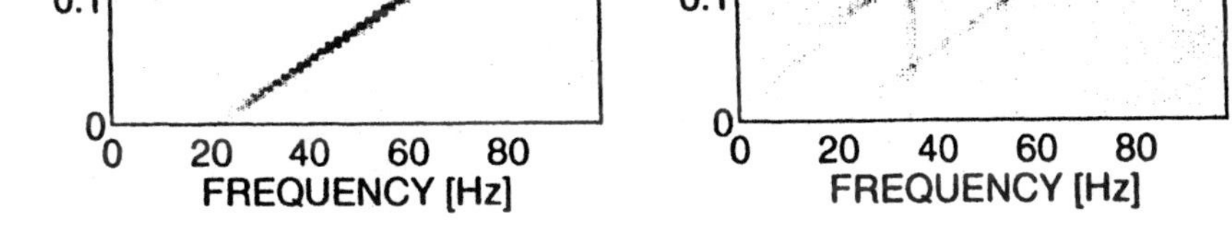

(a) (b)

Figure 14.Smoothed moment WVT in (a) the lag-reduced form; (b) the frequency-reduced form of a signal with two linear FMs with non-overlapping content in the time domain

Appendices

A Noise performance of the instantaneous frequency estimator for cubic FM signals

The following derivation was initiated by P. O'Shea [7], with some later modifications being done by B. Ristic and B. Boashash. Consider a constant amplitude second or third order polynomial FM signal, $z_s[n]$ embedded in white Gaussian noise. N samples of the observation are available, and the observed signal is given, after an amplitude normalisation, by:

$$z_r[n] = z_s[n] + z_w[n] = e^{j\phi[n]} + z_w[n] \tag{148}$$

where $\phi[n]$ is the time–varying phase function and $z_w[n]$ is complex white Gaussian noise of zero mean and variance $2\sigma^2$. Then the PWVD kernel defined in (98) is

$$\begin{aligned}
K_{z_r}[n,m] &= z_r^2[n+0.794m]\,(z_r^*[n-0.794m])^2\; z_r^*[n+m]z_r[n-m] \\
&= (z_s[n+0.794m]+z_w[n+0.794m])^2 \cdot \\
&\quad (z_s^*[n-0.794m]+z_w^*[n-0.794m])^2 \cdot \\
&\quad (z_s^*[n+m]+z_w^*[n+m])\cdot(z_s[n-m]+z_w[n-m]) \\
&= z_s^2[n+0.794m]\;(z_s^*[n-0.794m])^2\; z_s^*[n+m]\; z_s[n-m] \\
&\quad +z_s^2[n+0.794m]\;(z_s^*[n-0.794m])^2\; z_s^*[n+m]\; z_w[n-m] \\
&\quad +z_s^2[n+0.794m]\;(z_s^*[n-0.794m])^2\; z_s^*[n-m]\; z_w[n+m] \\
&\quad +2z_s[n+0.794m]\; z_w[n+0.794m]\;(z_s^*[n-0.794m])^2 \\
&\quad \cdot z_s[n-m]\; z_s^*[n+m] \\
&\quad +2z_s[n-0.794m]\; z_w[n-0.794m]\;(z_s^*[n+0.794m])^2 \\
&\quad \cdot z_s[n-m]\; z_s^*[n+m] \\
&\quad + \ldots\ldots \\
&\quad +z_w^2[n+0.794m]\;(z_w^*[n-0.794m])^2\; z_w^*[n+m]\; z_w[n-m]
\end{aligned} \tag{149}$$

The kernel expansion in (149) shows three types of terms. The first (on line 1) is due to the signal, the second (on line 2–6) is due to the cross–terms between signal and noise, and the third (on line 7) is due to the noise. The term due to the signal is simply the expression for the PWVD kernel of a noiseless complex exponential; this has been seen in Example 3. Since it has amplitude, A^6, it will have power A^{12}. The power of the term due to the noise (only) is given by

$$NP_{noise} = \mathcal{E}\left\{|z_w[n+0.794m]|^4 |z_w[n-0.794m]|^4\; |z_w[n+m]|^2\; |z_w[n-m]|^2\right\} \tag{150}$$

Since the noise is zero-mean Gaussian, it can be shown that the above expression reduces to:

$$NP_{noise} = \begin{cases} 256\sigma^{12} & \text{if } m \neq 0 \\ 46080\sigma^{12} & \text{if } m = 0 \end{cases} \tag{151}$$

that is, the power of noise is not stationary (with respect to the lag index m). Note that this noise is white for $m \neq 0$.

The power of the second type of terms in (149), due to the cross-components between signal and noise, can be expressed as:

$$NP_{cross-terms} = \begin{cases} 12A^{10}\sigma^2 + 68A^8\sigma^4 + 224A^6\sigma^6 + \dots & \text{if } m \neq 0 \\ 12A^{10}\sigma^2 + 120A^8\sigma^4 + 920A^6\sigma^6 + \dots & \text{if } m = 0 \end{cases} \tag{152}$$

At high SNRs ($A^2 >> \sigma^2$) the power of cross-terms reduces to

$$NP_{cross-terms} \approx 12A^{10}\sigma^2 \tag{153}$$

The total noise power in the kernel

$$NP_{kernel} = NP_{noise} + NP_{cross-terms} \tag{154}$$

at high SNRs reduces to:

$$NP_{kernel} \approx 12A^{10}\sigma^2 \tag{155}$$

Since the PWVD is the Fourier transform of the kernel, it is in fact the Fourier transform of a complex exponential of amplitude A^6, in white noise of variance given by (155). To determine the variance of the PWVD peak based IF estimate, one may follow the approach Rao and Taylor [10] [75] used for the conventional WVD estimate, that is, one can use the formula for the variance of the DFT peak based frequency estimate. This variance for white noise at high SNR, is given by

$$var_{DFT}(\hat{f}) = \frac{6}{(2\pi)^2(SNR)(N^2-1)} \tag{156}$$

where the "SNR" term in the above equation is the SNR in the DFT. Now since the PWVD kernel is conjugate symmetric, at most only $N/2$ samples can be independent. Thus the SNR in the PWVD spectrum at high SNR is

$$SNR_{PWVD} = \frac{A^{12}(N/2)}{12A^{10}\sigma^2} \tag{157}$$

[10] A small correction has to be made in eq.(4) of [75]. Namely, the noise of $z(n+k)z^*(n-k)$ term is expressed by σ_z^4 which is correct only for $k \neq 0$.

Substituting this expression for SNR in (156), and introducing a variance reduction factor of 0.85^2 to account for the overall 0.85 frequency axis scaling, the following result is obtained

$$var_{PWVD}(\hat{f}) = \frac{2 \cdot 6 \cdot 12A^{10}\sigma^2(0.85)^2}{(2\pi)^2A^{12}N(N^2-1)} = \frac{104.04\sigma^2}{(2\pi)^2A^2N(N^2-1)} \quad (158)$$

Now by comparison, the CR bound for a complex sinusoid in complex white Gaussian noise of variance, $2\sigma^2$, is given by:

$$var_{CR}(\hat{f}) = \frac{12\sigma^2}{(2\pi)^2A^2(N^2-1)} \quad (159)$$

It can be seen that the $PWVD_4$ based variance in (158) corresponds to 8.67 times the CR lower variance bound for estimating a sinusoid in white Gaussian noise. That is, it is approximately 9 dB higher than the stationary CR bound. Additionally, there will need to be some adjustment in the variance in (158) due to a small degree of interdependence between the $N/2$ samples of the half-kernel. Simulation has shown, however, that this adjustment is negligible.

Thus the variance for the PWVD peak based IF estimate for a fourth order polynomial phase law is seen to be 9 dB higher than the CR bound for estimating the frequency of a *stationary tone*. The problem of determining the CR bound on IF estimates for *polynomial phase signals* has been addressed in [69], [67]. For fourth order polynomial phase signals the CR bound may be shown to be 9 dB higher than for the stationary bounds, i.e exactly the same as the variance of the PWVD peak based IF estimate at high SNR. Thus the PWVD peak based IF estimator yields estimates which meet the CR bound.

Fig.5 shows the actual variance of the PWVD peak based IF estimator plotted against the CR bound for 64 points. The correspondence is seen to be quite close at high SNR.

The approximate SNR "threshold" for the PWVD based IF estimator may be determined by noting that for stationary frequency estimation, deterioration occurs when the SNR in the DFT is 15dB [84]. Thus the approximate threshold for the PWVD (at high SNR) is given by

$$\frac{A^{12}(N/2)}{12A^{10}\sigma^2} \approx 15dB \quad (160)$$

or

$$\frac{A^2(N)}{2\sigma^2} \approx 26dB \quad (161)$$

The threshold of 8 dB seen in Fig.5 is exactly as predicted by eq.(161).

B Group delay of the WVT

The local moment of the WVT with respect to time is by definition:

$$< t >_f = \frac{\int_{-\infty}^{\infty} t W_x^{(4)}(t,f)dt}{\int_{-\infty}^{\infty} W_x^{(4)}(t,f)dt} \tag{162}$$

Consider a deterministic signal, $x(t) = a(t)e^{j\phi(t)}$. Its Fourier transform can be represented by $X(f) = A(f)e^{j\varphi(f)}$. The WVT is defined as:

$$W_x^{(4)}(t,f) = \int_{-\infty}^{\infty} [x(t+\frac{\tau}{4})]^2 \, [x^*(t-\frac{\tau}{4})]^2 e^{-j2\pi f\tau} d\tau \tag{163}$$

If we substitute $y(t) = x^2(t)$ it is easy to show that

$$W_x^{(4)}(t,f) = 2\int_{-\infty}^{\infty} Y(2f+\frac{\theta}{2}) \, Y^*(2f-\frac{\theta}{2}) e^{j2\pi\theta t} d\theta \tag{164}$$

After several lines of mathematical manipulations based on the same derivation as for the WVD [9], one can show that:

$$< t >_f = -\frac{1}{2\pi}\Im m\left\{\frac{d}{df}\ln[X(2f) * X(2f)]\right\} \tag{165}$$

Now we can observe that the local moment in time of the WVT is equal to the *group delay* of the signal if and only if:

$$\arg\{X(2f) * X(2f)\} = \arg\{X(f)\} \tag{166}$$

The proof is given in [5]. Almost all practical signals do not satisfy condition (166).

C Properties of PWVDs

The PWVDs satisfy the following properties which were originally derived by B. Ristic, see [87]:

P-1. The PWVD is real for any signal $x(t)$:

$$\left[W_{\{x(t)\}}^{(k)}(t,f)\right]^* = W_{\{x(t)\}}^{(k)}(t,f) \tag{167}$$

Proof:

$$\begin{aligned}\left[W_{\{x(t)\}}^{(k)}(t,f)\right]^* &= \left[\int_{-\infty}^{\infty}\prod_{l=1}^{q/2}[x(t+c_l\tau)]^{b_l}\,[x^*(t+c_{-l}\tau)]^{-b_{-l}}\,e^{-j2\pi f\tau}d\tau\right]^* \\ &= \int_{-\infty}^{\infty}\prod_{l=1}^{q/2}[x^*(t+c_l\tau)]^{b_l}\,[x(t+c_{-l}\tau)]^{-b_{-l}}\,e^{+j2\pi f\tau}d\tau\end{aligned}$$

Substitution of τ by $-u$ yields:

$$\left[W^{(k)}_{\{x(t)\}}(t,f)\right]^* = -\int_{\infty}^{-\infty}\prod_{l=1}^{q/2}[x^*(t-c_l u)]^{b_l}\,[x(t-c_{-l}u)]^{-b_{-l}}\,e^{-j2\pi f u}\,du$$

Since coefficients b_i and c_i obey (100) and (101) we have:

$$\begin{aligned}\left[W^{(k)}_{\{x(t)\}}(t,f)\right]^* &= \int_{-\infty}^{\infty}\prod_{l=1}^{q/2}[x^*(t+c_{-l}u)]^{-b_{-l}}\,[x(t+c_l u)]^{b_l}\,e^{-j2\pi f u}\,du \\ &= W^{(k)}_{\{x(t)\}}(t,f) \quad \blacksquare \end{aligned} \tag{168}$$

P-2. The PWVD is an even function of frequency if $x(t)$ is real:

$$W^{(k)}_{\{x^*(t)\}}(t,-f) = W^{(k)}_{\{x(t)\}}(t,f) \tag{169}$$

Proof:

$$W^{(k)}_{\{x(t)\}}(t,-f) = \int_{-\infty}^{\infty}\prod_{l=1}^{q/2}[x(t+c_l\tau)]^{b_l}\,[x^*(t+c_{-l}\tau)]^{-b_{-l}}\,e^{-j2\pi(-f)\tau}d\tau$$

Substitution of τ by $-u$ yields:

$$\begin{aligned}W^{(k)}_{\{x(t)\}}(t,-f) &= -\int_{\infty}^{-\infty}\prod_{l=1}^{q/2}[x(t-c_l u)]^{b_l}\,[x^*(t-c_{-l}u)]^{-b_{-l}}\,e^{-j2\pi f u}du \\ &= W^{(k)}_{\{x(t)\}}(t,f)\end{aligned} \tag{170}$$

since coefficients b_i and c_i satisfy (100) and (101).$\blacksquare$

P-3. A shift in time by t_0 and in frequency by f_0 (i.e. modulation by $e^{j2\pi f_0 t}$) of signal $x(t)$ results in the same shift in time and frequency of the PWVD ($\forall t_0, f_0 \in R$):

$$W^{(k)}_{\{x(t-t_0)e^{j2\pi f_0(t-t_0)}\}}(t,f) = W^{(k)}_{\{x(t)\}}(t-t_0, f-f_0) \tag{171}$$

Proof:

$$W^{(k)}_{\{x(t-t_0)e^{j2\pi f_0(t-t_0)}\}}(t,f) =$$
$$\int_{-\infty}^{\infty}\prod_{l=1}^{q/2}[x(t-t_0+c_l\tau)]^{b_l}\,[x^*(t-t_0+c_{-l}\tau)]^{-b_{-l}}$$
$$\cdot\exp\{j2\pi f_0[t\cdot\textstyle\sum_{l=1}^{q/2}(b_l+b_{-l})+\tau\cdot\sum_{l=1}^{q/2}(c_l b_l+c_{-l}b_{-l})]\}\,e^{-j2\pi f\tau}d\tau$$

If

$$\sum_{l=1}^{q/2}(b_l+b_{-l}) = 0 \tag{172}$$

and

$$\sum_{l=1}^{q/2}(c_l b_l + c_{-l}b_{-l}) = 1 \tag{173}$$

then:

$$W^{(k)}_{\{x(t-t_0)e^{j2\pi f_0(t-t_0)}\}}(t,f) = W^{(k)}_{\{x(t)\}}(t-t_0, f-f_0) \tag{174}$$

Note that (172) and (173) are satisfied if conditions (100), (101) and (102) are valid.■

P-4. If $y(t) = w(t)z(t)$ then:

$$W^{(k)}_{\{y(t)\}}(t,f) = W^{(k)}_{\{w(t)\}}(t,f) *_f W^{(k)}_{\{z(t)\}}(t,f) \tag{175}$$

where $*_f$ denotes the convolution in frequency.

Proof:

$$\begin{aligned} K^{(k)}_{\{y(t)\}}(t,\tau) &= \prod_{l=1}^{q/2}[y(t+c_l\tau)]^{b_l}[y^*(t+c_{-l}\tau)]^{-b_{-l}} \\ &= K^{(k)}_{\{w(t)\}}(t,\tau)\cdot K^{(k)}_{\{z(t)\}}(t,\tau) \end{aligned} \tag{176}$$

Now, the FT of the PWVD kernel:

$$W^{(k)}_{\{y(t)\}}(t,f) = \underset{\tau\to f}{\mathcal{F}}\left[K^{(k)}_{\{w(t)\}}(t,\tau)\cdot K^{(k)}_{\{z(t)\}}(t,\tau)\right] \tag{177}$$

$$= W^{(k)}_{\{w(t)\}}(t,f) *_f W^{(k)}_{\{z(t)\}}(t,f) \quad ■ \tag{178}$$

P-5. Projection of the $W^{(k)}_{\{x(t)\}}(t,f)$ onto the time axis (time marginal):

$$\int_{-\infty}^{\infty} W^{(k)}_{\{x(t)\}}(t,f)df = |x(t)|^k \tag{179}$$

Proof:

$$\underset{f\to\tau}{\mathcal{F}^{-1}}\left[W^{(k)}_{\{x(t)\}}(t,f)\right] = \int_{-\infty}^{\infty} W^{(k)}_{\{x(t)\}}(t,f)e^{j2\pi f\tau}df \tag{180}$$

$$= K^{(k)}_{\{x(t)\}}(t,\tau) \tag{181}$$

For $\tau = 0$ it becomes:

$$\int_{-\infty}^{\infty} W^{(k)}_{\{x(t)\}}(t,f)df = K^{(k)}_{\{x(t)\}}(t,0) \tag{182}$$

$$= \prod_{l=1}^{q/2}[x(t)]^{b_l}\cdot[x^*(t)]^{-b_{-l}} \tag{183}$$

$$= |x(t)|^k \tag{184}$$

since coefficients b_l satisfy (100) and k is defined by (103).■

P-6. The local moment of the PWVD with respect to frequency gives the instantaneous frequency of the signal $x(t)$:

$$\frac{\int_{-\infty}^{\infty} f W_{\{x(t)\}}^{(k)}(t,f)df}{\int_{-\infty}^{\infty} W_{\{x(t)\}}^{(k)}(t,f)df} = \frac{1}{2\pi}\frac{d\phi(t)}{dt} \tag{185}$$

Proof: The local moment of the PWVD in the frequency is:

$$< f >_t = \frac{\int_{-\infty}^{\infty} f W_{\{x(t)\}}^{(k)}(t,f)df}{\int_{-\infty}^{\infty} W_{\{x(t)\}}^{(k)}(t,f)df} \tag{186}$$

$$= \frac{\frac{1}{2\pi j}\frac{\partial K_{\{x(t)\}}^{(k)}(t,\tau)}{\partial \tau}\Big|_{\tau=0}}{K_{\{x(t)\}}^{(k)}(t,\tau)\,|_{\tau=0}} \tag{187}$$

Since:

$$\frac{d\{\prod_{l=1}^{q/2}\Psi_l(\tau)\}}{d\tau} = \sum_{j=1}^{q/2}\left(\prod_{i=1,\, i\neq j}^{q/2}\Psi_i(\tau)\right)\cdot\frac{d\Psi_j(\tau)}{d\tau}$$

and assuming that coefficients b_i and c_i satisfy (100) and (101), it follows that:

$$\frac{\partial K_{\{x(t)\}}^{(k)}(t,\tau)}{\partial\tau}\Big|_{\tau=0} = [x'(t)x^*(t) - x(t)\,(x^*(t))']\cdot|x(t)|^{k-2}\cdot\sum_{j=1}^{q/2} c_j b_j \tag{188}$$

Thus we have:

$$< f >_t = \frac{1}{2}\frac{1}{2\pi j}\frac{x'(t)x^*(t) - x(t)\,(x^*(t))'}{x(t)\,x^*(t)} \tag{189}$$

The eq.(189) is identical to the corresponding one obtained for the conventional WVD [9]. Thus it is straightforward to show that:

$$< f >_t = \frac{1}{2\pi}\mathcal{I}m\left\{\frac{d}{dt}[\ln z(t)]\right\} \tag{190}$$

For a complex signal, $z(t) = A(t)e^{j\phi(t)}$, the average frequency of the PWVD at time instant, t, is:

$$< f >_t = \frac{1}{2\pi}\frac{d\phi(t)}{dt} \quad ■ \tag{191}$$

P-7. Time-frequency scaling: for $y(t) = \sqrt[k]{|a|}\cdot x(at)$

$$W^{(k)}_{\{y(t)\}}(t,f) = W^{(k)}_{\{x(t)\}}(at, \frac{f}{a}) \tag{192}$$

Proof:

$$W^{(k)}_{\{y(t)\}}(t,f) = \int_{-\infty}^{\infty} \prod_{l=1}^{q/2} [y(t+c_l\tau)]^{b_l} \, [y^*(t+c_{-l}\tau)]^{-b_{-l}} \, e^{-j2\pi f\tau} d\tau \tag{193}$$

$$= a \int_{-\infty}^{\infty} \prod_{l=1}^{q/2} [x(at+ac_l\tau)]^{b_l} \, [x^*(at+ac_{-l}\tau)]^{-b_{-l}} \, e^{-j2\pi f\tau} d\tau$$

Substitution of $a\tau$ by u yields:

$$W^{(k)}_{\{y(t)\}}(t,f) = \int_{-\infty}^{\infty} \prod_{l=1}^{q/2} [x(at+c_lu)]^{b_l} \, [x^*(at+c_{-l}u)]^{-b_{-l}} \, e^{-j2\pi \frac{f}{a} u} du$$

$$= W^{(k)}_{\{x(t)\}}(at, \frac{f}{a}) \quad \blacksquare \tag{194}$$

P-8. Finite time support: $W^{(k)}_{\{x(t)\}}(t,f) = 0$ for t outside $[t_1, t_2]$ if $x(t) = 0$ outside $[t_1, t_2]$.

Proof: Suppose $t < t_1$. Since coefficients c_l satisfy (101) we have,

$$W^{(k)}_{\{x(t)\}}(t,f) = \int_{-\infty}^{\infty} \prod_{l=1}^{q/2} [x(t+c_l\tau)]^{b_l} \, [x^*(t-c_l\tau)]^{-b_{-l}} \, e^{-j2\pi f\tau} d\tau \tag{195}$$

$$= \int_{-\infty}^{0} \prod_{l=1}^{q/2} [x(t+c_l\tau)]^{b_l} \, [x^*(t-c_l\tau)]^{-b_{-l}} \, e^{-j2\pi f\tau} d\tau$$

$$+ \int_{0}^{\infty} \prod_{l=1}^{q/2} [x(t+c_l\tau)]^{b_l} \, [x^*(t-c_l\tau)]^{-b_{-l}} \, e^{-j2\pi f\tau} d\tau$$

$$= I_1 + I_2 \tag{196}$$

Integral $I_1 = 0$ since $x(t+c_l\tau) = 0$; integral $I_2 = 0$ since $x^*(t - c_l\tau) = 0$. Therefore $W^{(k)}_{\{x(t)\}}(t,f) = 0$. Similarly, for $t > t_2$, it can be shown that $W^{(k)}_{\{x(t)\}}(t,f) = 0$. $\blacksquare$

References

[1] B. Boashash. *Methods and Applications of Time-Frequency Signal Analysis.* Longman Cheshire, ISBN No. 0-13-007444-6, Melbourne, Australia, 1991.

[2] B. Boashash. Time-frequency signal analysis. In S. Haykin, editor, *Advances in Spectral Estimation and Array Processing*, volume 1 of 2, chapter 9, pages 418–517. Prentice Hall, Englewood Cliffs, New Jersey, 1991.

[3] B. Boashash. Interpreting and estimating the instantaneous frequency of a signal – Part I: Fundamentals. *Proceedings of the IEEE*, pages 519–538, April 1992.

[4] B. Boashash. Interpreting and estimating the instantaneous frequency of a signal – Part II: Algorithms. *Proceedings of the IEEE*, pages 539–569, April 1992.

[5] B. Boashash and B. Ristich. Polynomial Wigner-Ville distributions and time-varying polyspectra. In B. Boashash, E. J. Powers, and A. M. Zoubir, editors, *Higher Order Statistical Signal Processing*. Longman Cheshire, Melbourne, Australia, 1993.

[6] G. Jones. *Time-frequency analysis and the analysis of multicomponent signals.* PhD Thesis, Queensland University of Technology, Australia, 1992.

[7] P.J. O'Shea. *Detection and estimation methods for non-stationary signals.* PhD Thesis, University of Queensland, 1991.

[8] B. Boashash, B. Escudie, and J. M. Komatitsch. Sur la possibilite d'utiliser la representation conjointe en temps et frequence dans l'analyse des signaux modules en frequence emis en vibrosismiques. In *7th Symposium on Signal Processing and its Applications*, pages 121–126, Nice, France, 1979. GRETSI. in French.

[9] T.A.C.M. Classen and W.F.G. Mecklenbrauker. The Wigner distribution - Part I. *Phillips Journal of Research*, 35:217–250, 1980.

[10] T.A.C.M. Classen and W.F.G. Mecklenbrauker. The Wigner distribution - Part II. *Phillips Journal of Research*, 35:276–300, 1980.

[11] T.A.C.M. Classen and W.F.G. Mecklenbrauker. The Wigner distribution - Part III. *Phillips Journal of Research*, 35:372–389, 1980.

[12] B. Boashash, P. Flandrin, B. Escudie, and J. Grea. Positivity of time-frequency distributions. *Compte Rendus Acad. des Sciences de Paris*, Series A(t288):307–309, January 1979.

[13] B. Boashash. Wigner analysis of time-varying signals - Its application in seismic prospecting. In *Proceedings of EUSIPCO*, pages 703–706, Nuernberg, West Germany, September 1983.

[14] D. Gabor. Theory of communication. *Journal of the IEE*, 93:429–457, 1946.

[15] R. Lerner. Representation of signals. In E. Baghdady, editor, *Lectures on Communications System Theory*, pages 203–242. McGraw-Hill, 1990.

[16] C. Helstrom. An expansion of a signal into gaussian elementary signals. *IEEE Trans. Information Theory*, 13:344–345, 1966.

[17] I. Daubeshies. The wavelet transform: A method for time-frequency localisation. In S. Haykin, editor, *Advances in Spectral Estimation and Array Processing*, volume 1 of 2. Prentice Hall, Englewood Cliffs, New Jersey, USA, 1990.

[18] C.H. Page. Instantaneous power spectra. *Journal of Applied Physics*, 23(1):103–106, 1953.

[19] C. Turner. On the concept of an instantaneous spectrum and its relation to the autocorrelation function. *Journal of Applied Puysics*, 25:1347–1351, 1954.

[20] M. Levin. Instantaneous spectra and ambiguity functions. *IEEE Transactions on Information Theory*, 13:95–97, 1967.

[21] A.W. Rihaczek. Signal energy distribution in time and frequency. *IEEE Transactions on Information Theory*, 14(3):369–374, 1968.

[22] J. Ville. Theorie et application de la notion de signal analytique. *Cables et Transmissions*, 2A(1):61–74, 1948.

[23] L. Cohen. Time-frequency distributions - A review. *Proceedings of the IEEE*, 77(7):941–981, July, 1989.

[24] B. Boashash. Note on the use of the Wigner distribution. *IEEE Transactions on Acoustics, Speech and Signal Processing*, 36(9):1518–1521, September 1988.

[25] E.P. Wigner. On the quantum correction for thermodynamic equilibrium. *Physics Review*, 40:748–759, 1932.

[26] B. Boashash. *Representation Temps-Frequence.* Dipl. de Docteur-Ingenieur these, University of Grenoble, France, 1982.

[27] B. Boashash. Note D'information sur la representation des signaux dans le domaine temps-frequence. Technical Report 135 81, Elf-Aquitaine Research Publication, 1981.

[28] B. Boashash. Representation conjointe en temps et en frequence des signaux d'energie finie. Technical Report 373 78, Elf-Aquitaine Research Publication, 1978.

[29] W. Martin. Time-frequency analysis of random signals. In *Proceedings of the IEEE International Conference on Acoustics, Speech and Signal Processing*, pages 1325–1328, Paris, France, April 1982.

[30] G.F. Boudreaux-Bartels. *Time-frequency signal processing algorithms: Analysis and synthesis using Wigner distributions.* PhD Thesis, Rice University, Houston, Texas, 1983.

[31] B. Boashash and P. J. Black. An efficient real-time implementation of the Wigner-Ville distribution. *IEEE Transactions on Acoustics, Speech and Signal Processing*, ASSP-35(11):1611–1618, November 1987.

[32] V.J. Kumar and C. Carroll. Performance of Wigner distribution function based detection methods. *Optical Engineering*, 23:732–737, 1984.

[33] S. Kay and G.F. Boudreaux-Bartels. On the optimality of the Wigner distribution for detection. In *Proceedings of the IEEE International Conference on Acoustics, Speech and Signal Processing*, pages 1263–1265, Tampa, Florida, USA, 1985.

[34] B. Boashash and F. Rodriguez. Recognition of time-varying signals in the time-frequency domain by means of the Wigner distribution. In *Proceedings of the IEEE International Conference on Acoustics, Speech and Signal Processing*, pages 22.5.1–22.5.4, San Diego, USA, April 1984.

[35] B. Boashash and P. J. O'Shea. A methodology for detection and classification of some underwater acoustic signals using time-frequency analysis techniques. *IEEE Transactions on Acoustics, Speech and Signal Processing*, 38(11):1829–1841, November 1990.

[36] B. Boashash and P. J. O'Shea. Signal detection and classification by time-frequency distributions. In B. Boashash, editor, *Methods and Applications of Time-Frequency Signal Analysis*, chapter 12. Longman Cheshire,, Melbourne, Australia, 1991.

[37] B. Boashash and H. J. Whitehouse. High resolution Wigner-Ville analysis. In *11th GRETSI Symposium on Signal Processing and its Applications*, pages 205–208, Nice, France, June 1987.

[38] H. J. Whitehouse, B. Boashash, and J. M. Speiser. High resolution processing techniques for temporal and spatial signals. In *High Resolution Techniques in Underwater Acoustics*. Springer-Verlag, 1990. Lecture Notes in Control and Information Science.

[39] N. Marinovic. *The Wigner distribution and the ambiguity function: generalisations, enhancement, compression and some applications*. PhD Thesis, City University of New York, 1986.

[40] A.J. Janssen. *Application of the Wigner distribution to harmonic analysis of generalised stochastic processes*. PhD Thesis, Amsterdam, 1990.

[41] M. Amin. Time-frequency spectrum analysis and estimation for non-stationary random processes. In B. Boashash, editor, *Methods and Applications of Time-Frequency Signal Analysis*, chapter 9. Longman Cheshire,, Melbourne, Australia, 1992.

[42] Y. Zhao, L.E. Atlas, and R.J. Marks II. The use of cone-shaped kernels for generalised time-frequency representation of non-stationary singals. *IEEE Trans. on Acoustics, Speech and Signal Processing*, 38(7), June 1990.

[43] I. Choi and W. Williams. Improved time-frequency representation of multicomponent signals using exponential kernels. *IEEE Transactions on Acoustics, Speech and Signal Processing*, 38(4):862–871, April 1990.

[44] P. Flandrin. Some features of time-frequency representations of multicomponent signals. In *Proceedings of the IEEE International Conference on Acoustics, Speech and Signal Processing*, pages 41B.1.4–41B.4.4, San Diego, USA, 1984.

[45] P. J. Kootsookos, B. C. Lovell, and B. Boashash. A unified approach to the STFT, TFD's and instantaneous frequency. *IEEE Transactions on Acoustics, Speech and Signal Processing*, August 1991.

[46] G. Jones and B. Boashash. Instantaneous quantities and uncertainty concepts for signal dependent time frequency distributions. In Franklin T. Luk, editor, *Advanced Signal Processing Algorithms, Architectures and Implementations*, San Diego, USA, July 1991. Proceedings of SPIE.

[47] H.H. Szu. Two-dimensional optical processing of one-dimensional acoustic data. *Optical Engineering*, 21(5):804–813, September/October 1982.

[48] P. J. Boles and B. Boashash. Application of the cross Wigner-Ville distribution to seismic surveying. In B. Boashash, editor, *Methods and Applications of Time-Frequency Signal Analysis*, chapter 20. Longman Cheshire, Melbourne, Australia, 1992.

[49] D. L. Jones and T. W. Parks. A high resolution data-adaptive time-frequency representaion. *IEEE Transactions on Acoustics, Speech and Signal Processing*, 38(12):2127–2135, December 1990.

[50] J. Bertrand and P. Bertrand. Time-frequency representations of broad-band signals. In *Proc. of Intern. Conf. on Acoust. Speech and Signal Processing*, pages 2196–2199, New York, USA, 1988.

[51] O. Rioul and P. Flandrin. Time-scale energy distributions: a general class extending wavelet transforms. *IEEE Transactions on Acoustics, Speech and Signal Processing*, pages 1746–1757, July 1992.

[52] R. Altes. Detection, estimation and classification with spectrograms. *Journal of the Acoustical Society of America*, 67:1232–1246, 1980.

[53] T.E. Posch. Kernels, wavelets and time-frequency distributions. *IEEE Transactions on Information Theory.* submitted.

[54] L. Cohen. Generalised phase space distributions. *Journal of Mathematical Physics*, 7:181–186, 1967.

[55] B. Boashash and A. P. Reilly. Algorithms for time-frequency signal analysis. In B. Boashash, editor, *Methods and Applications of Time-Frequency Signal Analysis*, chapter 7. Longman Cheshire,, Melbourne, Australia, 1991.

[56] B. Boashash and P. O'Shea. Time-varying higher order spectra. In Franklin T. Luk, editor, *Advanced Signal Processing Algorithms, Architectures and Implementations*, San Diego, USA, July 1991. Proceedings of SPIE.

[57] B. Boashash and P. J. O'Shea. Polynomial Wigner-Ville distributions and their relationship with time-varying higher spectra. *IEEE Transactions on Acoustics, Speech and Signal Processing*, 1993.

[58] B. Boashash and B. Ristich. Time-varying higher order spectra and the reduced Wigner trispectrum. In Franklin T. Luk, editor, *Advanced Signal Processing Algorithms, Architectures and Implementations*, volume 1770, pages 268–280, San Diego, USA, July 1992. Proceedings of SPIE.

[59] B. Boashash and B. Ristich. Analysis of FM signals affected by gaussian AM using the reduced Wigner-Ville trispectrum. In *Proc. of the Intern. Conf. Acoustic, Speech and Signal Processing*, Mineapolis, Minnesota, April, 1993.

[60] J.F. Randolf. *Basic Real and Abstract Analysis.* Academic Press, New York, 1968.

[61] S. M. Kay. *Modern Spectral Estimation: Theory and Application.* Prentice Hall, Englewood Cliffs, New Jersey, USA, 1987.

[62] R. Altes. Sonar for generalized target description and its similarity to animal echolocation systems. *J. Acoust. Soc. Am.*, 59(1):97–105, Jan. 1976.

[63] A.W. Rihaczek. *Principles of High-Resolution Radar.* Pensinsula Publishing, Los Altos, 1985.

[64] A. Dziewonski, S. Bloch, and M. Landisman. A technique for the anlysis of the transient signals. *Bull. Seismolog. Soc. Am.*, pages 427–449, Feb. 1969.

[65] B. Ferguson. A ground based narrow-band passive acoustic technique for estimating the altitude and speed of a propeler driven aircraft. *J. Acoust. Soc. Am.*, 92(3), September 1992.

[66] S. Peleg and B. Porat. Estimation and classification of polynomial phase signals. *IEEE Trans. Information Theory*, 37:422–429, March 1991.

[67] S. Peleg and Porat. The Cramer-Rao lower bound for signals with constant amplitude and polynomial phase. *IEEE Transactions on Signal Processing*, 39(3):749–752, March 1991.

[68] Z. Faraj and F. Castanie. Polynomial phase signal estimation. In *Signal Processing IV: Theories and Applications*, pages 795–798, Proc. EUSIPCO-92, August 1992.

[69] B. Boashash, P. J. O'Shea, and M. J. Arnold. Algorithms for instantaneous frequency estimation: A comparative study. In Franklin T. Luk, editor, *Advanced Signal Processing Algorithms, Architectures and Implementations*, pages 24–46, San Diego, USA, August 1990. Proceedings of SPIE 1348.

[70] M. Arnold. *Estimation of the instantaneous parameters of a signal.* Thesis, University of Queensland, Australia, 1991.

[71] T. Soderstrom. On the design of digital differentiating filters. Technical Report UPTEC 8017, University of Technology, Uppsala University, March 1980.

[72] M. J. Arnold and B. Boashash. The generalised theory of phase difference estimators. *IEEE Tran. Signal Processing*, 1993. Submitted.

[73] B. Boashash and B. Ristic. Time varying higher order spectra. In *Proceedings of 25th Asilomar Conference*, Pacific Grove, California, Nov. 1991.

[74] B. Boashash and B. Ristich. Application of cumulant tvhos to the analysis of composite fm signals in multiplicative and additive noise. In Franklin T. Luk, editor, *Advanced Signal Processing Algorithms, Architectures and Implementations*, San Diego, USA, July 1993. Proceedings of SPIE.

[75] P. Rao and F. J. Taylor. Estimation of instantaneous frequency using the Wigner distribution. *Electronics Letters*, 26:246–248, 1990.

[76] A. Dandawate and G. B. Giannakis. Consistent kth order time-frequency representations for (almost) cyclostationary processes. In *Proc. Ann. Conference on Information Sciences and Systems*, pages 976–984, Johns Hopkins University, March, 1991.

[77] J. R. Fonollosa and C. L. Nikias. General class of time-frequency higher-order spectra: Definitions, properties, computation and applications to transient signals. In *Proc. Int. Signal Processing Workshop on Higher-Order Statistics*, pages 132–135, Chamrousse, France, July 1991.

[78] A. Swami. Third-order Wigner distributions: definitions and properties. In *Proc. Int. Conf Acoustic, Speech and Signal Processing (ICASSP)*, pages 3081–4, Toronto, Canada, May, 1991.

[79] P. O. Amblard and J. L. Lacoume. Construction of fourth-order Cohen's class: A deductive approach. In *Proc. Int. Symp. Time-Frequency Time-Scale Analysis*, Victoria, Canada, October 1992.

[80] R. F. Dwyer. Fourth-order spectra of Gaussian amplitude-modulated sinusoids. *J. Acoust. Soc. Am.*, pages 919–926, August 1991.

[81] J.R. Fonollosa and C.L.Nikias. Analysis of transient signals using higher-order time- frequency distributions. In *Proc. Int. Conf. Acoustic, Speech and Signal Processing (ICASSP)*, pages V–197 - V–200, San Francisco, March, 1992.

[82] H. Van Trees. *Detection, Estimation and Modulation Theory: Part III.* John Wiley, New York, 1971.

[83] B. Boashash, P. J. O'Shea, and B. Ristic. A statistical/computational comparison of some algorithms for instantaneous frequency estimation. In *ICASSP*, Toronto, May, 1991.

[84] D.C. Rife and R.R. Boorstyn. Single tone parameter estimation from discrete-time observations. *IEEE Transactions on Information Theory*, 20(5):591–598, 1974.

[85] F. Hlawatsch and G. F. Boudreaux-Bartels. Linear and quadratic time-frequency signal representations. *IEEE Signal Processing Magazine*, 9(2):21–67, April 1991.

[86] B. Ristic and B. Boashash. Kernel design for time-frequency analysis using Radon transform. *IEEE Transactions on Signal Processing*, 41(5):1996–2008, May 1993.

[87] B. Ristic. *Adaptive and higher-order time-frequency analysis methods for nonstationary signals.* PhD Thesis, Queensland University of Technology, Australia, (to appear).

Fundamentals of Higher-Order s-to-z Mapping Functions and their Application to Digital Signal Processing

Dale Groutage
David Taylor Research Center
Detachment Puget Sound, Bremerton, WA 98314-5215

Alan M. Schneider
Department of Applied Mechanics and Engineering Sciences
University of California at San Diego, La Jolla, CA 92093-0411

John Tadashi Kaneshige
Mantech NSI Technology Services Corp., Sunnyvale, CA 94089

Abstract - The principal advantage of using higher-order mapping functions is increased accuracy in digitizing linear, time-invariant, continuous-time filters for real-time applications. A family of higher-order numerical integration formulas and their corresponding s-to-z mapping functions are presented. Two of the main problems are stability and handling discontinuous inputs. The stability question is resolved by analyzing the stability regions of the mapping functions. Sources of error in the accuracy of the output of digitized filters relative to their continuous-time counterparts are explored. Techniques for digitizing continuous-time filters, using the mapping functions, are developed for reducing different sources of error, including error resulting from discontinuous inputs. Performance improvement of digital filters derived from higher-order s-to-z mapping functions, as compared to those derived from linear mapping functions, is demonstrated through the use of examples. Analysis to demonstrate improvement is carried out in both the time and frequency domains.

Based on "Higher-Order s-to-z Mapping Functions and Their Application in Digitizing Continuous-Time Filters" by A. M. Schneider, J. T. Kaneshige, and F. D. Groutage which appeared in Proceedings of the IEEE, Vol. 79, No. 11, pp. 1661-1674; Nov. 1991.

I. INTRODUCTION

Higher-order s-to-z mapping functions were derived in Schneider *et al.*, [1]. This paper outlines the derivation of these higher-order mapping functions and demonstrates their performance improvement over linear mapping functions in both the time and frequency domains. A Figure-of-Merit is developed which provides a measure of the comparative performance between two digital filters derived using different mapping functions. A Figure-of-Merit for evaluating comparative performance between digital filters is applied to frequency-domain analysis data.

The preferred classical technique for converting a linear, time-invariant, continuous-time filter to a discrete-time filter with a fixed sample-time is through the use of the so-called linear s-to-z mapping functions. Of the linear mapping functions, the most popular is the bilinear transformation, also known as Tustin's rule. Derived from trapezoidal integration, it converts a transfer function F(s) in the s-domain to another transfer function $F_D(z)$ in the z-domain by the mapping function

$$s = \frac{2}{T} \cdot \frac{(z-1)}{(z+1)} = f(z) \quad , \tag{1}$$

where T is the time interval between samples of the discrete-time system. The procedure consists of using the mapping function to replace every s in F(s) by the function of z, to obtain $F_D(z)$.

Until now, higher-order mapping from the s-domain to the z-domain was not practical because of the stability limitations associated with conventional mapping functions of higher-order numerical integration methods. This is pointed out by Kuo [2] who states:

> "In general, higher-order and more complex numerical integration methods such as the Simpson's rules are available. However, these schemes usually generate higher-order transfer functions which cause serious stability problems in the simulation models. Therefore, in control system applications, these higher-order integration methods are seldom used."

The essence of Schneider *et al.*, [1] was the derivation of higher-order mapping functions that do not suffer a stability problem.

In a typical real-time application, a continuous-time signal u(t) enters the continuous-time filter F(s) and produces the continuous-time output y(t), as shown in the top path of Fig. 1. If the system design calls for digital implementation, then the discrete-time "equivalent" filter $F_D(z)$ must be found, where the subscript D stands for "digital". The input u(t) is sampled every T seconds, producing $u(kT) = u_k$, the kth sample of which is processed at time $kT = t_k$ by $F_D(z)$, to produce the output sample $y(kT) = y_k$ as in the bottom path of Fig. 1. It should be clearly understood that $F_D(z)$ is not the z-transform of F(s);

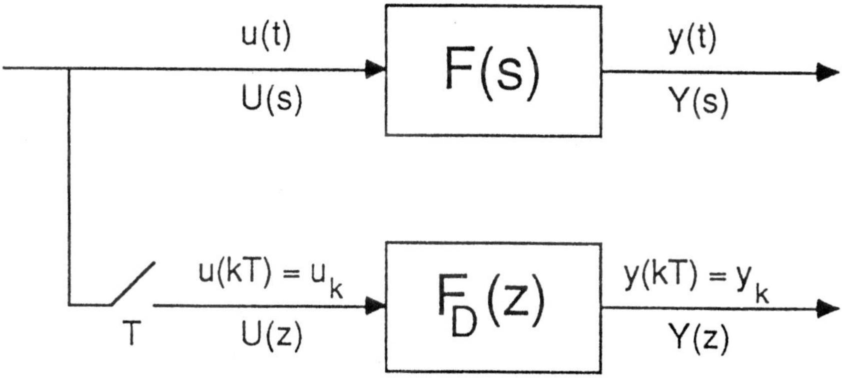

Fig. 1 A continuous-time filter F(s) and its digital "equivalent" $F_D(z)$.

rather, it is the pulse transfer function of the discrete-time filter which is intended to reproduce, as closely as possible, the behavior of the continuous-time filter F(s). The accuracy of the approximation is dependent on how F(s) is converted to $F_D(z)$, and how frequently the input samples arrive. References [3] and [4] present Groutage's algorithm, a general method for automating the transformation from F(s) to $F_D(z)$. In [4], Groutage further suggested that improved accuracy in the response of the digital filter could be obtained by using higher-order s-to-z mapping functions. This is an analysis and development of that suggestion [5].

II. MAPPING FUNCTIONS

The continuous-time filters which are considered are described by linear, constant-coefficient ordinary differential equations with time as the running variable, subject to an independent input u(t). The general transfer function representation is of the form

$$F(s) = \frac{Y(s)}{U(s)} = \frac{B_0 s^m + B_1 s^{m-1} + \ldots + B_m}{A_0 s^n + A_1 s^{n-1} + \ldots + A_n} \qquad m < n. \tag{2}$$

The state-variable representation of this continuous-time filter is

$$\dot{\underline{x}}(t) = \mathbf{A}\underline{x}(t) + \underline{b}u(t) \tag{3a}$$

$$y(t) = \underline{c}^T \underline{x}(t), \tag{3b}$$

and can be represented, for example, in phase-variable form ,where A_0 in (2) must be normalized to unity, and where

$$\mathbf{A} = \begin{bmatrix} 0 & 1 & 0 & \cdots & 0 & 0 \\ 0 & 0 & 1 & \cdots & 0 & 0 \\ \vdots & \vdots & \vdots & \ddots & \vdots & \vdots \\ 0 & 0 & 0 & \cdots & 1 & 0 \\ 0 & 0 & 0 & \cdots & 0 & 1 \\ -A_n & -A_{n-1} & -A_{n-2} & \cdots & -A_2 & -A_1 \end{bmatrix} \tag{4a}$$

$$\underline{b}^T = [\,0\,0\,0 \cdots 0\,1\,] \tag{4b}$$

$$\underline{c}^T = [\,B_m\,B_{m-1}\,\cdots\,B_0\,0\cdots 0\,]. \tag{4c}$$

The corresponding elementary block diagram of this filter is displayed in Fig. 2, and demonstrates one way by which an nth-order filter can be implemented using n integrators [6].

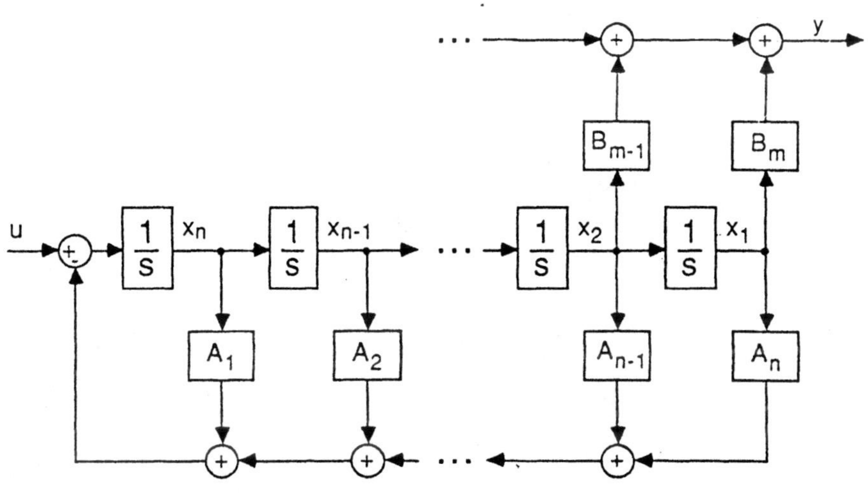

Fig. 2 An nth-order filter implemented with n integrators.

In time-domain analysis, the continuous-time integrator, $\frac{1}{s}$, can be approximated by different numerical integration formulas. Thus numerical integration provides a natural method for digitizing continuous-time filters. In frequency-domain analysis, the corresponding mapping functions can be used to map s-domain filters into z-domain filters; the resulting discrete-time filters can be implemented in the time-domain through difference equations.

For stability reasons to be outlined later, the Adams-Moulton family of numerical integration formulas (5a) - (5e) was chosen, from which we generate the corresponding family of mapping functions (6a) - (6e).

Order of Integration	Adams-Moulton Numerical Integration Formulas	
2	$x_k = x_{k-1} + \frac{T}{2} \cdot (\dot{x}_k + \dot{x}_{k-1})$	(5a)
3	$x_k = x_{k-1} + \frac{T}{12} \cdot (5\dot{x}_k + 8\dot{x}_{k-1} - \dot{x}_{k-2})$	(5b)
4	$x_k = x_{k-1} + \frac{T}{24} \cdot (9\dot{x}_k + 19\dot{x}_{k-1} - 5\dot{x}_{k-2} + \dot{x}_{k-3})$	(5c)
5	$x_k = x_{k-1} + \frac{T}{720} \cdot (251\dot{x}_k + 646\dot{x}_{k-1} - 264\dot{x}_{k-2} + 106\dot{x}_{k-3} - 19\dot{x}_{k-4})$	(5d)
6	$x_k = x_{k-1} + \frac{T}{1440} \cdot (475\dot{x}_k + 1427\dot{x}_{k-1} - 798\dot{x}_{k-2} + 482\dot{x}_{k-3} - 173\dot{x}_{k-4} + 27\dot{x}_{k-5})$	(5e)

Order of Mapping Function	Corresponding Mapping Functions	
1	$s = \frac{2}{T} \cdot \frac{z-1}{z+1}$	(6a)
2	$s = \frac{12}{T} \cdot \frac{z^2 - z}{5z^2 + 8z - 1}$	(6b)
3	$s = \frac{24}{T} \cdot \frac{z^3 - z^2}{9z^3 + 19z^2 - 5z + 1}$	(6c)
4	$s = \frac{720}{T} \cdot \frac{z^4 - z^3}{251z^4 + 646z^3 - 264z^2 + 106z - 19}$	(6d)
5	$s = \frac{1440}{T} \cdot \left(\frac{z^5 - z^4}{475z^5 + 1427z^4 - 798z^3 + 482z^2 - 173z + 27} \right)$	(6e)

Note that the notation of the Adams-Moulton formulas for numerical solution of differential equations has been changed from that appearing in [7] to apply to the input and output of an integrator, thereby turning these into formulas for numerical integration. The indices have also been lowered by one to correspond with the engineering practice of calculating the present output given the present input. We define the order of the mapping function to be the number of its z-plane poles. The second-order numerical integration formula, (5a), represents trapezoidal integration, which corresponds to the first-order mapping function, (6a), Tustin's rule.

The unification of this family originates from the fact that all of these formulas approximate the incremental area, $\Delta x_k = x_k - x_{k-1}$, under the derivative curve, $\dot{x}(t)$, over the single time-interval, t_{k-1} to t_k. Eq. (5a) approximates the derivative curve by fitting a straight line to the two derivative values $\dot{x}_{k-1}$ and $\dot{x}_k$, (5b) fits a parabola to the three derivative values $\dot{x}_{k-2}$, $\dot{x}_{k-1}$, and $\dot{x}_k$, (5c) fits a cubic to the four derivative values $\dot{x}_{k-3}$, $\dot{x}_{k-2}$, $\dot{x}_{k-1}$, and $\dot{x}_k$, and so forth. A particular Adams-Moulton numerical integration formula is said to be of order p if the one-step, or local, truncation error is $O(T^{p+1})$ [8]. Thus, the third-order integration formula (5b) has a one-step truncation error of $O(T^4)$.

The digitized state-variable equations,

$$\underline{x}_k = f(\underline{x}_{k-1}, \dot{\underline{x}}_i) \quad \text{for } i = k - r\cdot n, \ldots, k \tag{7a}$$

where

f = linear function corresponding to a particular mapping function (7b)

$$\dot{\underline{x}}_i = \mathbf{A}\underline{x}_i + \underline{b}u \tag{7c}$$

$$y_i = \underline{c}^T \underline{x} \tag{7d}$$

r = order of the mapping function (7e)

n = order of the continuous-time filter, (7f)

are equivalent to the discrete-time filter,

$$F_D(z) = \frac{B'_0 z^{r \cdot n} + B'_1 z^{r \cdot n - 1} + \ldots + B'_{r \cdot n}}{A'_0 z^{r \cdot n} + A'_1 z^{r \cdot n - 1} + \ldots + A'_{r \cdot n}}. \tag{8}$$

Eq. (7c) demonstrates how the state-derivatives, $\dot{\underline{x}}_k$, are dependent upon the states, $\underline{x}_k$. Therefore $\underline{x}_k$ appears on both sides of equation (7a), i.e. implicitly on the right side. However one can still solve for $\underline{x}_k$, when limited to linear constant-coefficient differential equations. This implies that predictor-corrector pairs, typically used when numerically solving nonlinear or time-varying differential equations, are not necessary; our mapping functions use the Adams-Moulton family of interpolation formulas, ("correctors"), but not the Adams-Bashford extrapolation formulas, ("predictors").

The parabolic and cubic numerical integration formulas, (5b) and (5c), are used to generate Schneider's rule and the Schneider-Kaneshige-Groutage (SKG) rule, (6b) and (6c), which are derived in Appendix A. These are the principal higher-order mapping functions whose stability properties and applications are analyzed throughout this paper. Extension of the principles described herein, to the higher-order mapping functions (6d), (6e), and so forth, is straight-forward.

While methods of numerical integration are inherently applicable in the time domain, mapping functions can be analyzed in the frequency domain. Mapping functions can be viewed as frequency-domain descriptions of numerical integration formulas. As a result of performing frequency-domain analysis, one can determine the stability of the resulting discrete-time filter by analyzing the z-plane map of the s-plane poles.

Simpson's numerical integration formula,

$$x_k = x_{k-2} + \frac{T}{3} \cdot (\dot{x}_k + 4\dot{x}_{k-1} + \dot{x}_{k-2}), \tag{9}$$

approximates the area under the derivative curve over two time-steps. When analyzed in the frequency-domain, it is easily shown (Appendix B) that the corresponding Simpson's rule,

$$s = \frac{3}{T} \cdot \frac{z^2 - 1}{z^2 + 4z + 1}, \tag{10}$$

maps any stable filter s-plane pole to an unstable z-plane pole, and is dropped from further consideration.

The Runge-Kutta method is commonly used to approximate integration. An advantage of this method is its self-starting characteristic. However, the intermediate time-steps and the pseudo-iterative calculations inherent in this method, result in its being limited to the time-domain.

III. STABILITY REGIONS

The stability region of a mapping function refers to the portion of the left-half s-plane which maps into the interior of the unit circle in the z-plane. A stable filter s-plane pole inside the stability region will map into r stable z-plane poles, where r is the order of the mapping function. For $F_D(z)$ to be stable, all of its poles must lie inside the unit circle of the complex z-plane. It has been shown [9] that the bilinear transformation, or Tustin's rule (6a), maps the primary strip of the s-plane into the unit circle of the z-plane. The primary strip is the portion of the left-half of the s-plane which extends from $j\omega = -j\frac{\omega_s}{2}$ to

$j\frac{\omega_s}{2}$. The s-plane can be divided into an infinite number of these periodic strips, each of which is mapped into the interior of the unit circle of the z-plane by Tustin's rule. Thus Tustin's rule has a stability region which covers the entire left-half of the s-plane; any stable continuous-time filter, F(s), will be converted into a stable discrete-time filter, $F_D(z)$, using this mapping function.

Since higher-order mapping functions are derived from higher-order numerical integration formulas, it seems logical that higher-order mapping functions would result in greater accuracy. Stability comes into question since the stability regions of the higher-order mapping functions, including those generated by the Adams-Moulton numerical integration formulas, do not contain the entire left-half of the s-plane. However since the stability regions are well defined, stability can easily be determined for a particular filter at various sampling frequencies.

The s-plane stability regions of the mapping functions are found to be inversely proportional to the sampling time, T. Figure 3 shows an s-plane which has been non-dimensionalized by the sampling period T, so that the axes are σT and $j\omega T$. Non-dimensionalizing permits a single curve to represent the stability boundary for each higher-order rule.* Figure 3 contains the boundaries of the stability regions for Schneider's rule and for the SKG rule. Since these plots are symmetrical about the real axis, only the top half is shown. The boundaries of the stability regions were calculated by setting z equal to points

---

* In the usual s-plane, where $s = \sigma + j\omega$, the stability boundary for a given higher-order rule changes in size inversely proportional to T; if T is halved, then the size of the stability boundary doubles. It would be cumbersome to replot the irregularly-shaped Schneider or SKG stability boundaries for each new value of T. It is easier to change the scale on the axes of the s-plane, and to replot the poles of F(s). That is why the s-plane of Fig. 3 has been non-dimensionalized.

along the unit circle of the z-plane and solving for s, using (6b) and (6c) respectively.

There are two methods for determining stability. The first is graphical. It uses the stability boundaries of Fig. 3. If, for a given sampling period T, all of the poles of F(s) lie inside the heavy solid contour of Fig. 3, then the $F_D(z)$ obtained by Schneider's rule will be stable for that value of T. A similar statement holds for the SKG rule, using the heavy dashed curve of Fig. 3.

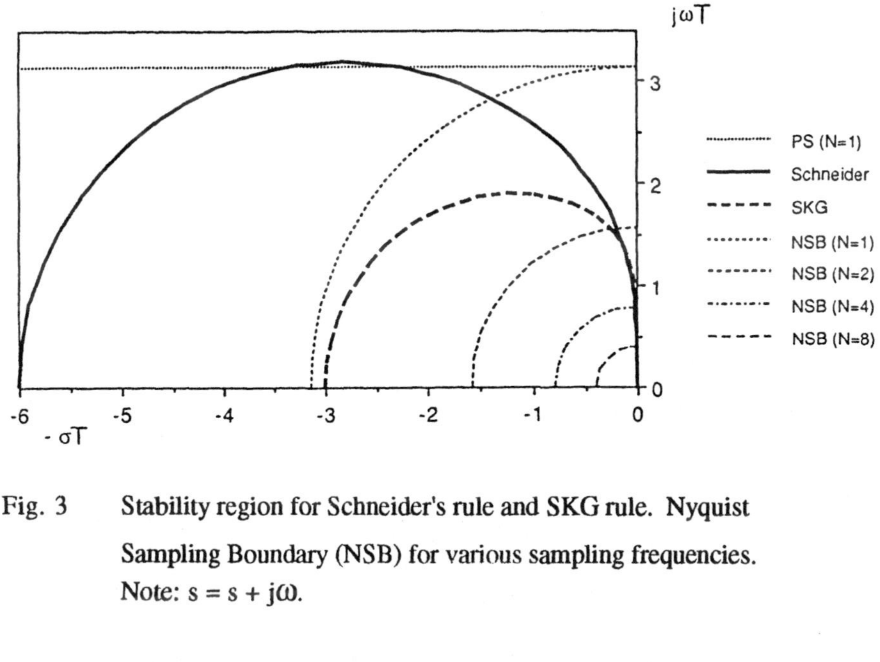

Fig. 3 Stability region for Schneider's rule and SKG rule. Nyquist Sampling Boundary (NSB) for various sampling frequencies. Note: $s = s + j\omega$.

For a stable filter F(s), there typically is a maximum sample time for which $F_D(z)$ will be stable, using a selected higher-order rule. For any sample time less than this maximum, the resulting $F_D(z)$ will be stable. As an example, suppose one wishes to convert the filter $F(s) = \dfrac{80}{(s+8)^2+4^2} = \dfrac{80}{s^2+16s+80}$ to discrete form by Schneider's rule.

The poles of F(s) are at $s = -8 \pm j4$. Select $T = 1$ as a trial sampling period.

Then the upper-half plane pole of F(s) plots onto the non-dimensional axes of Fig. 3 at (-8 + j4). This point lies outside the stability boundary for Schneider's rule – the resulting $F_D(z)$ will be unstable. Now change the sample time to T = 1/2. The upper-half-plane pole of F(s) now plots onto the axes of Fig. 3 at (-4 + j2). This is inside the stability boundary – the $F_D(z)$ produced by Schneider's rule will be stable. It will also be stable for any T less than 1/2 second.

The second method for determining stability is analytic. It is particularly applicable in situations where the pole locations are close to the stability boundary curves of Fig. 3, where it may be difficult to assess stability graphically. Define a single complex pole (for physical systems, complex poles occur in conjugate pairs) of F(s) as follows:

$$CP(s) = \frac{1}{s + a + jb} \tag{11}$$

This complex pole is mapped to the z-plane domain by a particular mapping function f(z). Thus, the complex poles in the z-plane are defined by the expression

$$CP(z) = \frac{1}{f(z) + a + jb} \tag{12}$$

where we note that higher-order mapping functions map a single s-plane pole into multiple z-plane poles. If all of the poles of CP(z) lie within the unit circle, the mapping function f(z) is a stable mapping function.

As an example, consider the higher-order mapping function

$$f(z) = \frac{12}{T} \frac{z^2 - z}{5z^2 + 8z - 1} \tag{13}$$

Thus, for a pole located at $s = a + jb$

$$CP(z) = \frac{A}{z^2 + Bz + C} \tag{14}$$

where

$$A = \frac{T(5z^2 + 8z - 1)}{12 + 5T(a + jb)}$$

$$B = \frac{-12 + 8T(a + jb)}{12 + 5T(a + jb)}$$

and

$$C = \frac{-T(a + jb)}{12 + 5T(a + jb)}$$

Note that A, B, and C will be complex numbers. The poles of CP(z), equal to the zeroes of $z^2 + Bz + C$, determine stability.

To apply this technique to a specific s-plane transfer function, F(s), the following steps would be carried out.

1. Factor the denominator of F(s).
2. Isolate the pole or pair of complex-conjugate poles that are most likely to result in an unstable z-plane pole. One can start with that real or pair of complex poles located furthermost from the $j\omega$ axis.
3. Map this pole (either a real or one of the complex-conjugate pair) to the z-plane via a specific higher-order mapping function f(z).

4. Determine the location of CP(z). Note that for a real pole, CP(z) is evaluated with b = 0 in the above equations for A,B, and C.

The numerical evaluation of equation (14) can be accomplished quite easily using the MATLAB R computer program, which has a routine "roots" for solving polynomial equations with complex coefficients.*

5. Repeat steps 1-4 above for any other poles of F(s) which may possibly lead to unstable z-plane poles, such as any lying in a narrow wedge along the $j\omega$ -axis.

6. The mapping function f(z) and selected sampling period T lead to an unstable $F_D(z)$ if any poles tested lie outside the unit circle of the z-plane. In this case, make T smaller and repeat the test.

Figure 3 also presents what we call the Nyquist Stability Boundary for several values of N, N being the ratio of the actual sampling frequency to the minimum sampling frequency satisfying the Nyquist Sampling Criteria. These curves are not needed to determine the stability of a given filter; the point of these curves is to provide an intuitive grasp of the interrelationship between typical sampling frequencies used in practice and the stability boundaries. Specifically, the point of these curves is to show, for sampling frequencies which satisfy the Nyquist Sampling Criterion with a margin N, where N is typically 2,4, or more, that almost all continuous-time filters F(s) converted by either Schneider's rule or the SKG rule will lead to a stable $F_D(z)$. The possible exceptions are those F(s) filters with poles lying close to the $j\omega$ - axis, i.e. having a very low damping ratio.

---

* MATLAB is a trademark of the Math Works, Inc.

We now define these terms. The Nyquist Sampling Criterion states that, in order to avoid aliasing,

$$\omega_s \geq 2 \, . \, \omega_N, \tag{15}$$

where ω_S is the sampling frequency,

$$\omega_s = \frac{2\pi}{T}, \tag{16}$$

and ω_N is the Nyquist frequency, which is the highest frequency appearing in the sampled signal. When digitizing continuous-time filters, we redefine ω_N a s follows.

$$\omega_N \equiv \max\left\{\omega_{n(fastest)}, \omega_{0(fastest)}\right\}, \tag{17}$$

where

$\omega_{n(fastest)}$ = the highest frequency in the filter F(s) (18a)

(equal to the distance in the s-plane of the furthest filter pole from the origin)

$\omega_{0fastest)}$ = the highest frequency in the input signal (18b)

The reason for this redefinition is that transients in the input will excite the natural modes of the filter, which will then appear in the response. The lowest natural frequency of the filter, $\omega_{n(slowest)}$, is the distance from the origin of the closest s-plane filter-pole.

The minimum sampling frequency which satisfies the Nyquist Sampling Criterion is defined as

$$\omega_{s(min)} \equiv 2 \cdot \omega_N . \tag{19}$$

By the definition of $\omega_{s(min)}$, all s-plane poles of the stable filter F(s) will lie on or inside the left-half of the circle which is centered at the origin and has a radius of $\frac{\omega_{s(min)}}{2}$, since

$$\frac{\omega_{s(min)}}{2} \geq \omega_{n(fastest)}. \tag{20}$$

This left-half circle will be referred to as the Nyquist Sampling Boundary (NSB). We next define the ratio of the sampling frequency to the minimum allowable frequency satisfyiing the Nyquist Sampling Criterion to be the Nyquist Sampling Ratio, N:

$$N \equiv \frac{\omega_s}{\omega_{s(min)}} \tag{21}$$

Figure 3 displays the Nyquist Sampling Boundary (NSB) for N = 1,2,4 and 8, and also the boundary of the primary strip (PS) for N = 1. For example, consider sampling at 4 times the minimum sampling frequency satisfying the Nyquist Sampling Criterion. Then all poles of F(s) will lie on or inside the small circle shown with dot-dash (radius = $\pi/4$ = .79). Referring now to the stability boundary for, say, Schneider's rule, it is seen that, except for a filter having a pole in a very slim wedge along the $j\omega$ -axis, all filter poles of F(s) lie inside the stability region, and hence Schneider's rule will produce stable poles in $F_D(z)$. Similar conclusions hold for the SKG rule.

IV. SOURCES OF ERROR

The accuracy in digitizing a continuous-time filter is defined by the error in the output of the discrete-time equivalent filter relative to its continuous-time counterpart, for a given input. The error resulting from the digitized filter can be separated into truncation error, roundoff error, and startup error.

Truncation error arises from the digital approximation of continuous-time integration. Truncation error depends heavily on the order of the mapping function by which the discrete-time equivalent filter is obtained. As the order of the mapping function increases, the resulting truncation error decreases. The local truncation error, i.e. the error over one time interval, is of the order of T^{r+2}, and the global truncation error, i.e. the error after a fixed length of time, is of the order of T^{r+1}, where r is the order of the mapping function [10].

Roundoff error occurs since digital computers have finite precision in real-number representation. Roundoff error can be generated when obtaining the discrete-time filter coefficients, and also during the real-time processing of the discrete-time filter. When generating discrete-time filter coefficients, the magnitude of the roundoff error will depend on the order and coefficients of the continuous-time filter, the order and coefficients of the mapping function, and the digitizing technique used. The stability and accuracy of a digitized filter has often been found to be surprisingly sensitive to the roundoff error in the discrete-time filter coefficients. In order to reduce this roundoff error, different digitizing techniques have been created and will be discussed later.

Startup error refers to the error caused by initialization and discontinuous inputs. In many instances, startup error will tend to dominate truncation and roundoff error for a period of time. Startup error is demonstrated through the example of a pure integrator. Assume that the integrator is on-line, ready to act when any input comes along. Prior to time $t = t_0 = 0$, the input is zero, and the integrator output is also zero. Then suppose a unit step, $\mu(t)$, arrives at t_0. The continuous-time output of the integrator is a unit ramp starting at t_0. Now consider the output of various discrete-time equivalent integrators, with sample-time T.

At time t_0, trapezoidal integration, (5a), will approximate $\mu(t)$ over the interval [-T,0] by fitting a line to the points $\mu(-T) = 0$ and $\mu(0) = 1$. The area under this approximation establishes an erroneous non-zero output at t_0. At each additional time step, trapezoidal integration will obtain the correct approximation of $\mu(t)$ and the correct increment to be added to the past value of the integrator's output. However, the startup error has already been embedded in the integrator's output, and in the case of a pure integrator, will persist forever. In the case of a stable filter, error introduced during startup will eventually decay to zero.

Parabolic integration, (5b), will result in startup error occurring over the first two time steps. At time t_0, the input to the integrator will be incorrectly approximated over the interval [-T,0] by fitting a parabola to the three points $\mu(-2T) = 0$, $\mu(-T) = 0$, and $\mu(0) = 1$. At time t_1, $\mu(t)$ will be incorrectly approximated over the interval [0,T] by fitting a parabola to the three points $\mu(-T) = 0$, $\mu(0) = 1$, and $\mu(T) = 1$. After the initial two time steps, parabolic integration will obtain the correct approximation of the input, but once again, an error has already been introduced into the integrator's output and will persist forever. Cubic integration, (5c), and other higher-order numerical integration formulas will introduce similar startup error every time there is a discontinuity in the continuous-time input.

V. DIGITIZING TECHNIQUES

Numerous techniques exist in the literature for finding the coefficients of $F_D(z)$ from those of F(s) when using Tustin's rule. Equivalent techniques using higher-order mapping functions like Schneider's rule and the SKG rule are not so well known. Two digitizing methods are presented below that address this situation. Round-off error in these techniques is of key importance. Digitizing an nth-order continuous-time filter, Eq. (2), by the use of an rth-order mapping function, produces an $n \cdot r^{th}$-order discrete-time equivalent filter, Eq. (8). The discrete-time filter can be implemented in the time-domain by the single multi-order difference equation

$$y_k = \frac{1}{A'_0} \cdot \left[-\sum_{i=1}^{n \cdot r} A'_i \, y_{k-i} + \sum_{i=0}^{n \cdot r} B'_i u_{k-i} \right], \tag{22}$$

or by the multiple first-order difference equations

$$\begin{bmatrix} x_1([k+1]T) \\ x_2([k+1]T) \\ \vdots \\ x_{n \cdot r-1}([k+1]T) \\ x_{n \cdot r}([k+1]T) \end{bmatrix} = \begin{bmatrix} 0 & 1 & 0 & \cdots & 0 \\ 0 & 0 & 1 & \cdots & 0 \\ \vdots & \vdots & \vdots & \ddots & \vdots \\ 0 & 0 & 0 & \cdots & 1 \\ -A'_{n \cdot r} & -A'_{n \cdot r-1} & -A'_{n \cdot r-2} & \cdots & -A'_1 \end{bmatrix} \cdot \begin{bmatrix} x_1(kT) \\ x_2(kT) \\ \vdots \\ x_{n \cdot r-1}(kT) \\ x_{n \cdot r}(kT) \end{bmatrix} + \begin{bmatrix} 0 \\ 0 \\ \vdots \\ 0 \\ 1 \end{bmatrix} \cdot u(kT) \tag{23}$$

and

$$y(kT) = B'_0 \cdot u(kT) + \sum_{l=1}^{n \cdot r} (B'_{n \cdot r - l + 1} - A'_{n \cdot r - l + 1} \cdot B'_0) \cdot x_l\,(kT) \tag{24}$$

where A'_0 in (8) must be normalized to unity [9], or by any other customary method for writing a pulse transfer function in state-variable form. All tests were implemented using (22), since it can be performed using only one division, while (23-24) requires 2r n +1 divisions resulting from the normalization.

The first digitizing technique, the Plug-In-Expansion (PIE) method (derived in Appendix C), takes advantage of the integer-coefficient property of the mapping functions to reduce roundoff error by requiring only multiplications and additions of integers until the final steps of the algorithm. The second technique, Groutage's algorithm, is based on a set of simultaneous linear equations that possess unique features which allow a solution through the use of the Inverse Discrete Fourier Transform (IDFT). Appendix D derives Groutage's algorithm using this approach and illustrates the procedure with a fifth-order numerical example.

In order to prevent startup error caused by discontinuous inputs, a time-domain processing approach was developed, which has the capability of an "aware" processing mode. Once a discontinuous input has been received, the algorithm uses a special procedure to compensate for the discontinuity.

Aware processing can be implemented only in systems in which there is knowledge that a discontinuity in the input has occurred. This is a small but definitely non-trivial class of systems. Systems in which there is a definite startup procedure, such as closing a switch or turning a key, can utilize this method. For example, a digital autopilot, used to control the relatively short burn of a high-thrust chemical rocket for changing a spacecraft's orbit, can be sent a signal indicating thrust turn-on and turn-off. In process control, there may well be situations in which changing the process or changing a set point is information that can be made available. A digital controller used in the operating room can be notified when the bottle containing the drug being infused into the patient is about to be changed [11]. Additionally, there are sophisticated on-line methods by which a system can itself detect a change [12]. If the sample-rate of the change-detecting process is much higher than that of the digital control process, then it may be possible to use aware processing in the slower loop. And finally, not to be overlooked, is digital simulation, that is, the numerical integration of ordinary differential equations in initial-value problems. Here the simulationist supplies the information that initializes the process.

The time-domain processing method is derived from the state-variable representation of the continuous-time filter. The numerical integration formulas are used to perform numerical integration at the state-space level. The purpose for developing this time-domain processing method is to enable the application of aware-processing-mode compensation.

Time-domain processing can be visualized by letting the derivatives of the states, $\dot{\underline{x}}$, be calculated using the equation

$$\dot{\underline{x}}_k = \mathbf{A}\underline{x}_k + \underline{b}u_k. \tag{25}$$

These derivatives enter n parallel integrators, each of which is approximated by a numerical integration formula. The outputs of these approximate integrators represent the states $\underline{x}_k$. Recall that when using the Adams-Moulton family of numerical integration formulas on linear time-invariant filters, $\underline{x}_k$ can be solved for even though it appears on both sides of the equation, when integrating Eq. (25). Using the trapezoidal integration formula, (5a), results in the trapezoidal time-domain processing formula

$$\underline{x}_k = \left[I - \frac{T}{2}\cdot A\right]^{-1} \cdot \left\{\left[I + \frac{T}{2}\cdot A\right]\cdot \underline{x}_{k-1} + \frac{T}{2}\cdot \underline{b}\cdot (u_k + u_{k-1})\right\}, \tag{26}$$

derived in Appendix E. Using the parabolic integration formula, (5b), results in the parabolic time-domain processing formula

$$\underline{x}_k = \left[I - \frac{5T}{12}\cdot A\right]^{-1} \cdot \left\{\left[I - \frac{2T}{3}\cdot A\right]\cdot \underline{x}_{k-1} - \frac{T}{12}\cdot A\cdot \underline{x}_{k-2} + \frac{T}{12}\cdot \underline{b}\cdot (5u_k + 8u_{k-1} - u_{k-2})\right\}, \tag{27}$$

derived in Appendix F.

The scenario of aware-processing-mode compensation, at the arrival of a discontinuous input, may be visualized as follows. Assume that the input stream has a discontinuity at time $t = t_0 = 0$. This causes a discontinuity in the derivatives of the states, $\dot{\underline{x}}_0$. Let u_{0-} represent the extrapolated input signal prior to the discontinuity, and let u_{0+} represent the actual value of the input signal directly after the discontinuity. These two different input values correspond to two different state-variable derivative values, $\dot{\underline{x}}_{0-}$ and $\dot{\underline{x}}_{0+}$ from (25). Note that u_{0-} can be determined by a prediction, or extrapolation, algorithm. In the case of a unit step input, u_{0-} is known to be zero.

To prevent startup error when using the trapezoidal time-domain processing formula (26), the state vector $\underline{x}_0$ must be computed by setting $\dot{\underline{x}}_0 = \dot{\underline{x}}_{0-}$, which corresponds to setting $u_0 = u_{0-}$. That is, at the time t_0 of the discontinuity, $\underline{x}_0$ is computed from (26), with $k = 0$, by setting $u_0 = u_{0-}$. Since trapezoidal integration requires only one past value of the derivative, the state-variables $\underline{x}_1$ can be computed from (26) in the normal manner with $k = 1$ and u_0 equal to u_{0+}. This procedure will be referred to as trapezoidal aware-processing-mode compensation.

In order to prevent startup error when using the parabolic time-domain processing formula (27) with k = 0, the state vector $\underline{x}_0$ must be computed by setting $\dot{\underline{x}}_0 = \dot{\underline{x}}_{0-}$, which corresponds to setting $u_0 = u_{0-}$. However at time t_1, $\underline{x}_1$ cannot be computed from (27) with k = 1, since u_{-1}, u_0, and u_1 do not come from a single continuous input function. Recall that fitting a smooth curve, like a trapezoid or a parabola, to an input function with a discontinuity results in startup error. One way around this obstacle is to wait until time t_2 before attempting to obtain $\underline{x}_1$. Then, $\underline{x}_1$ can be computed with k = 1, using the parabolic aware-processing-mode compensation formula

$$\underline{x}_k = \left[8\mathbf{A}^2T^2 - 24\mathbf{A}T + 24\mathbf{I}\right]^{-1} \cdot \left\{ \begin{array}{c} \left[-4\mathbf{A}^2T^2 + 24\mathbf{I}\right] \cdot \underline{x}_k \\ -\, 2\underline{b}T \cdot u_{k+1} \\ +\left[-8\mathbf{A}\underline{b}T^2 + 16\underline{b}T\right] \cdot u_k \\ +\left[-4\mathbf{A}\underline{b}T^2 + 10\underline{b}T\right] \cdot u_{(k-1)+} \end{array} \right\} \tag{28}$$

derived in Appendix G. Another way around this obstacle is to compute $\underline{x}_1$ by using the trapezoidal time-domain processing formula (26) at time t_1, with k = 1 and $u_0 = u_{0+}$. This has the advantage of processing the states $\underline{x}_1$, and producing the output y_1, in real time without delay. However, the disadvantage is the use of lower-order integration, which results in less accuracy. In either case, the states $\underline{x}_2$ can be computed at time t_2, using (27) with k = 2 and $u_0 = u_{0+}$. This procedure will be referred to as parabolic aware-processing-mode compensation.

Once the discontinuity has been bridged by the appropriate use of (26) or (27) and (28), from then on, the state vectors can be propagated once each sample time by (26) for trapezoidal integration or (27) for parabolic integration. Alternatively, (26) can be reconfigured in the standard state-variable form

$$\underline{w}\,(k+1) = F\,\underline{w}\,(k) + G\,u\,(k) \tag{29}$$

$$y(k) = H\underline{w}(k) + Ju(k) \tag{30}$$

The relation of F, G, H, J, and $\underline{w}$ to A, $\underline{b}$, $\underline{c}$ and $\underline{x}$ are defined for trapezoidal integration in reference [13]. Presumably (27) can be converted to the form (29, 30) as well; however, it will result in a state vector of twice the dimension of $\underline{x}$, since (27) is a second-order equation in $\underline{x}$.

The formula corresponding to (27) for cubic time-domain processing can be derived in a similar manner. In such case, the aware-processing-mode compensation would consist of computing x_0 by using $u_0 = u_{0-}$. The states x_1 and x_2 can be computed by either waiting until the input at time t_3 is received, or by using (26) and (27) at times t_1 and t_2 respectively.

VI. RESULTS

A. TIME-DOMAIN EVALUATION

The transformations were applied to two separate transfer functions for evaluation purposes. The first transfer function is a fifth-order filter with real poles, two real zeroes, and a dc gain of unity:

$$F(s) = \frac{1152s^2 + 2304s + 864}{s^5 + 27.5s^4 + 261.5s^3 + 1039s^2 + 1668s + 864} \tag{31a}$$

$$F(s) = \frac{1152(s+0.5)(s+1.5)}{(s+1)(s+2)(s+4.5)(s+8)(s+12)} \tag{31b}$$

This is the normalized version of the filter which was analyzed in [3], [4], and [14]. The second filter is a simple band-pass filter:

$$F(s) = \frac{s}{s^2 + s + 25} \tag{32}$$

This filter is analyzed in [15]. Evaluations were conducted in both the time and frequency domains. In the time domain, only the fifth-order filter was evaluated, whereas in the frequency domain both filters were evaluated.

All tests were performed on an IBM-compatible Personal Computer using double precision. Every effort was made to reduce roundoff error, since the focus of these tests was to study truncation, startup error, and stability. The numerical coefficients of the discrete-time filters in this section were obtained with the PIE digitizing technique, unless otherwise specified. The fifth-order filter has a fastest natural frequency, $\omega_{n(fastest)}$, of 12 radians per second, and a slowest natural frequency $\omega_{n(slowest)}$, of 1 radian per second. The input frequencies used in sinusoidal testing were taken to be strictly slower than $\omega_{n(fastest)}$. Therefore the minimum allowable sampling frequency satisfying the Nyquist Sampling Criterion, $\omega_{s(min)}$, was twice $\omega_{n(fastest)}$. The sampling frequencies, ω_s, were varied over the values in Table I.

Table I

$\log_2(N)$	N	ω_s	T
0.0	1.000	24.000	0.261799
0.5	1.414	33.941	0.185120
1.0	2.000	48.000	0.130900
1.5	2.828	67.882	0.092560
2.0	4.000	96.000	0.065450
2.5	5.657	135.765	0.046280
3.0	8.000	192.000	0.032725
3.5	11.314	271.529	0.023140
4.0	16.000	384.000	0.016362

This filter was transformed into different discrete-time filters using Tustin's rule, Schneider's rule, and the SKG rule, for the several sampling frequencies. Fig. 4 displays (in the sT-plane) the fastest pole of F(s), s = -12, relative to the stability regions of Schneider's rule and the SKG rule, as N increases. This demonstrates that while Schneider's rule results in stable filters for all frequencies satisfying the Nyquist Sampling Criterion, the SKG rule results in an unstable pole for N = 1. This pole is stable for N ≥1.047.

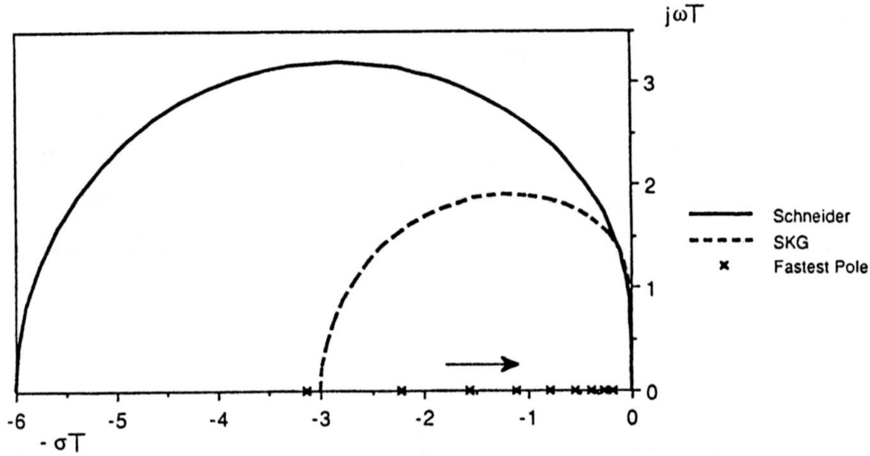

Fig. 4 Location in sT-plane of the fastest pole of F(s) for nine increasing values of N from 1 to 16.

The first test analyzed the accuracy of the digitized filters using the sinusoidal input $u(t) = \sin(1.5t)$ The objective of this test was to examine the truncation error resulting from the different mapping functions. To eliminate the starting transient, the input was run for several cycles until the output had reached steady-state. The outputs of the different discrete-time filters were compared with the exact solution, obtained analytically for the continuous-time filter. The root-mean-square error was computed over one cycle for each of the different filters at each of the different sampling frequencies. Fig. 5 contains a logarithmic plot of rms error vs sampling frequency, for all of the stable discrete-time filters. In order to highlight the proportionality relationship between the global truncation error and T^{r+1}, the sampling-frequency axis has been scaled as $\log_2(N)$. As the sampling frequency doubles, the rms error decreases by 0.5^{r+1}, which plots linearly on the scales chosen. Fig. 5 confirms this linear relationship and also demonstrates that, as the order of the mapping function increases, the magnitude of the downward slope of the error-vs-frequency curve also increases.

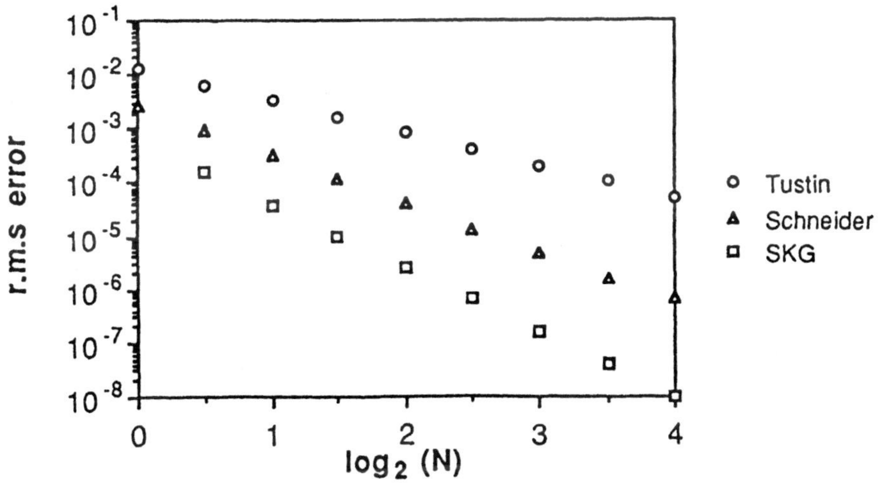

Fig. 5 RMS error in steady-state output for sinusoidal input as a function of sampling frequency for three discrete equivalent filters of a fifth-order F(s). Primarily truncation error. F(s) has DC gain = 1.

The results of the digitized filters using higher-order mapping functions can be compared with Fig. 6, the results of the Boxer-Thaler and Madwed integrator methods covered in reference [14], for the continuous-time filter

(31a,b). Through stability analysis, the Boxer-Thaler integrator was found to result in unstable discrete-time filters for the sampling frequencies corresponding to N = 1 and N = 1.414. The Madwed integrator was found to result in an unstable discrete-time filter for N = 1. Fig. 6 demonstrates that although there is an improvement in the global truncation error resulting from the Boxer-Thaler and Madwed integrators, relative to Tustin's rule, they are of the same order, since the slope of the error-vs-frequency curves are the same. In comparison, since the slope resulting from the mapping functions increases as r increases, higher-order mapping functions will always result in truncation error which improves at a faster rate than the Boxer-Thaler and Madwed integrators, as the sampling frequency is increased. Thus, when sampling fast enough, higher-order mapping functions can always result in smaller truncation error.

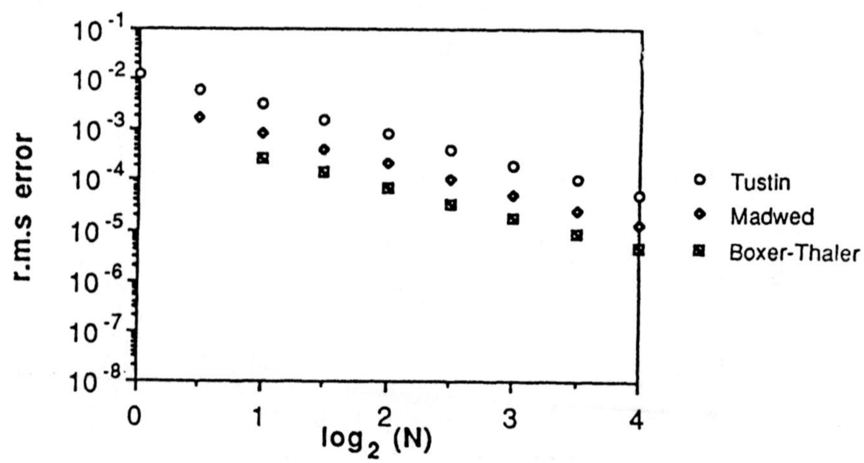

Fig. 6 Same F(s) as Fig. 5 with results from two additional equivalent filters.

The second test analyzed the accuracy of the digitized filters using a unit step input. The objective of this test was to examine startup error during discontinuous inputs. The rms errors were computed from the time of the discontinuity, to five times the filter time-constant, computed as $\frac{1}{\omega_{n(slowest)}}$. Fig. 7a contains the results using Tustin's rule and Schneider's rule in the on-line mode, plus trapezoidal and parabolic time-domain processing in the aware-processing mode. The top two curves of this figure represent the rms errors resulting from Tustin's rule and Schneider's rule . Note that the rms errors are nearly equal for the two filters. The lower two curves of the figure represent the rms errors of the corresponding trapezoidal and parabolic time-domain processing formulas using aware-processing-mode compensation. A plot of the total instantaneous error at each sample vs time is given in Fig. 7b. Three curves are plotted— one each for the Tustin-rule filter, the Schneider-rule filter in the on-line mode, and the Schneider-rule filter in the aware-processing mode. The sample time was .065450 seconds (N = 4) for all three curves. First, note that aware-processing-mode compensation produces a substantial improvement over the corresponding mapping function's on-line ("unaware") processing. This demonstrates that compensating for discontinuous inputs eliminates startup error. Second, with aware-processing-mode compensation, parabolic time-domain processing results in substantially smaller error than trapezoidal time-domain processing. This demonstrates that parabolic integration results in smaller truncation error than trapezoidal.

To demonstrate the importance of satisfying the Nyquist Sampling Criterion, trapezoidal time-domain processing was used in the aware-processing mode, for sampling frequencies for which $N < 1$. A unit step input was used to generate transients by exciting the natural modes of the filter. Trapezoidal integration was used since it is stable for all sampling frequencies, and the aware-processing mode was used in order to prevent startup error. Note that since all of

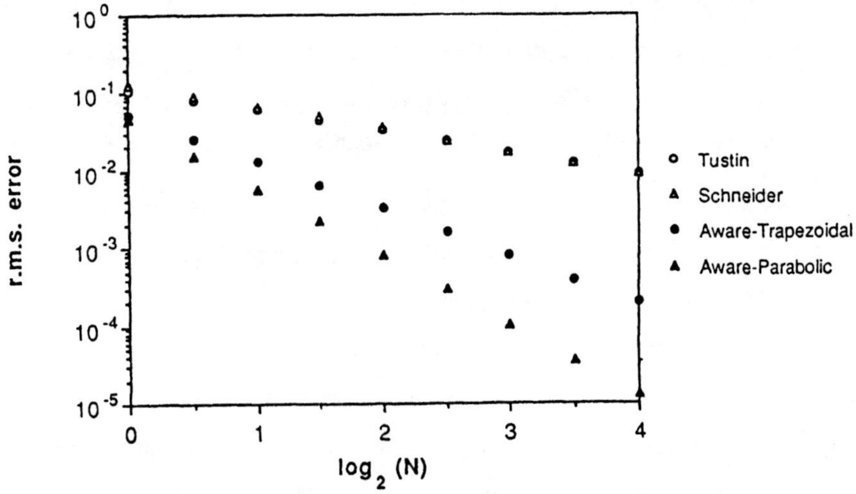

Fig. 7a RMS error results for same F(s) as Fig. 5 but now using unit step input. Aware and unaware ("on-line") processing are compared.

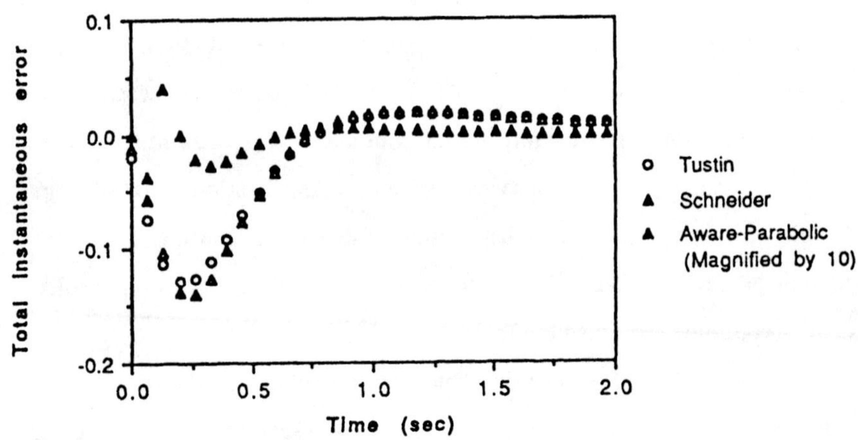

Fig. 7b Time response.

the s-plane poles are real, they lie inside the primary strip regardless of the sampling frequency. Fig. 8 contains the rms errors, which demonstrate that there is no uniform reduction in error with increasing N for sampling frequencies which do not satisfy the Nyquist Sampling Criterion. Since, for the sampling frequencies that were tested, the Nyquist Sampling Criterion was not satisfied primarily as a result of the transients generated from the faster filter-poles, the rms errors were computed over the interval from 0 to .5 seconds or 1 sample, whichever is longer, in order to highlight the effect of aliasing.

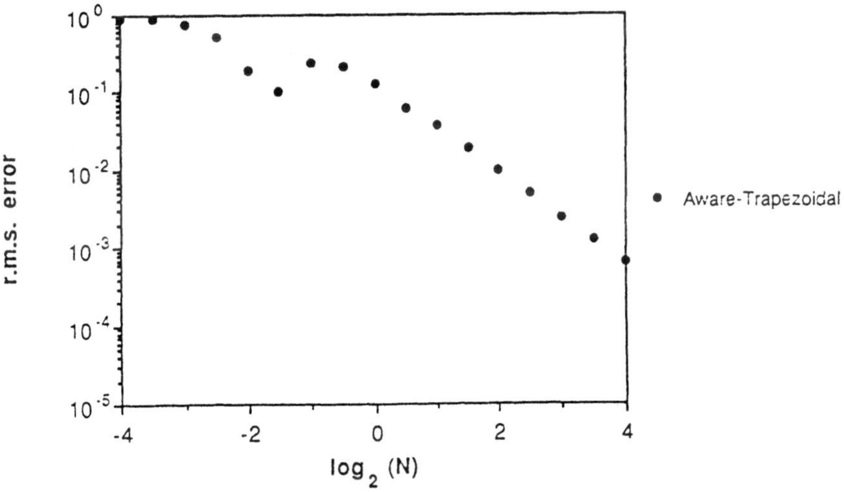

Fig. 8 Effect on error of sampling at a frequency lower than the Nyquist Sampling Criterion.

References [3], [4], and [14] considered the following unnormalized version of filter (31a,b)

$$F(s) = \frac{s^2 + 2s + 0.75}{s^5 + 27.5s^4 + 261.5s^3 + 1039s^2 + 1668s + 864} \tag{33a}$$

$$= \frac{(s+0.5)(s+1.5)}{(s+1)(s+2)(s+4.5)(s+8)(s+12)}, \tag{33b}$$

which has a DC gain of $\frac{1}{1152}$. This filter was tested with a unit step input and a sampling period of T = 0.01 seconds. The discrete-time filter coefficients resulting from Tustin's rule and Schneider's rule are listed in Table II.

Table II

	Tustin's rule	Schneider's rule
B'_0	1.26252343750000014E-07	7.29417212215470783E-08
B'_1	1.28761718750000011E-07	2.05809039834104971E-07
B'_2	-2.47476562500000014E-07	-1.05011551468460664E-07
B'_3	-2.52476562500000019E-07	-5.20858651620370421E-07
B'_4	1.21261718750000003E-07	8.77540418836805781E-08
B'_5	1.23752343750000011E-07	3.74249725115740820E-07
B'_6		-1.28926965784143536E-07
B'_7		1.46656539351851887E-08
B'_8		-5.47060366030092715E-10
B'_9		-9.52449845679012609E-13
B'_{10}		-3.01408179012345785E-16
A'_0	1.14416842019999998E+00	1.11919892686631939
A'_1	-5.41890448400000047E+00	-5.27387884476273161
A'_2	1.02616673620000007E+01	9.91098259555844940
A'_3	-9.71218680800000023E+00	-9.26048848877314867
A'_4	4.59416426099999953E+00	4.26763188433449070
A'_5	-8.68908664799999950E -01	-0.743678472985648 15
A'_6		-1.96006569432870398E-02
A'_7		-1.66281162037037057E-04
A'_8		-5.74941550925926069E-07
A'_9		-7.90509259259259456E-10
A'_{10}		-3.47222222222222319E-13

In order to stress the importance of roundoff error in the filter coefficients, it was found that the Tustin's-rule filter becomes unstable when its coefficients are rounded to 7 significant figures.

The resulting rms errors, using the different methods are listed in Table III.

Table III

Method	rms Error
Boxer-Thaler	4.70034725296039265E-06
Madwed	4.68114761718508622E-06
Tustin's Rule	4.64400700511313123E-06
Schneider's Rule	4.68337204603581546E-06
Aware Trapezoidal	6.15693246289917751E-08
Aware Parabolic	2.73096449871258854E-09

Boxer-Thaler and Madwed filter coefficients were taken from reference [14]. Note that the error resulting from the Boxer-Thaler and Madwed filters are relatively close to the error resulting from Tustin's rule and Schneider's rule. Also note that aware-processing-mode compensation results in a significant improvement in both the trapezoidal and parabolic cases; the observed error is primarily that due to truncation, start-up error having been eliminated. These results demonstrate that the errors, resulting from methods described in reference [14], were dominated primarily by startup error, with truncation error effectively masked.

In general, errors associated with the design of digital filters can occur in several stages of the design process. We have discussed errors associated with mapping a rational function in the complex s-domain to a rational function in the complex z-domain. These errors are attributed to the method that is used to map from the s- to z-domain. Another major source of error is associated with the method of realizing the digital filter from the rational function $F_D(z)$. This subject has received considerable attention in the literature. Ogata [9] notes the error due to quantization of the coefficients of the pulse transfer function. He states that this type of error, which can be destabilizing, can be reduced by mathematically decomposing a higher-order pulse transfer function into a combination of lower-order blocks. He goes on to describe the series or cascade, parallel, and ladder programming methods. (The mapping functions presented here are compatible with all of these programming methods. One can decompose F(s) into simple blocks, and transform each to discrete form by the selected mapping function. Alternatively, one can convert F(s) to $F_D(z)$ by the selected mapping function, and decompose the latter into the desired form. In either case, and no matter which programming method is selected, the resulting overall pulse transfer function from input to output will be the same, a condition not generally true for other mapping methods, such as Boxer - Thaler.) Additional insight on digital filter realization can be found in Johnson [16], Antoniou [17], Strum and Kirk [18], and Rabiner and Gold [19].

B. FREQUENCY DOMAIN EVALUATION

Frequency domain evaluations were carried out in terms of comparing the analog filter frequency response, F(s) for $s = j\omega$, to the corresponding frequency response for digital filters derived from Tustin's and Schneider's rule, $F_T(z)$ and $F_S(z)$ for $z = e^{j\omega T}$. Fig. 9 presents plots for the analog filter and corresponding digital filters derived by both Tustin's and Schneider's rules. The dotted-line curves (magnitude and phase) represent the analog filter response, the dashed curves represent the Schneider digital filter response, and the solid line curves represent the Tustin digital filter response. The sampling time interval for the filters represented by the response curves of Fig. 9 is 0.126 seconds. Note also that the range of the frequency response is over two-thirds of the Nyquist frequency range. (The Nyquist frequency range for a low-pass filter $F_D(z)$ is defined to be $0 < \omega < \omega_{s/2}$.) Fig. 10 presents magnitude and phase plots of the complex error, as a function of frequency, for digital filters derived using Tustin's and Schneider's rules. The complex error functions for these two filters are defined as

$$E_T(j\omega) = F(j\omega) - F_T(e^{j\omega T}) \tag{34}$$

and

$$E_S(j\omega) = F(j\omega) - F_S(e^{j\omega T}) \tag{35}$$

In the above equation, $E_T(j\omega)$ is the error associated with the digital filter using Tustin's rule and $E_S(j\omega)$ is the error associated with the digital filter using Schneider's rule, $F(j\omega)$ is the response function for the analog filter and $F_T(e^{j\omega T})$ and $F_S(e^{j\omega T})$ are the frequency response functions of the respective digital filters. The dashed curves of Fig. 10 are for the Schneider digital filter representation, whereas the solid curves are for the Tusin digital filter representation. A Figure-of-Merit (FOM) was established for determining the performance level of a digital filter derived by a specific mapping function as a function of the sampling frequency. The Figure-of-Merit (FOM) is defined as

$$\text{FOM} = \left(\frac{1}{L}\sum_{i=1}^{L} |E(j\omega_i)|^2\right)^{\frac{1}{2}} \tag{36}$$

where $E(j\omega_i)$ is the complex error at L discrete frequencies, ω_i. The range of discrete frequencies for which the Figure-of-Merit is calculated is somewhat

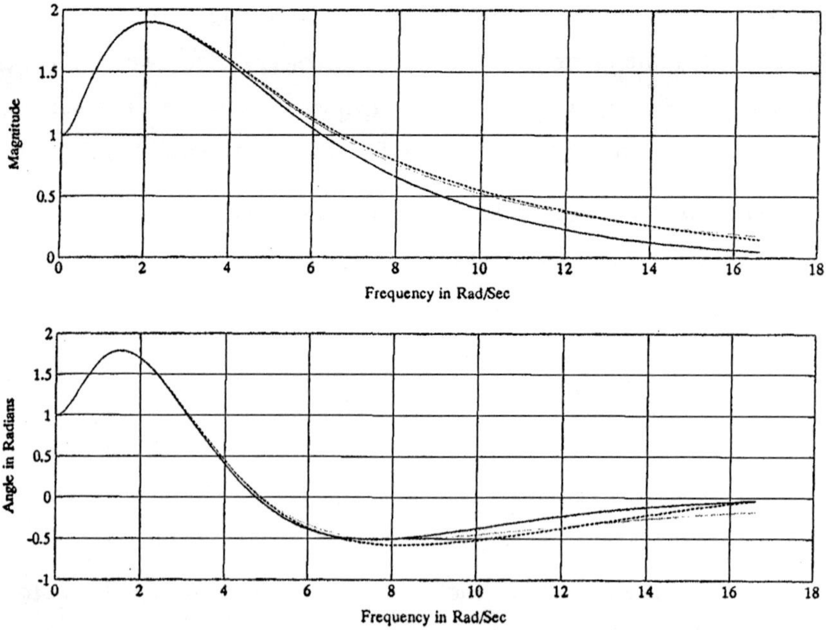

Fig. 9 Fifth-order F(s) frequency responses: analog filter, Schneider digital filter, Tustin Digital Filter.

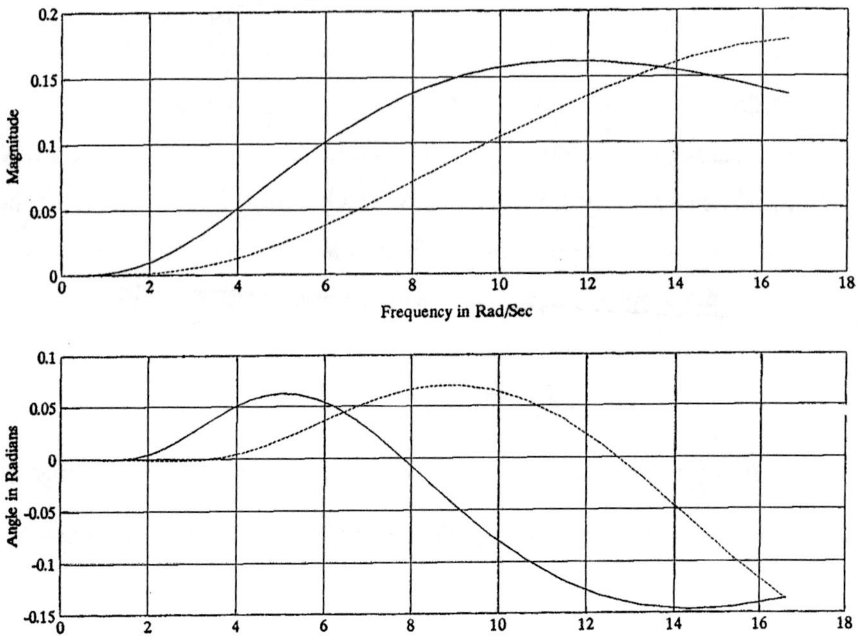

Fig. 10 Error as a function of frequency for Schneider and Tustin digital filters (fifth-order F(s)).

arbitrary. As a rule-of-thumb, this could be the Nyquist frequency range, the band-pass frequency range of the filter, or some other range of interest.

The fifth-order filter was evaluated using the Figure-of-Merit, as described above. One hundred equally-spaced points (L = 100) over a logarithmic frequency scale defined by the range from 0.01 to two-thirds of the high limit of the Nyquist frequency range, $\frac{2}{3} \cdot \frac{\omega_S}{2} = \omega_S/3$, were used to evaluate the Figure-of-Merit. Table IV presents Figure-of-Merit data for selected sampling time intervals for digital filters derived using Tustin's and Schneider's rules.

Table IV

SAMPLING TIME INTERVAL-SEC	FOM_T	FOM_S
0.0100	0.00044367	0.00006679
0.0390	0.00685940	0.00278154
0.0680	0.01986582	0.01131582
0.0970	0.03753012	0.02580793
0.1260	0.05786764	0.04489918
0.1550	0.07917394	0.06672674
0.1840	0.10015588	0.08949417
0.2130	0.11993386	0.11175256
0.2420	0.13797713	0.13248408
0.2710	0.15401902	0.15107094

Performance of the digital filters derived using Schneider's rule for selective sampling intervals, as compared to the digital filters derived using Tustin's rule, is presented in Fig. 11. Comparative performance is in terms of percent improvement, where percent improvement is defined as

$$\%Imp = \left(\frac{FOM_T - FOM_S}{FOM_S}\right) \times 100 \tag{37}$$

For this evaluation, Schneider's mapping function outperformed the linear mapping (Tustin's) function over the entire range of sampling intervals. However, it is interesting to note that the greatest margin of improvement is obtained at the higher sampling frequencies (smaller sampling intervals), and that improvement falls off as the sampling interval is increased.

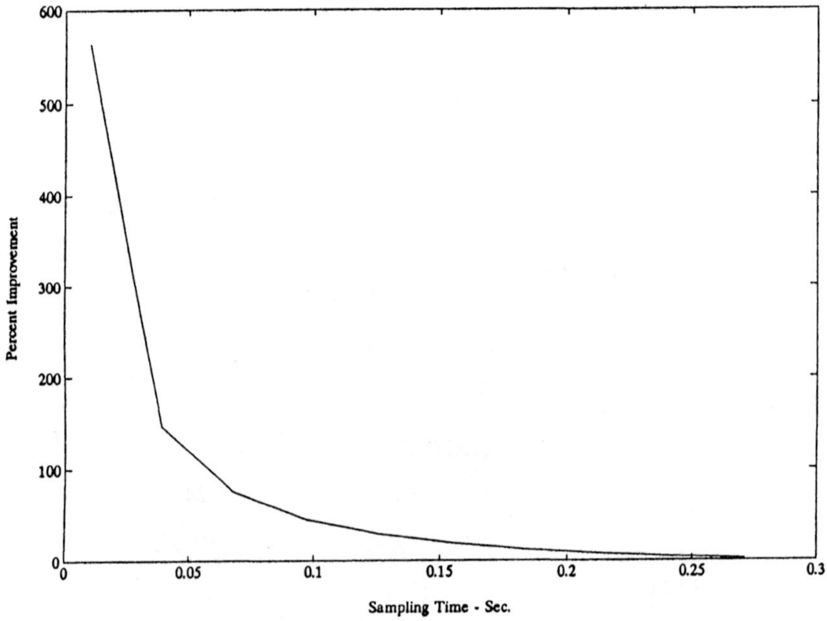

Fig. 11 Percent improvement of Higher-Order Mapping over Linear Mapping as a function of sampling time.

A similar performance evaluation was carried out for the band-pass filter. Fig. 12 presents frequency response information in terms of magnitude and phase plots for the analog representation of this filter. Fig. 13 presents magnitude and phase plots for the error functions $E_T(j\omega)$ and $E_S(j\omega)$ associated

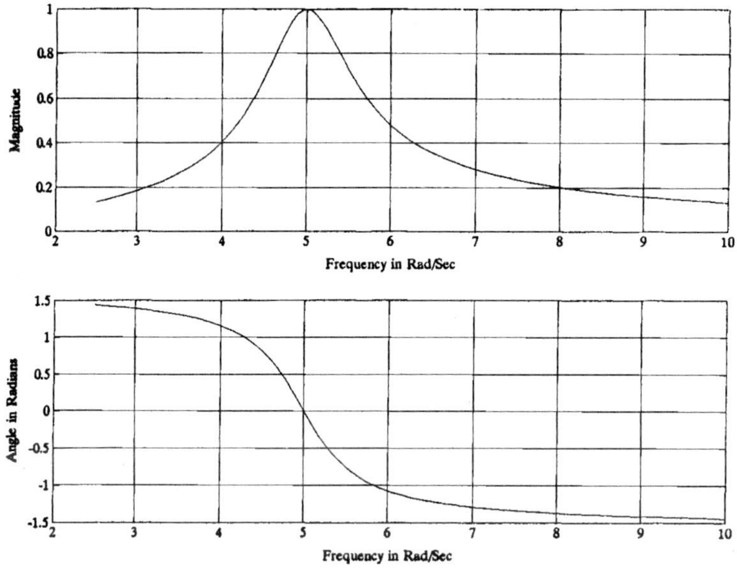

Fig. 12 **Band-pass filter response: analog filter.**

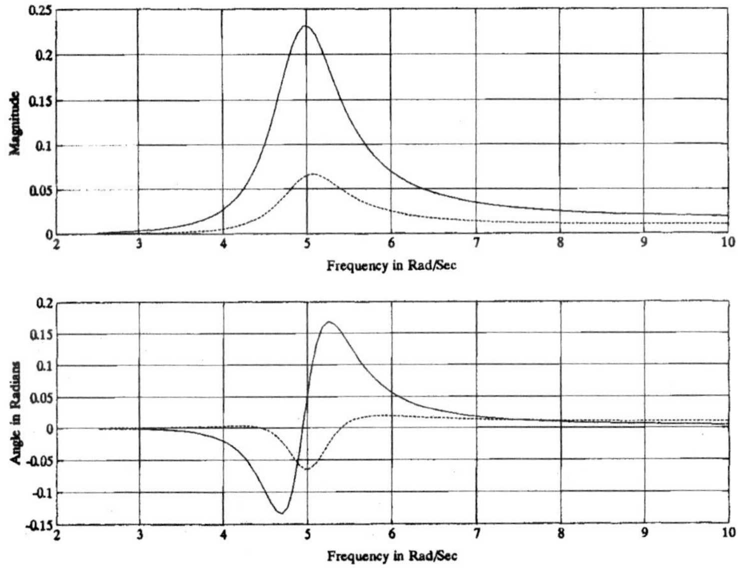

Fig. 13 **Band-pass filter error for Schneider and Tustin digital filter representations.**

with the digital filter for the band-pass example. The dashed curves of Fig. 13 are for Schneider digital filter representation, whereas the solid curves are for Tustin digital filter representation. The sample time for the filter response of Fig. 13 is 0.106 seconds. For this evaluation, the Figure-of-Merit was calculated using one hundred (L = 100) equally-spaced points over the logarithmic frequency scale defined by the upper and lower limits

$$F_{low} = \frac{CF}{2} \tag{38}$$

and

$$F_{high} = 2CF \tag{39}$$

where CF is the center frequency of the band-pass filter. Table V presents Figure-of-Merit data for this example.

Table V

SAMPLING TIME INTERVAL-SEC	FOM_T	FOM_S
0.0100	0.00072459	0.00001913
0.0340	0.00836010	0.00075265
0.0580	0.02422157	0.00375336
0.0820	0.04804772	0.01072336
0.1060	0.07933866	0.02363077
0.1300	0.11725235	0.04499489
0.1540	0.16051583	0.07836798
0.1780	0.20740427	0.12922418
0.2020	0.25583856	0.20680980
0.2260	0.30361501	0.32874937

The performance improvement of the digital filter derived using Schneider's rule, as compared to the digital filter derived using Tustin's rule, is presented in Fig. 14. Again, just as for the fifth-order example, the filter derived using a higher-order mapping function outperforms the equivalent filter derived using linear mapping function for all of the sampling intervals listed in Table V, except for the largest sampling interval (T = 0.2260 seconds). This sampling interval of 0.2260 seconds is close to the sampling time for which the higher-order filter approaches the stability boundary. Also note that the performance improvement for smaller sampling intervals, as is the case with the fifth-order filter, is significantly better than for larger sampling intervals.

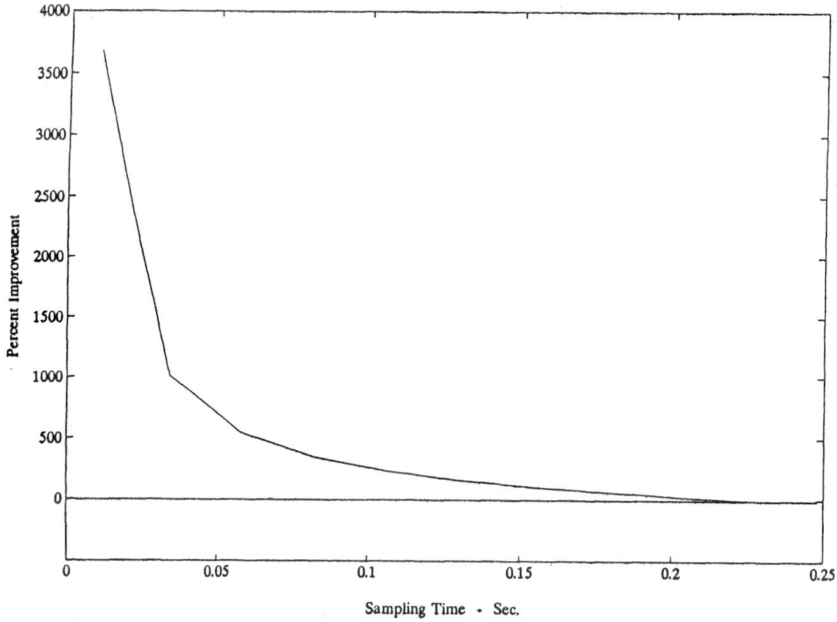

Fig. 14 Percent improvement in the band-pass filter.

VII. CONCLUSION

Higher-order mapping functions can be used in digitizing continuous-time filters in order to achieve increased accuracy. The question of stability is resolved in practical application; when the sampling frequency is chosen high enough to satisfy the Nyquist Sampling Criterion, then the Schneider-rule and SKG-rule filter are almost always stable. Stability of higher-order filters can be achieved by sampling fast enough. The stability boundaries of Fig. 3 provide a graphical test for determining stability of Schneider-rule and SKG-rule filters. Equations (11)-(14) present an analytical procedure for testing stability.

The difficulty in handling discontinuous inputs is approached through the introduction of aware-processing mode-compensation.

Results demonstrate that without aware-processing-mode compensation, higher-order mapping functions can produce digitized filters with significantly greater accuracy when handling smooth inputs, and approximately equivalent accuracy during the transient stages after a discontinuous input, relative to filters obtained using Tustin's rule. With aware-processing-mode compensation, the higher-order numerical integration formulas can result in increased accuracy for both smooth and discontinuous inputs. A digital filter derived using higher-order mapping functions demonstrates improved performance over the filter's frequency pass-band, when compared to a similar digital filter derived using a linear mapping function. The level of improvement is better at the smaller sampling intervals and falls off as the sampling frequency approaches the Nyquist sampling rate.

APPENDIX A
DERIVATION OF SCHNEIDER'S RULE AND THE SKG RULE

Here we derive Schneider's rule and the SKG rule, (6b) and (6c), from the parabolic and cubic numerical integration formulas, (5b) and (5c). Suppose we have the continuous-time filter displayed in Fig. 15, consisting of a pure.

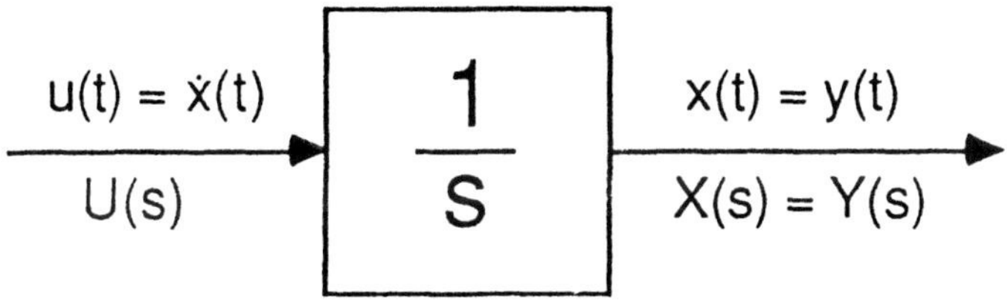

Fig. 15 Input-output relations for a pure integrator.

integrator with input u(t), output y(t), and state x(t). The continuous-time transfer function of the resulting filter,

$$\frac{Y(s)}{U(s)} = \frac{1}{s}, \tag{A.1}$$

represents an integrator which can be implemented digitally using the numerical integration formulas. Note from Fig. 15 that $\dot{x} = u$ and $x = y$. Changing the variable-notation of (5b) and (5c) results in:

$$y_k = y_{k-1} + \frac{T}{12} \cdot (5u_k + 8u_{k-1} - u_{k-2}) \tag{A.2}$$

$$y_k = y_{k-1} + \frac{T}{24} \cdot (9u_k + 19u_{k-1} - 5u_{k-2} + u_{k-3}). \tag{A.3}$$

Taking the z-transform of both sides of each equation and multiplying through by z^2 and z^3 respectively, leads to:

$$(z^2 - z)Y(z) = \frac{T}{12} \cdot (5z^2 + 8z - 1)U(z) \tag{A.4}$$

$$(z^3 - z^2)Y(z) = \frac{T}{24} \cdot (9z^3 + 19z^2 - 5z + 1)U(z). \tag{A.5}$$

Cross-dividing both equations results in the discrete-time transfer functions,

$$\frac{Y(z)}{U(z)} = \frac{T}{12} \cdot \frac{5z^2 + 8z - 1}{z^2 - z} \tag{A.6}$$

$$\frac{Y(z)}{U(z)} = \frac{T}{24} \cdot \frac{9z^3 + 19z^2 - 5z + 1}{z^3 - z^2}. \tag{A.7}$$

Comparing the discrete-time transfer functions, (A.6) and (A.7), with the continuous-time transfer function (A.1), leads to the s-to-z mapping function relationships representing Schneider's rule and the Schneider-Kaneshige-Groutage (SKG) rule respectively,

$$s = \frac{12}{T} \cdot \frac{z^2 - z}{5z^2 + 8z - 1} \tag{A.8}$$

$$s = \frac{24}{T} \cdot \frac{z^3 - z^2}{9z^3 + 19z^2 - 5z + 1}. \tag{A.9}$$

APPENDIX B

PROOF OF INSTABILITY OF SIMPSON'S RULE

Consider the pure integrator system, Fig. 15, where the input to the integrator is u, and the output of the integrator is y. Changing the variable notation of Simpson's numerical integration formula, (9), results in

$$y_k = y_{k-2} + \frac{T}{3} \cdot (u_k + 4u_{k-1} + u_{k-2}). \tag{B.1}$$

Taking the z-transform of both sides of (B.1) and multiplying through by z^2, results in:

$$(z^2 - 1)Y(z) = \frac{T}{3} \cdot (z^2 + 4z + 1)U(z). \tag{B.2}$$

Cross-dividing to obtain the discrete-time transfer function,

$$\frac{Y(z)}{U(z)} = \frac{T}{3} \cdot \frac{z^2 + 4z + 1}{z^2 - 1}, \tag{B.3}$$

and comparing with the continuous-time transfer function (A.1), results in Simpson's rule

$$s = \frac{3}{T} \cdot \frac{z^2 - 1}{z^2 + 4z + 1}. \tag{B.4}$$

Now consider the stable continuous-time filter

$$F(s) = \frac{a}{s+a} \qquad a > 0. \tag{B.5}$$

Using Simpson's rule to map from s to z results in the discrete-time filter

$$F_D(z) = \frac{z^2 + 4z + 1}{(1 + \frac{3}{aT})z^2 + 4z + (1 - \frac{3}{aT})}. \tag{B.6}$$

The two poles of $F_D(z)$ are obtained by setting the denominator equal to zero, and solving for z. It is readily found that there is one pole lying outside of the unit circle for all positive values of the product aT. Therefore Simpson's rule leads to an unstable filter, since it maps any stable continuous-time filter-pole to at least one unstable z-plane pole.

APPENDIX C
DERIVATION OF PLUG-IN-EXPANSION (PIE) METHOD

This appendix shows one method for computing the coefficients of the discrete-time filter from the coefficients of the continuous-time filter and the mapping function. Let $f_D(z)$ be a mapping function of order r, where the polynomials C(z) and D(z) have integer coefficients. Then

$$f_D(z) = \frac{\alpha}{T} \cdot \frac{C(z)}{D(z)} = \frac{1}{\beta} \cdot \frac{C(z)}{D(z)}, \tag{C.1}$$

where α is an integer and $\beta = \frac{T}{\alpha}$. Recall the mapping functions

$$f_{D(Tustin)}(z) \quad s = \frac{2}{T} \cdot \frac{z - 1}{z + 1} \tag{C.2}$$

$$f_{D(Schneider)}(z) \quad s = \frac{12}{T} \cdot \frac{z^2 - z}{5z^2 + 8z - 1} \tag{C.3}$$

$$f_{D(SGK)}(z) \quad s = \frac{24}{T} \cdot \frac{z^3 - z^2}{9z^3 + 19z^2 - 5z + 1}. \tag{C.4}$$

Given a continuous-time filter

$$F(s) = \frac{B_0 s^m + B_1 s^{m-1} + \ldots + B_m}{A_0 s^n + A_1 s^{n-1} + \ldots + A_n} \quad m \le n, \tag{C.5}$$

and substituting the selected mapping function for s yields

$$F_D(z) = \frac{\dfrac{B_0 \cdot C^m(z)}{\beta^m \cdot D^m(z)} + \cdots + \dfrac{B_m \cdot C^0(z)}{\beta^0 \cdot D^0(z)}}{\dfrac{A_0 \cdot C^n(z)}{\beta^n \cdot D^n(z)} + \cdots + \dfrac{A_n \cdot C^0(z)}{\beta^0 \cdot D^0(z)}} \cdot \tag{C.6}$$

Multiplying through by β^n and $D^n(z)$ results in

$$F_D(z) = \frac{B_0 \cdot \beta^{n-m} \cdot C^m(z) \cdot D^{n-m}(z) + \cdots + B_m \cdot \beta^n \cdot C^0(z) \cdot D^n(z)}{A_0 \cdot \beta^0 \cdot C^n(z) \cdot D^0(z) + \cdots + A_n \cdot \beta^n \cdot C^0(z) \cdot D^n(z)} . \tag{C.7}$$

Note that each term is of the order $z^{r \cdot n}$. Define the expansion of $C^k(z)$ and $D^k(z)$ by

$$C^k(z) = c_{k,0} z^{r \cdot k} + c_{k,1} z^{r \cdot k - 1} + \cdots + c_{k,r \cdot k} \quad k = 0, 1, \ldots\ldots n \tag{C.8}$$

$$D^k(z) = d_{k,0} z^{r \cdot k} + d_{k,1} z^{r \cdot k - 1} + \cdots + d_{k,r \cdot k}, \quad k = 0, 1, \ldots\ldots n \tag{C.9}$$

where

$$c_{0,0} = d_{0,0} = 1 \tag{C.10}$$

$$c_{k,j} = 0 \qquad j > rk \tag{C.11}$$

$$d_{k,j} = 0 \qquad j > rk \tag{C.12}$$

and where $c_{k,j}$ and $d_{k,j}$ are integers and can be calculated by integer arithmetic. The resulting discrete-time filter,

$$F_D(z) = \frac{B'_0 z^{r \cdot n} + B'_1 z^{r \cdot n - 1} + \ldots + B'_{r \cdot n}}{A'_0 z^{r \cdot n} + A'_1 z^{r \cdot n - 1} + \ldots + A'_{r \cdot n}}, \tag{C.13}$$

has coefficients which can be computed by

$$B'_k = \sum_{i=0}^{m} \left[B_i \cdot \beta^{n-m+i} \sum_{j=0}^{k} c_{m-i,\, k-j}\, d_{n-m+i,\, j} \right] \quad 0 \le k \le nr \tag{C.14}$$

$$A'_k = \sum_{i=0}^{n} \left[A_i \cdot \beta^{i} \sum_{j=0}^{k} c_{n-i,\, k-j}\, d_{i,\, j} \right] \quad 0 \le k \le nr \tag{C.15}$$

Note that non-integers appear for the first time in (C.14) and (C.15). Also note that, except for the calculation of β, no divisions are performed. These features preserve numerical accuracy.

APPENDIX D
DERIVATION OF THE DISCRETE FOURIER TRANSFORM (DFT) METHOD

The continuous-time filter F(s), Eq. (C.5), is transformed to a discrete-time filter $F_D(z)$, Eq. (C.7), by a given mapping function $f_D(z)$, Eq. (C.1). Substituting the right-hand side of (C.1) for s in (C.5) leads to $F_D(z)$ as in (C.7), which can be expressed as a ratio of polynomials in z as in (C.13). Equate coefficients of like powers of z in (C.7) and (C.13), taking the numerator and denominator separately. Letting z take on the successive values of the nr + 1 roots of unity,

$$\omega_k = \exp\left[i\, 2\pi k / (rn + 1)\right] \quad k = 0, 1, \ldots .nr \qquad i = \sqrt{-1} \tag{D.1}$$

leads to the linear simultaneous equations

$$\begin{bmatrix} 1 & 1 & \cdots & 1 \\ \omega_1^{nr} & \omega_1^{nr-1} & \cdots & \omega_1^{0} \\ \cdots & \cdots & \cdots & \cdots \\ \cdots & \cdots & \cdots & \cdots \\ \cdots & \cdots & \cdots & \cdots \\ \omega_{nr}^{nr} & \omega_{nr}^{nr-1} & \cdots & \omega_{nr}^{0} \end{bmatrix} \begin{bmatrix} B'_0 \\ B'_1 \\ \cdots \\ \cdots \\ \cdots \\ B'_{nr} \end{bmatrix} = \begin{bmatrix} \sum_{i=0}^{m} B_i \beta^{n-m+i} D^{n-m+i}(z)\, C^{m-i}(z) \big|_{z=\omega_0} \\ \sum_{i=0}^{m} B_i \beta^{n-m+i} D^{n-m+i}(z)\, C^{m-i}(z) \big|_{z=\omega_1} \\ \cdots \\ \cdots \\ \sum_{i=0}^{m} B_i \beta^{n-m+i} D^{n-m+i}(z)\, C^{m-i}(z) \big|_{z=\omega_{nr}} \end{bmatrix} \tag{D.2}$$

and

$$\begin{bmatrix} 1 & 1 & \cdots & 1 \\ \omega_1^{nr} & \omega_1^{nr-1} & \cdots & \omega_1^{0} \\ \cdots & \cdots & \cdots & \cdots \\ \cdots & \cdots & \cdots & \cdots \\ \cdots & \cdots & \cdots & \cdots \\ \omega_{nr}^{nr} & \omega_{nr}^{nr-1} & \cdots & \omega_{nr}^{0} \end{bmatrix} \begin{bmatrix} A'_0 \\ A'_1 \\ \cdots \\ \cdots \\ \cdots \\ A'_{nr} \end{bmatrix} =$$

$$= \begin{bmatrix} \sum_{i=0}^{n} A_i \beta^i D^i(z) C^{n-i}(z) \big|_{z=\omega_0} \\ \sum_{i=0}^{n} A_i \beta^i D^i(z) C^{n-i}(z) \big|_{z=\omega_1} \\ \cdots \\ \cdots \\ \sum_{i=0}^{n} A_i \beta^i D^i(z) C^{n-i}(z) \big|_{z=\omega_{nr}} \end{bmatrix} \quad \text{(D.3)}$$

Equations (D.2) and (D.3) can be abbreviated to matrix equations

$$F_{nr}\, b_{nr} = \Sigma_b \quad \text{(D.4)}$$

and

$$F_{nr}\, a_{nr} = \Sigma_a \quad \text{(D.5)}$$

where F_{nr} is the (nr + 1) x (nr + 1) Fourier matrix. The solutions of (D.4) and (D.5) are obtained by taking the Inverse Discrete Fourier Transform (IDFT) of the vectors Σ_b and Σ_a. Thus, the nr x 1 vector of B'_i coefficients, b_{nr}, is obtained by

$$b_{nr} = \text{IDFT}(\Sigma_b) \tag{D.6}$$

and the nr x 1 vector of A'_i coefficients, a_{nr}, is obtained by

$$a_{nr} = \text{IDFT}(\Sigma_a) \tag{D.7}$$

Consider the fifth-order example, (33). Coefficients for this example, obtained by the DFT method for both Tustin's rule and Schneider's rule, as implemented with MATLAB, are listed in Table VI.

Table VI

	Tustin's rule	Schneider's rule
B'_0	1.26252343750000E-07	0.07294172122155E-6
B'_1	1.28761718750000E-07	0.20580903983411E-6
B'_2	-2.47476562500000E-07	-0.10501155146846E-6
B'_3	-2.52476562500000E-07	-0.52085865162037E-6
B'_4	1.21261718750000E-07	0.08775404188368E-6
B'_5	1.23752343750000E-07	0.37424972511574E-6
B'_6		-0.12892696578414E-6
B'_7		0.01466565393518E-6
B'_8		-0.00054706036603E-6
B'_9		-0.00000095244985E-6
B'_{10}		-0.00000000030141E-6
A'_0	1.14416842020000E+00	1.11919892686631
A'_1	-5.41890448400000E+00	-5.27387884476273
A'_2	1.02616673620000E+01	9.91098259555845
A'_3	-9.71218680800000E+00	-9.26048848877315
A'_4	4.59416426100000E+00	4.26763188433450

A'_5	-8.68908664800001E -01	-0.74367847298565
A'_6		-0.01960065694328
A'_7		-0.00016628116204
A'_8		-0.00000057494154
A'_9		-0.00000000079052
A'_{10}		-0.00000000000034

The MATLAB* * program for the DFT method using Schneider's rule is:

```
n = [0 1 2 3 4 5 6 7 8 9 10];
z = exp ( (i*n*2*pi) / 11);
sn = z .* (z-1);
sd = (.01/12) .* (5 .* z.^2 + 8 .* z -1);
y = (1 .* sn.^2 .* sd.^3 + 2 .* sn.^1 .* sd.^4 + .75 .* sd.^5);
N = ifft(y);
x = (1 .* sn.^5 + 27.5 .*sn.^4 .* sd.^1 + 261.5 .* sn.^3 .*sd.^2 . . .
      + 1039 .* sn.^2 .*sd.^3 + 1668 .*sn.^1 .*sd.^4 + 864 .*sd.^5);
D = ifft(x);
```

APPENDIX E
DERIVATION OF THE TRAPEZOIDAL TIME-DOMAIN PROCESSING FORMULA

Consider the state-variable formulation of the continuous-time filter F(s)

* MATLAB is a trademark for The Math Works, Inc.

$$\dot{\underline{x}}(t) = \mathbf{A}\underline{x}(t) + \underline{b}u(t). \tag{E.1}$$

where A is not necessarily in the phase-variable form of Eq. (4a).
Letting $\dot{\underline{x}}(t) = \dot{\underline{x}}(kT) = \dot{\underline{x}}_k$ implies that

$$\dot{\underline{x}}_k = \mathbf{A}\underline{x}_k + \underline{b}u_k \tag{E.2}$$

$$\dot{\underline{x}}_{k-1} = \mathbf{A}\underline{x}_{k-1} + \underline{b}u_{k-1}. \tag{E.3}$$

Substituting (E.4) and (E.3) for the derivatives into the trapezoidal numerical integration formula:

$$\underline{x}_k = \underline{x}_{k-1} + \frac{T}{2}\cdot(\dot{\underline{x}}_k + \dot{\underline{x}}_{k-1}), \tag{E.4}$$

results in

$$\underline{x}_k = \underline{x}_{k-1} + \frac{T}{2}\cdot(\mathbf{A}\underline{x}_k + \underline{b}u_k + \mathbf{A}\underline{x}_{k-1} + \underline{b}u_{k-1}). \tag{E.5}$$

Solving for $\underline{x}_k$ yields the trapezoidal time-domain processing formula

$$\underline{x}_k = \left[\mathbf{I} - \frac{T}{2}\cdot\mathbf{A}\right]^{-1}\cdot\left\{\left[\mathbf{I} + \frac{T}{2}\cdot\mathbf{A}\right]\cdot\underline{x}_{k-1} + \frac{T}{2}\cdot\underline{b}\cdot(u_k + u_{k-1})\right\}, \tag{E.6}$$

for the continuous-time filter (E.1).

APPENDIX F

DERIVATION OF THE PARABOLIC TIME-DOMAIN PROCESSING FORMULA

Consider the state-variable formulation of the continuous-time filter F(s) represented in (E.1). Letting $\dot{\underline{x}}(t) = \dot{\underline{x}}(kT) = \dot{\underline{x}}_k$ implies that

$$\dot{\underline{x}}_k = \mathbf{A}\underline{x}_k + \underline{b}u_k \tag{F.1}$$

$$\dot{\underline{x}}_{k-1} = \mathbf{A}\underline{x}_{k-1} + \underline{b}u_{k-1} \tag{F.2}$$

$$\dot{\underline{x}}_{k-2} = \mathbf{A}\underline{x}_{k-2} + \underline{b}u_{k-2}. \tag{F.3}$$

Substituting (F.1 - 3) for the derivatives into the parabolic numerical integration formula

$$\underline{x}_k = \underline{x}_{k-1} + \frac{T}{12}\cdot(5\dot{\underline{x}}_k + 8\dot{\underline{x}}_{k-1} - \dot{\underline{x}}_{k-2}), \tag{F.4}$$

and solving for $\underline{x}_k$, yields the parabolic time-domain processing formula

$$\underline{x}_k = \left[\mathbf{I} - \frac{5T}{12}\cdot\mathbf{A}\right]^{-1}\cdot\left\{\left[\mathbf{I} + \frac{2T}{3}\cdot\mathbf{A}\right]\cdot\underline{x}_{k-1} - \frac{T}{12}\cdot\mathbf{A}\cdot\underline{x}_{k-2} + \frac{T}{12}\cdot\underline{b}\cdot(5u_k + 8u_{k-1} - u_{k-2})\right\} \tag{F.5}$$

for the continuous-time filter (E.1).

APPENDIX G

DERIVATION OF THE PARABOLIC AWARE-PROCESSING-MODE COMPENSATION FORMULA

Consider the state-variable formulation of the nth-order continuous-time filter represented in (E.1). Consider an input, u(t), with a discontinuity at time t_0. The parabolic aware-processing-mode compensation formula calculates $\underline{x}_1$, given $\underline{x}_0$, u_{0+}, u_1, and u_2. The n parabolas connecting the points $\dot{\underline{x}}_0$, $\dot{\underline{x}}_1$, and $\dot{\underline{x}}_2$, can be expressed in the form

$$\underline{P}(t) = \underline{\alpha}t^2 + \underline{\beta}t + \underline{\gamma}. \tag{G.1}$$

Note that $\underline{x}_1$ can be calculated by

$$\underline{x}_1 = \underline{x}_0 + \underline{\text{Area}}_{[0,1]}, \tag{G.2}$$

where $\underline{\text{Area}}_{[0,1]}$ denotes the area under the parabolas in the interval [0,T]. Using the fact that the derivatives can be expressed as values of the vector set of parabolas $\underline{P}(t)$ at various times, we find that

$$\dot{\underline{x}}_0 = \underline{\gamma} \tag{G.3}$$

$$\dot{\underline{x}}_1 = \underline{\alpha}T^2 + \underline{\beta}T + \underline{\gamma} \tag{G.4}$$

$$\dot{\underline{x}}_2 = 4\underline{\alpha}T^2 + 2\underline{\beta}T + \underline{\gamma}, \tag{G.5}$$

results in the parabolas having the form

$$\underline{P}(t) = \underline{\alpha}(\dot{\underline{x}}_0,\dot{\underline{x}}_1,\dot{\underline{x}}_2)t^2 + \underline{\beta}(\dot{\underline{x}}_0,\dot{\underline{x}}_1,\dot{\underline{x}}_2)t + \dot{\underline{x}}_0. \tag{G.6}$$

The fact that the derivatives can also be expressed as

$$\dot{\underline{x}}_0 = \mathbf{A}\underline{x}_0 + \underline{b}u_{0+} \tag{G.7}$$

$$\dot{\underline{x}}_1 = \mathbf{A}\underline{x}_1 + \underline{b}u_1 \tag{G.8}$$

$$\dot{\underline{x}}_2 = \mathbf{A}\underline{x}_2 + \underline{b}u_2, \tag{G.9}$$

results in the parabolas being functions of $\dot{\underline{x}}_0$, $\dot{\underline{x}}_1$, $\dot{\underline{x}}_2$, u_{0+}, u_1, and u_2. The resulting parabolas can be expressed as

$$\underline{P}(t) = f(\underline{x}_0, \underline{x}_1, \underline{x}_2, u_{0+}, u_1, u_2, t). \tag{G.10}$$

In order to calculate the areas under the different parabolas, the derivatives $\dot{\underline{x}}(t) = \underline{P}(t)$ are passed through a set of n parallel integrators. The areas under the parabolas over the intervals [0,T] and [0,2T] can be calculated using the formulas

$$\underline{\text{Area}}_{[0,1]} = \frac{T}{12}\cdot\left(-\dot{\underline{x}}_2 + 8\dot{\underline{x}}_1 + 5\dot{\underline{x}}_0\right) \tag{G.11}$$

$$\underline{\text{Area}}_{[0,2]} = \frac{T}{3}\cdot\left(\dot{\underline{x}}_2 + 4\dot{\underline{x}}_1 + \dot{\underline{x}}_0\right). \tag{G.12}$$

The resulting values of $\underline{x}_1$ and $\underline{x}_2$

$$\underline{x}_1 = \underline{x}_0 + \underline{\text{Area}}_{[0,1]} \tag{G.13}$$

$$\underline{x}_2 = \underline{x}_0 + \underline{\text{Area}}_{[0,2]}, \tag{G.14}$$

are both functions of $\underline{x}_0$, $\underline{x}_1$, $\underline{x}_2$, u_{0+}, u_1, and u_2. Since $\underline{x}_0$, u_{0+}, u_1, and u_2 are known, $\underline{x}_1$ and $\underline{x}_2$ can be solved for simultaneously. The solution for $\underline{x}_1$ yields the parabolic aware-processing-mode compensation formula

$$\underline{x}_1 = \left[8\mathbf{A}^2T^2 - 24\mathbf{A}T + 24\mathbf{I}\right]^{-1} \cdot \left\{ \begin{matrix} \left[-4\mathbf{A}^2T^2 + 24\mathbf{I}\right] \cdot \underline{x}_1 \\ -2\underline{b}T \cdot u_2 \\ +\left[-8\mathbf{A}\underline{b}T^2 + 16\underline{b}T\right] \cdot u_1 \\ +\left[-4\mathbf{A}\underline{b}T^2 + 10\underline{b}T\right] \cdot u_{0+} \end{matrix} \right\}, \tag{G.15}$$

which can be generalized to apply to a discontinuity which occurs at time t_{k-1},

$$\underline{x}_k = \left[8\mathbf{A}^2T^2 - 24\mathbf{A}T + 24\mathbf{I}\right]^{-1} \cdot \left\{ \begin{matrix} \left[-4\mathbf{A}^2T^2 + 24\mathbf{I}\right] \cdot \underline{x}_k \\ -2\underline{b}T \cdot u_{k+1} \\ +\left[-8\mathbf{A}\underline{b}T^2 + 16\underline{b}T\right] \cdot u_k \\ +\left[-4\mathbf{A}\underline{b}T^2 + 10\underline{b}T\right] \cdot u_{(k-1)+} \end{matrix} \right\}. \tag{G.16}$$

REFERENCES

1. Alan M. Schneider, John Tadashi Kaneshige, and F. Dale Groutage, "Higher Order s-to-z Mapping Functions and Application in Digitizing Continuous-time Filters," Proc. of the IEEE, **Vol. 79, No. 11**, pp. 1661-1674, (November 1991).

2. Benjamin C. Kuo, Digital Control Systems. Holt, Rinehart, and Winston, Inc., New York, NY, p.315. (1980).

3. F. Dale Groutage, Leonid E. Volfson, and Alan M. Schneider, "S-Plane to Z-Plane Mapping Using a Simultaneous Equation Algorithm Based on the Bilinear Transformation," IEEE Transactions on Automatic Control, **AC-32**, pp. 365-637, (July 1987).

4. F. Dale Groutage, Leonid E. Volfson, and Alan M. Schneider, "S-Plane to Z-Plane Mapping, The General Case," IEEE Transactions on Automatic Control, **33**, pp. 1087-1088, (November 1988).

5. John T. Kaneshige, "Higher-Order S-Plane to Z-Plane Mapping Functions and Their Application in Digitizing Continuous-Time Filters," MS Thesis, Department of Applied Mechanics and Engineering Sciences, University of California at San Diego (June 1990).

6. J.L. Melsa and D.G. Schultz, Linear Control Systems, McGraw-Hill, New York, NY; pp. 41-63 (1969).

7. T. S. Parker and L. O. Chua, Practical Numerical Algorithms for Chaotic Systems, Springer-Verlag, New York, NY; pp. 90-101 (1989).

8. C. William Gear, Numerical Initial Value Problems in Ordinary Differential Equations, Prentice-Hall, Engelwood Cliffs, NJ; pp. 106 (1971).

9. Katsuhiko Ogata, Discrete-Time Control Systems, Prentice-Hall, Engelwood Cliffs, NJ; pp. 217-218, 234-235, 483-485 (1987).

10. Kendall E. Atkinson, An Introduction to Numerical Analysis, Wiley, New York, NY; Ch. 6 (1989).

11. James F. Martin, Alan M. Schneider, and N. Ty Smith, "Multiple-Model Adaptive Control of Blood Pressure Using Sodium Nitroprusside," IEEE Transactions on Biomedical Engineering, **Vol. BME-34**, No.8 (August 1987).

12. Leonid Volfson, "Pattern Recognition and Classification and its Applications to Pre-Processing Arterial Pressure Waveform Signals and Image Processing Problems," Ph.D. Thesis, Department of Applied Mechanics and Engineering Sciences, University of California at San Diego (September 1990).

13. Gene F. Franklin, J. David Powell, and Michael L. Workman, Digital Control of Dynamic Systems, 2nd edition, Addison-Wesley, Reading, MA; pp.147-149 (1990).

14. Chi-Hsu Wang, Mon-Yih Lin and Ching-Cheng Teng, "On the Nature of the Boxer-Thaler and Madwed Integrators and Their Application in Digitizing a Continuous-Time System," IEEE Transactions on Automatic Control, (Oct. 1990).

15. Gene F. Franklin and J. David Powell, Digital Control of Dynamic Systems, 1st edition, Addison Wesley, Reading, MA; pp. 65-66 (1981).

16 Johnny R. Johnson, Introduction to Digital Signal Processing, Prentice-Hall, Englewood Cliffs, NJ; (1975).

17. Andreas Antoniou, Digital Filters: Analysis and Design, McGraw-Hill, New York; McGraw-Hill, (1979).

18. Robert D. Strum and Donald E. Kirk, First Principles of Discrete Systems and Digital Signal Processing, Addison-Wesley, New York; (1989).

19. Lawrence R. Rabiner and Bernard Gold, Theory and Application of Digital Signal Processing, Prentice-Hall, Englewood Cliffs, NJ; (1975).

DESIGN OF 2-DIMENSIONAL RECURSIVE DIGITAL FILTERS

M. Ahmadi
Department of Electrical Engineering
University of Windsor
Windsor, Ontario, N9B 3P4, Canada

I. INTRODUCTION

Two-dimensional (2-D) digital filters are computational algorithms that transform a 2-D input sequence of numbers into a 2-D output sequence of numbers according to pre-specified rules, hence yielding some desired modifications to the characteristics of the input sequences.

Applications of 2-D digital filters cover a wide spectrum. The objective is usually being either enhancement of an image to make it more acceptable to the human eye, or removal of the effects of some degradation mechanisms, or separation of features to facilitate identification and measurement by human or machine.

II. CHARACTERIZATION OF 2-D DIGITAL FILTERS

Linear digital filters are generally classified into two groups, namely non-recursive, that is known as Finite Impulse Response (FIR) and recursive, which is known as Infinite Impulse Response (IIR) digital filter. The emphasis of this chapter will be on the design of linear causal 2-D IIR digital filters.

CONTROL AND DYNAMICS SYSTEMS, VOL. 78

A 2-D recursive digital filter can be characterized by its difference equation as

$$y(m,n) = \sum_{i=0}^{M_1}\sum_{j=0}^{N_1} a_{ij}x(m-i,n-j) - \sum_{i=0}^{M_2}\sum_{\substack{j=0\\ i+j\neq 0}}^{N_2} b_{ij}y(m-i,n-j) \tag{1}$$

or by its transfer function as

$$H(z_1, z_2) = \frac{A(z_1,z_2)}{B(z_1,z_2)} = \frac{\sum_{i=0}^{M_1}\sum_{j=0}^{N_1} a_{ij}\, z_1^{-i}\, z_2^{-j}}{\sum_{i=0}^{M_2}\sum_{j=0}^{N_2} b_{ij}\, z_1^{-i}\, z_2^{-j}} \tag{2}$$

As can be seen from eqn.(1) output of the recursive filter is a function of the present and the past inputs as well as the past outputs of the filter. Therefore, it is possible for the output to become increasingly large despite the size of the input signal. As a result, these filters can become unstable. It should be noted that stability means the response to a bounded input data array should be a bounded array. This is commonly known as BIBO stability. To ensure stability, the denominator polynomial must satisfy the following constraints [1]

$$B(z_1, z_2) \neq 0 \quad \text{for} \bigcap_{i=1}^{2} |z_i| \geq 1 \tag{3}$$

Assuming that factorization of $A(z_1, z_2)$ and/or $B(z_1, z_2)$ is possible, one may define three different subclasses of 2-D recursive digital filters as follows:

II. 1. Separable Product

In this subclass of 2-D recursive digital filter, it is assumed that factorization of both $A(z_1, z_2)$ and $B(z_1, z_2)$ into polynomials consist of z_1 and z_2 variables is possible. Therefore, the transfer function of the 2-D filter becomes

$$H(z_1, z_2) = \frac{A_1(z_1)A_2(z_2)}{B_1(z_1)B_2(z_2)} = H_1(z_1)H_2(z_2) \tag{4}$$

where

$$H_i(z_i) = \frac{A_i(z_i)}{B_i(z_i)} \quad \text{for } i = 1, 2 \tag{5}$$

and

$$A_i(z_i) = \sum_{m=0}^{M_i} a_{im}\, z_i^{-m} \tag{6a}$$

$$B_i(z_i) = \sum_{n=0}^{N_i} b_{in}\, z_i^{-n} \quad \text{for } i = 1, 2 \tag{6b}$$

The advantages of this subclass of 2-D filters are:

(a) The stability problem is reduced to that of the 1-D filter.

(b) The problem of 2-D filter design will be reduced to that of the known design procedures of 1-D filters.

The disadvantage of this subclass is the shape of cutoff boundary, which is restricted to a rectangular one.

II. 2 Separable Numerator Non-Separable Denominator 2-D Recursive Filters

The transfer function of this class of 2-D filter is driven by assuming only the reparability of $A(z_1, z_2)$ exists. In this case we obtain

$$H(z_1, z_2) = \frac{A_1(z_1)A_2(z_2)}{B(z_1, z_2)} \tag{7}$$

where $A_i(z_i)$ for i = 1, 2 is the same as eqn.(6a) while

$$B(z_1, z_2) = \sum_{i=0}^{M_2} \sum_{j=0}^{N_2} b_{ij}\, z_1^{-i}\, z_2^{-j} \tag{8}$$

This class of 1-D filters presents the same difficulties regarding the stability problem as of the general class of eqn.(2), since there are no advantages in using them we will omit any further discussion on their design.

II. 3 Separable Denominator Non-Separable Numerator 2-D Recursive Filters

The transfer function of this subclass of filters is

$$H(z_1, z_2) = \frac{A_1(z_1)A_2(z_2)}{B(z_1, z_2)} = \frac{\sum_{i=0}^{M_1}\sum_{i=0}^{N_1} a_{ij} z_1^{-i} z_2^{-j}}{\left[\sum_{i=0}^{M_2} b_{1i} z_1^{-i}\right]\left[\sum_{j=0}^{N_2} b_{2j} z_2^{-j}\right]} \tag{9}$$

This sub-class of filters has the advantages of the separable product filters but not their disadvantage, i.e. circular, elliptical and in general non-symmetrical cutoff boundary 2-D filters can be designed using this sub-class of 2-D filters [2] - [10].

The transfer function of eqn.(9) represents a 2-D filter with central symmetry. This means that the magnitude response of the 2-D filter is the same in the first $(\omega_1, \omega_2 \geq 0)$ and the third $(\omega_1, \omega_2 \leq 0)$ quadrants; similarly for the second $(\omega_1 \leq 0,\ \omega_2 \geq 0)$ and the fourth $(\omega_1 \geq 0,\ \omega_2 \leq 0)$ quadrants.

A quadrantal symmetric 2-D filter is a filter with identical magnitude response in all quadrants. A quadrantal symmetric 2-D filter [2], [11] is o btained if the following constraints are imposed on the transfer function of central symmetric 2-D filter defined in eqn.(9).

$$a_{ij} = a_{M-i,j} = a_{i,M-j} = a_{M-i,\,M-j} \tag{10}$$

$$b_{1i} = b_{2j} \tag{11}$$

and considering $M_1 = M_2 = N_1 = N_2 = M$. In this case eqn.(9) can be rewritten as

$$H(z_1, z_2) = \frac{z_1^{-M/2} z_2^{-M/2} \sum_{i=0}^{M/2} \sum_{j=0}^{M/2} a_{ij} \cos(i w_1) \cos(j w_2)}{\left[\sum_{i=0}^{M} b_i z_1^{-i}\right]\left[\sum_{j=0}^{M} b_j z_2^{-j}\right]} \tag{12}$$

where $\cos \omega_k = \left(z_k^{-1} + z_k\right) / 2$ for $k = 1, 2$ or alternatively as

$$H\left(z_1, z_2\right) = \frac{z_1^{-M/2} \, z_2^{-M/2} \sum_{i=0}^{M/2} \sum_{j=0}^{M/2} a'_{ij} \cos\left(w_1\right)^i \cos\left(w_2\right)^j}{\left[\sum_{i=0}^{M} b_i \, z_1^{-i}\right]\left[\sum_{j=0}^{M} b_j \, z_2^{-j}\right]} \tag{13}$$

It has been shown in [2] that the design of a circular symmetric 2-D filter is only possible through this type of transfer function.

If in addition to the above constraints for the quadrantal symmetric 2-D filter, we include an additional constraint

$$a_{ij} = a_{ji} \tag{14}$$

then an octagonal symmetric 2-D filter is realized [2]. The magnitude response of this sub-class of 2-D filters is similar in all octants.

In the subsequent section, methods for the design of various sub-class of 2-D filters will be presented.

III. DESIGN OF SEPARABLE PRODUCT 2-D FILTERS

The design of 2-D digital filters can be simplified considerably if the transfer function of the 2-D filter can be factored into a product of two 1-D transfer function as in eqn.(4) [12]

$$H(z_1, z_2) = H_1(z_1)\, H_2(z_2) \tag{15}$$

This requires design of two 1-D filters using any of the known techniques presented in [13], [14]. It should be noted that filters designed using this technique will have rectangular cut-off boundary.

Let us assume that the 2-D filter specification in terms of passband gain and stopband attenuation is given. We would like to derive the constraints that must be imposed on the two 1-D filters so that such specifications are satisfied.

It is desired to design a 2-D filter with the following magnitude specification.

$$|H(\omega_1, \omega_2)| = \begin{cases} 1 \pm \Delta p & \text{for } 0 \le |\omega_i| \le \omega_{pi} \\ \Delta a & \text{for } \omega_{ai} \le |\omega_i| \le \frac{\omega_{si}}{2} \end{cases} \tag{16}$$

where ω_{pi} and ω_{ai}, $i = 1, 2$, are passband and stop-band edges along the ω_1 and ω_2 axes, respectively, and Δp and Δa are passband ripple and stopband losses.

Let us assume that the two 1-D filters which are cascaded to form a 2-D filter are specified by ω_{pi}, ω_{ai}, δ_{pi} and δ_{ai} (i, 1,2) as the pass-band edge, stop-band edge, pass-band ripple, and stop-band loss, respectively. Then from equation (15) we can deduct

$$\text{Max}\left\{M(\omega_1,\omega_2)\right\} = \max\left\{M_1(\omega_1)\right\}\max\left\{M_2(\omega_2)\right\}$$

and

$$\text{Min}\left\{M(\omega_1,\omega_2)\right\} = \min\left\{M_1(\omega_1)\right\}\min\left\{M_2(\omega_2)\right\}$$

where $M(\omega_1,\omega_2)$, $M_1(\omega_1)$ and $M_2(\omega_2)$ are magnitude responses of 2-D filter, and the two 1-D filter along ω_1 and ω_2 axes, respectively. It is obvious from the above that the derived 2-D filter will satisfy the specification of equation (16) if the following relationships are met.

$$\left(1+\delta_{p1}\right)\left(1+\delta_{p2}\right) \leq 1 + \Delta p \tag{17}$$

$$\left(1-\delta_{p1}\right)\left(1-\delta_{p2}\right) \geq 1 - \Delta p \tag{18}$$

$$\left(1+\delta_{p1}\right)\delta_{a2} \leq \Delta a \tag{19}$$

$$\left(1+\delta_{p2}\right)\delta_{a1} \leq \Delta a \tag{20}$$

$$\delta_{a1}\,\delta_{a2} \leq \Delta a \tag{21}$$

Constraints (17) and (18) can be expressed respectively in the alternative forms as follows

$$\delta_{p1}+\delta_{p2}+\delta_{p1}\,\delta_{p2} \leq \Delta_p \tag{22}$$

and

$$\delta_{p1}+\delta_{p2}-\delta_{p1}\,\delta_{p2} \leq \Delta_p \tag{23}$$

Therefore, if the constraint (22) is satisfied, eqn.(23) is also satisfied. One can also deduct that the constraints (19) - (21) will be satisfied if

$$\max\left\{\left(1+\delta_{p1}\right)\delta_{a2}\,,\,\left(1+\delta_{p2}\right)\delta_{a1}\right\} \le \Delta a \tag{24}$$

since

$$\left(1+\delta_{p1}\right) >> \delta_{a1} \quad \text{and} \quad \left(1+\delta_{p2}\right) >> \delta_{a2} \tag{25}$$

Now assuming that

$$\delta_{p1} = \delta_{p2} = \delta_p \tag{26}$$

and

$$\delta_{a1} = \delta_{a2} = \delta_a \tag{27}$$

One can write

$$\delta_p = (1+\Delta p)^{1/2} - 1 \tag{28}$$

and

$$\delta_a = \Delta a \;/\; (1+\Delta p)^{1/2} \tag{29}$$

so as to satisfy constraints (17) - (21). Now with the knowledge of the pass-band ripple and stop-band loss of the two 1-D filters using eqns.(28) and (29), one can employ any of the 1-D design technique reported in [13]-[14] to design them.

This technique is also capable of designing 2-D bandpass, highpass as well as band-stop filters with rectangular cut-off boundary.

IV. DESIGN OF NON-SEPARABLE NUMERATOR, SEPARABLE DENOMINATOR 2-D FILTERS

In this section several techniques to calculate the parameters of the separable denominator and non-separable numerator 2-D transfer functions with quadrantal or octagonal symmetric characteristics, and constant group delay responses are described. It is worth noting that these sub-classes of 2-D filters, unlike the separable product filters, are capable of providing any arbitrary cut-off boundaries. On the other hand, their stability problem is reduced to that of 1-D filters.

IV. 1 Method I [8]

In this method, without loss of generality, we assume that $M_1 = N_1 = M_2 = N_2 = M$ in eqn.(9). We also assume that

$$a_{ij} = a_{(M-i)j} = a_{i(M-j)} = a_{(M-i)(M-j)} \tag{30}$$

with

$$B_1(z_1) = B(z_1) \qquad B_2(z_2) = B(z_2) \tag{31}$$

to obtain quadrantal symmetric magnitude response.

In this case, the transfer function will be in the form of eqn. (12) which is

$$H(z_1,z_2)=\frac{z_1^{-M/2}z_2^{-M/2}\sum_{i=0}^{M/2}\sum_{j=0}^{M/2}a'_{ij}\cos(i\omega_1)\cos(j\omega_2)}{\left[\sum_{i=0}^{M}b_i z_1^{-i}\right]\left[\sum_{i=0}^{M}b_i z_2^{-i}\right]} \tag{32}$$

where $\cos\omega_k=\left(z_k^{-1}+z_k\right)/2$ for $k = 1, 2$.

Octagonal symmetric 2-D filter is obtained if

$$a'_{ij}=a'_{ji} \tag{33}$$

Designing a 2-D filter means the calculation of the coefficients of the filter transfer function (32) in such a way that the magnitude response and/or phase response of the designed 2-D filter approximates the desired characteristics while maintaining the stability. The latter requires that the roots of the 1-D polynomials eqn.(32) be calculated at the end of each design process. If any of the roots of 1-D polynomials is found to be outside the unit circle in the z_1 and z_2 plane, hence instability, they should be replaced by their mirror image with respect to their corresponding unit circle in the z_1 and z_2 plane to stabilize the filter.

This stabilization procedure unfortunately changes the group delay characteristics of the designed filter, i.e., if the designed 2-D filter has constant group delay responses (linear-phase), at the end of this process the group delay responses will no longer remain constant.

In this method, based on the properties of the positive definite and semi-definite matrices, we generate a polynomial which has all its zeros inside the unit circle and assign it to the denominator of eqn. (32). The new coefficients of the derived transfer function are then used as the parameters of the optimization so that the desired magnitude and phase responses are obtained.

IV. 1.1 Generation of 1-D Hurwitz polynomials

Any positive definite matrix can be decomposed as

$$\Delta_1 = D\Gamma D^T s + G \tag{34}$$

where "s" is a complex variable, "D" is an upper triangular matrix with unity elements in its diagonal, "D^T" is the transpose of matrix "D", "Γ" is a diagonal matrix with non-negative elements and "G" is a skewed symmetric matrix as follows

$$D = \begin{bmatrix} 1 & d_{12} & d_{13} & \cdots & d_{1n} \\ 0 & 1 & d_{23} & \cdots & d_{2n} \\ . & . & . & \cdots & . \\ . & . & . & \cdots & . \\ . & . & . & \cdots & . \\ 0 & 0 & 0 & \cdots & 1 \end{bmatrix} \tag{35}$$

$$\Gamma = \begin{bmatrix} \gamma_1^2 & 0 & 0 & \cdots & 0 \\ 0 & \gamma_2^2 & 0 & \cdots & 0 \\ . & . & . & \cdots & . \\ . & . & . & \cdots & . \\ . & . & . & \cdots & . \\ 0 & 0 & 0 & \cdots & \gamma_n^2 \end{bmatrix} \tag{36}$$

$$G = \begin{bmatrix} 0 & g_{12} & g_{13} & \cdots & g_{1n} \\ -g_{12} & 0 & g_{23} & \cdots & g_{2n} \\ -g_{13} & -g_{23} & 0 & \cdots & g_{3n} \\ \cdot & \cdot & \cdot & \cdots & \cdot \\ \cdot & \cdot & \cdot & \cdots & \cdot \\ -g_{1n} & -g_{2n} & -g_{3n} & \cdots & 0 \end{bmatrix} \tag{37}$$

It is known that Δ_1 is always physically realizable [15]. Therefore determinant of Δ_1 constitutes the even or odd part of a Hurwitz polynomial in "s". In this case

$$B(s) = \det \Delta_1 + k_1 \frac{\partial \det \Delta_1}{\partial s} \tag{38}$$

is a Hurwitz polynomial (HP) in s where k is a positive number. The order of B(s) is equal to the rank of matrix Δ_1. For example to generate a second order HP by using the above method, one may write

$$\Delta_1 = \begin{bmatrix} 1 & d \\ 0 & 1 \end{bmatrix} \begin{bmatrix} \gamma_1^2 & 0 \\ 0 & \gamma_2^2 \end{bmatrix} \begin{bmatrix} 1 & 0 \\ d & 1 \end{bmatrix} s + \begin{bmatrix} 0 & g \\ -g & 0 \end{bmatrix}$$

$$\det \Delta_1 = \gamma_1^2 \gamma_2^2 \ s^2 + g^2 \tag{39}$$

$$B(s) = \gamma_1^2 \gamma_2^2 \ s^2 + 2k \gamma_1^2 \gamma_2^2 \ s + g^2 \tag{40}$$

Assuming k= 2 gives

$$B(s) = \gamma_1^2 \gamma_2^2 \ s^2 + 4 \gamma_1^2 \gamma_2^2 \ s + g^2 \tag{41}$$

Higher order HP's can be obtained by either choosing Δ_1 with higher order rank or by cascading an appropriate number of lower order HP's. To obtain the discrete version of the above polynomial, one may use bilinear transformation.

Modification can be made to the above technique in order to alleviate the cumbersome process of calculating partial derivatives. It has been shown in [9] that addition of resistive matrices to eqn.(34) can give

$$\Delta_2 = D\Gamma D^T s + g + R\Sigma R^T \tag{42}$$

where "D", "Γ" and "G" are as described in eqn. (35) - (37) while R is an upper triangular matrix of the following form

$$R = \begin{bmatrix} 1 & r_{12} & r_{13} & \cdots & r_{1n} \\ 0 & 1 & r_{23} & \cdots & r_{2n} \\ 0 & 0 & 1 & \cdots & r_{3n} \\ \cdot & & & & \\ \cdot & & & & \\ \cdot & & & & \\ 0 & 0 & 0 & & 1 \end{bmatrix} \tag{43}$$

and Σ is a diagonal matrix with non-negative elements as shown below

$$\Sigma = \begin{bmatrix} \delta_1^2 & 0 & 0 & 0 \dots & 0 \\ 0 & \delta_2^2 & 0 & 0 \dots & 0 \\ \cdot & & & & \\ \cdot & & & & \\ \cdot & & & & \\ 0 & 0 & 0 & 0 & \delta_n^2 \end{bmatrix} \tag{44}$$

One can easily show that the determinant of Δ_2 in eqn.(42) constitutes a Hurwitz Polynomial. For example, a second order HP by setting the rank of Δ_2 to be equal to 2 is generated as follows

$$\Delta_2 = \begin{bmatrix} 1 & d \\ 0 & 1 \end{bmatrix} \begin{bmatrix} \gamma_1^2 & 0 \\ 0 & \gamma_2^2 \end{bmatrix} \begin{bmatrix} 1 & 0 \\ d & 1 \end{bmatrix} s + \begin{bmatrix} 0 & g \\ -g & 0 \end{bmatrix}$$

$$+ \begin{bmatrix} 1 & r \\ 0 & 1 \end{bmatrix} \begin{bmatrix} \delta_1^2 & 0 \\ 0 & \delta_2^2 \end{bmatrix} \begin{bmatrix} 1 & 0 \\ r & 1 \end{bmatrix} \tag{45}$$

The determinant of Δ_2 in eqn.(45) yields

$$B(s) = \det(\Delta_2) = \gamma_1^2 \gamma_2^2 s^2 + [\sigma_1^2 \gamma_2^2 + \sigma_1^2 \gamma_1^2 + (d-r)^2 \sigma_2^2 \gamma_2^2] s$$
$$+ (\sigma_1^2 \sigma_2^2 + g^2) \tag{46}$$

which is a second-order HP. Higher order HP's can be generated by simply raising the order of the matrices in eqn.(42) or just by cascading lower order HP's. However, it should be noted that the latter approach is suboptimal.

IV. 2 Formulation of The Design Problem

In this design method, a 1-D HP is generated using any of the two techniques presented earlier. The discrete version of the derived HP polynomial is obtained by the application of the bilinear transformation. Note that, this would yield a rational function with all its zeros inside the unit circle, and the poles at "-1". The numerator of this rational function, which is a stable 1-D polynomial in z, is assigned to the denominator of eqn.(13). Now the coefficients of this 2-D transfer function can be used as the parameters of optimization to meet the desired magnitude and phase responses.

A look at the transfer function in (13) would reveal that the numerator can either be considered as a linear phase or a zero phase polynomial in variable z_1 and z_2 depending whether "$z_1^{-M/2}\ z_2^{-M/2}$ " term is included in the transfer function or not and as a result the numerator can have no effect on the overall phase response of the transfer function. In fact, it is easy to show that the phase response is generated through the two 1-D denominator polynomials in z_1 and z_2. The obvious approach to calculate the parameters of the filter transfer function is to separately use the coefficients of the denominator polynomials for phase specification only and then the numerator coefficients to achieve the overall magnitude response of the 2-D filter. This approach, however, has a drawback, that is through the phase approximation step, the magnitude of the two 1-D all pole transfer function may generate spikes which cannot be compensated by the numerator which is a 2-variable zero-phase (or linear phase) polynomial. Therefore the following modification to the above approach has been added.

Assume that the two 1-D polynomials in eqn. (13) are identical except for the variable. This is true in octagonal and quadrantal symmetric 2-D filters. We calculate the error between the magnitude response of the ideal and the designed 1-D filter as follows

$$E_{mag}(\omega_{in}, \Psi) = \left|H_I(e^{j\omega_{in}T})\right| - \left|H_D(e^{j\omega_{in}T})\right| \text{ for } i = 1,2 \qquad (47)$$

where "n" is the number of discrete frequency points along the axes, "ψ" is the coefficient vector, $\left|H_I\left(e^{j\omega_i T}\right)\right|$ is the ideal magnitude response of the 2-D filter $\left(\left|H_I\left(e^{j\omega_1 T}, e^{j\omega_2 T}\right)\right|\right)$ along ω_i axis for i = 1, 2 and $|H_D|$ is the magnitude response of the designed 1-D all-pole filter defined as

$$\left|H_D(e^{j\omega_{in}T})\right| = \left|\frac{1}{\sum_{j=0}^{M} b_{ij} z_i^{-j}}\right|_{z_i = e^{j\omega_{in}T}} \quad \text{for } i = 1,2 \qquad (48)$$

The error between the group-delay response of the ideal and the designed filter is defined as

$$E_\tau(\omega_{in}, \psi) = \tau_I T - \tau(\omega_{in}) \qquad i = 1,\ 2 \qquad (49)$$

where τ_I is the ideal group-delay response of the 2-D filter $(\tau_I(\omega_1, \omega_2))$ along ω_i axis for i = 1, 2 while $\tau(\omega_i)$ is the group-delay response of the designed filter. The objective function is defined as the general least mean square error and is calculated using the following relationship:

$$E_g(\omega_{in}, \psi) = \sum_{n \in I_{ps}} E^2_{mag}(\omega_{in}, \underline{\psi}) + \sum_{n \in I_p} E^2_\tau(\omega_{in}, \psi) \qquad \text{for } i = 1, 2 \tag{50}$$

where I_{ps} is a set of all discrete frequency points along ω_i, i = 1, 2 axis in the passband and the stopband of the filter, and I_p is a set of all passband discrete frequency points along ω_i, i = 1, 2.

Now the coefficient vector "ψ" can be calculated by minimizing E_g in eqn. (50). This is a simple non-linear optimization problem and can be solved by using any suitable unconstrained non-linear optimization technique. After calculation of the coefficients of the two 1-D polynomial, in the denominator of eqn.(13) we employ another objective function for the calculation of the coefficients of the numerator of the transfer-functions using the following relationship

$$E_{mag}(j\omega_{1m}, j\omega_{1m}, \Psi) = \left|H_I(e^{j\omega_{1m}T}, e^{j\omega_{2n}T})\right| - \left|H_D(e^{j\omega_{1m}T}, e^{j\omega_{2n}T})\right| \tag{51}$$

where "ψ" is the coefficient vector (the coefficients of the numerator polynomials in (13)) and $|H_D|$ is the magnitude response of the ideal 2-D filter, while $|H_D|$ is the magnitude response of the designed filter. The least mean square error is calculated using the relationship

$$E_{\ell_2}(j\omega_{1m}, j\omega_{2n}, \Psi) = \sum_{m,n \in I_{ps}} \sum E^2_{mag}(j\omega_{1m}, j\omega_{2n}, \Psi) \quad (52)$$

where I_{ps} is a set of all discrete frequency points along ω_1, ω_2 axis covering both the pass-band and the stopband of the 2-D filter. By minimizing E_{ℓ_2} in eqn.(52) using any of the non-linear or linear unconstrained optimization technique, coefficients of the numerator of the transfer function can be determined so that the overall magnitude response is obtained. This technique though suboptimal but is extremely fast and efficient and offers a considerable reduction in the computation costs as compared to methods of [2], [5].

IV. 3 Design Example

To test the utility of the described design method we design an octagonal symmetry 2-D filter with the following magnitude specification and constant group-delay responses.

$$\left|H_I\left(e^{j\omega_{1m}T}, e^{j\omega_{2n}T}\right)\right| = \begin{cases} 1 & 0 \le \sqrt{\omega_{1m}^2 + \omega_{2n}^2} \le 1. \text{ rad/sec} \\ 0 & 2.5 \le \sqrt{\omega_{1m}^2 + \omega_{2n}^2} \le 5 \text{ rad/sec} \end{cases}$$

ω_s, the sampling frequency, is chosen to be equal to 10 rad/sec. Table (1) shows the coefficients of the designed 2-D filter while Fig. 1 (a-c) show the magnitude and group-delay responses of the designed 2-D filter with respect, respectively.

TABLE 1

Values of the Coefficients of the Designed 2-D Filters

denominator coefficients (eqn.(41))	numerator coefficients (eqn.(32))
$\gamma_{11} = \gamma_{12} = 2.5092$	$a'_{00} = -0.1757$
$\gamma_{21} = \gamma_{22} = 0.1831$	$a'_{01} = a'_{10} = 0.4312$
$g_1 = g_2 = 1.0153$	$a'_{11} = 0.4427$

In the design method presented earlier, both steps of the optimization techniques used for the determination of the coefficients of the 2-D filter eqn.(13) were non-linear. In the method that will be presented later a modification is made to the above technique to obtain a better local minimum. Thus modification uses non-linear optimization technique is the first step similar to that of [8] but in the second step linear programming approach is utilized to calculate the coefficients of the numerator of the 2-D transfer function. Details of the proposed modified design technique is presented in the next section.

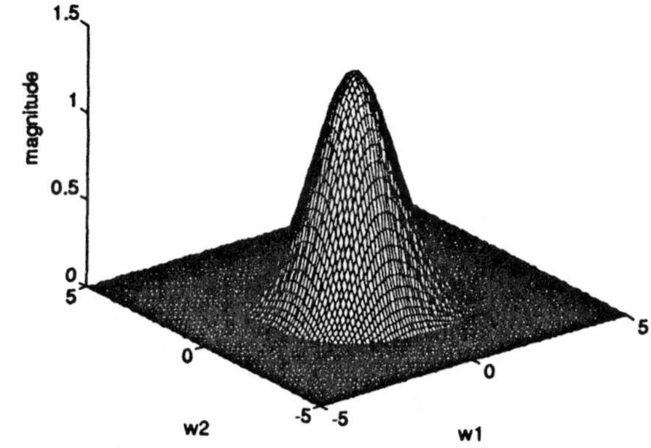

(a)

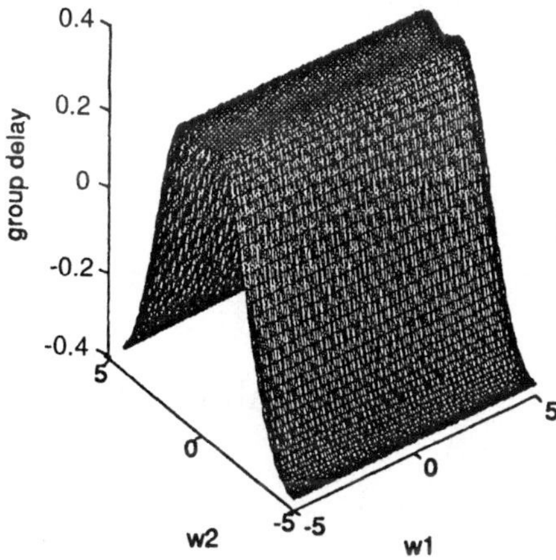

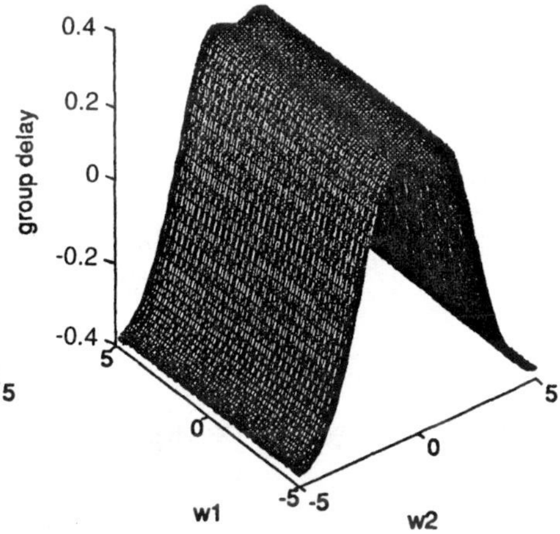

(b) (c)

Fig. 1 (a) magnitude response (b) group delay with respect to ω_1 (c) group delay with respect to ω_2

IV. 3 Modified Design Method for Quadrantal/Octagonal Symmetric 2-D Filter

In the modified design method presented here, the following steps are taken:

(i) Design two 1-D digital filters satisfying the magnitude specification of the designed 2-D filter along the ω_1 and ω_2 axes with or without constant group-delay responses using the technique given in [16]. Note that if the phase specification is not of any concern, two 1-D analog filters (Butterworth, Chebyshev or Elliptic) could be designed and then discretized using bilinear transformation. In this step

$$H_1(z_1, z_2) = H(z_1)\,H(z_2) \tag{53}$$

is a separable product 2-D filter with a rectangular cut-off boundary.

(ii) Cascade the designed $H_1(z_1, z_2)$ with a 2-D non-recursive digital filter of the form

$$H_2(z_1, z_2) = \sum_{i=0}^{M/2} \sum_{j=0}^{M/2} a'_{ij} \cos i\omega_1 \cos j\omega_2 \tag{54}$$

or

$$H_2(z_1, z_2) = \sum_{i=0}^{M/2} \sum_{j=0}^{M/2} \hat{a}_{ij} (\cos \omega_1)^i (\cos \omega_2)^j \tag{55}$$

Note that cascading H_1 by either of H_2 in eqns. (54) and (55) yields a 2-D filter with the same phase characteristics of $H_1(z_1, z_2)$ since $H_2(z_1, z_2)$ can only have either zero or linear phase characteristics.

(iii) Calculate the error of the magnitude response as

$$E_{mag}(j\omega_{im}, j\omega_{2n}, \psi) = \left|H_I\left(e^{j\omega_{1m}T}, e^{j\omega_{2n}T}\right)\right|$$

$$- \left|H_1\left(e^{j\omega_{1m}T}, e^{j\omega_{2n}T}\right)\right| \left|H_2\left(e^{j\omega_{1m}T}, e^{j\omega_{2n}T}\right)\right| \le \varepsilon \tag{56}$$

or

$$\left|H_2(e^{j\omega_{1m}T}, e^{j\omega_{2n}T})\right| \ge \frac{\left|H_I(e^{j\omega_{1m}T}, e^{j\omega_{2n}T})\right|}{\left|H_1(e^{j\omega_{1m}T}, e^{j\omega_{2n}T})\right|} - \frac{\varepsilon}{\left|H_1(e^{j\omega_{1m}T}, e^{j\omega_{2n}T})\right|} \tag{57}$$

or alternatively one can write

$$\begin{cases} H_2\left(e^{jw_{1m}T}, e^{jw_{2n}T}\right) + \dfrac{\varepsilon}{\left|H_1\left(e^{jw_{1m}T}, e^{jw_{2n}T}\right)\right|} \ge \dfrac{\left|H_I\left(e^{jw_{1m}T}, e^{jw_{2n}T}\right)\right|}{\left|H_1\left(e^{jw_{1m}T}, e^{jw_{2n}T}\right)\right|} \\ H_2\left(e^{jw_{1m}T}, e^{jw_{2n}T}\right) - \dfrac{\varepsilon}{\left|H_1(e^{jw_{1m}T}, e^{jw_{2n}T})\right|} \le \dfrac{\left|H_I\left(e^{jw_{1m}T}, e^{jw_{2n}T}\right)\right|}{\left|H_1\left(e^{jw_{1m}T}, e^{jw_{2n}T}\right)\right|} \end{cases} \tag{58}$$

where in the above equation ε is the error tolerance to be minimized, T is the sampling period.

(iv) Calculate a'_{ij} in eqn.(54) or $\hat{a}_{ij}$ in eqn.(55) by utilization of linear programming optimization technique, i.e.

$$\begin{cases} \text{Minimize } \varepsilon \text{ in equation (57)} \\ \text{Subject to constraints of eqn. (58)} \end{cases}$$

The revised simplex [17] method can be used for the determination of the coefficients of the $H_2(z_1, z_2)$.

To illustrate the usefulness of the proposed technique a 2-D lowpass filter with the following magnitude specifications is designed,

$$\left|H_I(e^{jw_{1m}T}, e^{jw_{2n}T})\right| = \begin{cases} e^{-W} & \text{for } 0 \le W \le 0.8 \text{ rad/sec} \\ e^{-6W} & \text{for } 0.8 \le W \le \pi \text{ rad/sec} \end{cases} \tag{59}$$

where $W = \sqrt{\omega_{1m}^2 + \omega_{2n}^2}$, and $\omega_s = 2\pi$ rad/sec.

In this example, the method of [16] was used to derive two 1-D all-pole filters in z_1 and z_2 variables. Coefficients of the two identical 1-D filters are shown in Table (2).

TABLE 2

The Coefficients of $B_i(z_i)$ for i =1, 2 (eqn. (32))

$b_{10} = b_{20} = 5.563914$

$b_{11} = b_{21} = -11.47563$

$b_{12} = b_{22} = 11.87721$

$b_{13} = b_{23} = -6.616661$

$b_{14} = b_{24} = 1.651168$

$H_2(z_1, z_2)$ is chosen to be,

$$H_2(z_1, z_2) = \sum_{i=0}^{2}\sum_{j=0}^{2} \hat{a}_{ij} \cos(\omega_1)^i \cos(\omega_2)^j \qquad (60)$$

Table 3 shows the values of the coefficients of $H_2(z_1, z_2)$ for $\alpha = 6$ in eqn. (59), while Fig. (2a) and (2b) show the 3-D plot and contour map of the designed 2-D filter respectively.

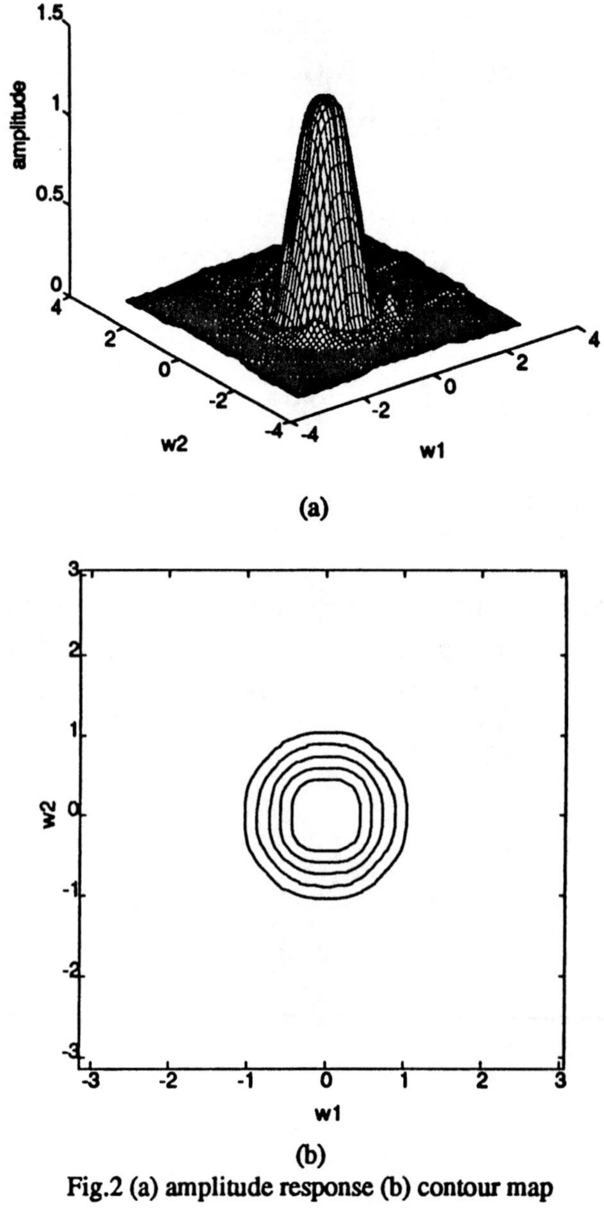

Fig.2 (a) amplitude response (b) contour map

TABLE 3

The Coefficients of $H_2(z_1, z_2)$ (eqn.(32))

$\hat{a}_{00}$ = 0.25633241

$\hat{a}_{01}$ = -0.96096868

$\hat{a}_{02}$ = 0.52468431

$\hat{a}_{10}$ = -0.96101382

$\hat{a}_{11}$ = 0.57964734

$\hat{a}_{12}$ = 0.55679245

$\hat{a}_{20}$ = 0.52472945

$\hat{a}_{21}$ = 0.55674732

$\hat{a}_{22}$ = 0.10711711

V. DESIGN OF GENERAL CLASS OF 2-D RECURSIVE DIGITAL FILTERS

In this class of 2-D filters numerator and denominator of the transfer function are non-separable 2-variable polynomials in z_1 and z_2 variables as follows

$$H(z_1, z_2) = \frac{A(z_1, z_2)}{B(z_1, z_2)} = \frac{\sum_{i=0}^{M} \sum_{j=0}^{M} a_{ij} z_1^{-i} z_2^{-j}}{\sum_{i=0}^{M} \sum_{j=0}^{M} b_{ij} z_1^{-i} z_2^{-j}} \tag{61}$$

The iterative techniques can be used to determine the coefficients of the 2-D transfer function by the means of minimizing an objective function, but they do not guarantee the stability of the designed 2-D filter. This usually requires carrying out a difficult stability test. If the designed filter is found to be unstable, optimization process has to be repeated by using a different set of initial values. Since to date there is no viable stabilization technique for 2-D filters is reported in literature. For this reason, researchers in the area of 2-D recursive digital filter design relied quite heavily on the transformation of a 2-variable passive analog network using double bilinear transformation [18]-[21]. However, Goodman [22] pointed out the possibility of instability of 2-D digital filters that have been derived through bilinear transformation applied to their analog 2-D counterpart. This is due to the presence of non-essential singularities of the second kind which are not present in the 1-D domain. Rajan et al [23] have studied the properties of Hurwitz Polynomials without any non-essential singularities of the second kind and termed them Very Strict Hurwitz Polynomials (VSHP). Ramachandran and Ahmadi [24] have used the properties of derivatives of the even or the odd parts of a 2-variable Hurwitz Polynomial to generate VSHPs. In another method, Ahmadi and Ramachandran [25] have used the properties of positive-definite and positive-semi-definite matrices to generate VSHPs. Both the above techniques require calculations of partial derivatives at some point in the process. They have also shown that if the derived 2-variable VSHP is discretized using double bilinear transformations and then assigned to the denominator of the 2-D filter transfer function in eqn.(61), the designed 2-D filter after the optimization process is guaranteed to be stable.

In the next section a method for generation of VSHPs is presented. This method does not require the cumbersome calculation of partial derivatives. It is also shown how this 2-variable VSHP can be used to design a stable 2-D recursive digital filter.

V. 1. Generation of 2-Variable VSHP's

As mentioned in [24]-[25], a symmetric positive-definite or a positive semi-definite matrix can always be realized physically. Any positive definite matrix "P" can be decomposed as a product of two upper or lower triangular matrices [26] as

$$P = QQ^T \tag{62}$$

where Q^T is the transpose of Q.

Consider the matrix D_m defined by

$$\begin{aligned} D_m &= AA^T s_1 + BB^T s_2 + CC^T + G \\ &= A_1 + B_1 + C_1 + G \end{aligned} \tag{63}$$

Matrices "A", "B", "C" are of order "m"as follows

$$A = \begin{bmatrix} a_{11} & a_{12} & a_{13} & a_{1m} \\ 0 & a_{22} & a_{23} & a_{2m} \\ 0 & 0 & a_{33} & a_{3m} \\ . & & & \\ . & & & \\ . & & & \\ 0 & 0 & 0 & a_{mm} \end{bmatrix} \tag{64}$$

$$B = \begin{bmatrix} b_{11} & b_{12} & b_{13} & b_{1m} \\ 0 & b_{22} & b_{23} & b_{2m} \\ 0 & 0 & b_{33} & b_{3m} \\ . & & & \\ . & & & \\ . & & & \\ 0 & 0 & 0 & b_{mm} \end{bmatrix} \quad (65)$$

$$C = \begin{bmatrix} c_{11} & c_{12} & c_{13} & c_{1m} \\ 0 & c_{22} & c_{23} & c_{2m} \\ 0 & 0 & c_{33} & c_{3m} \\ . & & & \\ . & & & \\ . & & & \\ 0 & 0 & 0 & c_{mm} \end{bmatrix} \quad (66)$$

where "a_{ii}", "b_{ii}" and "c_{ii}" are non-negative elements for $i = 1, 2, ..., m$.

The matrix "G" is skew-symmetric and is of the form

$$G = \begin{bmatrix} 0 & g_{12} & g_{13} & g_{1m} \\ -g_{12} & 0 & g_{23} & g_{2m} \\ -g_{13} & -g_{23} & 0 & g_{3m} \\ . & & & \\ . & & & \\ -g_{1m} & -g_{2m} & -g_{3m} & 0 \end{bmatrix} \quad (67)$$

If "D_m" is regarded as an impedance matrix, "A_1" (called s_1-matrix) can be realized by inductors in the variable "s_1". "B_1" (called s_2-matrix) can be realized by inductors in the variable "s_2", "C_1" can be realized by resistors (called resistor-matrix) and "G" can be realized by a m-port gyrator. Alternatively, if "D_m" is regarded as an admittance matrix, "A_1" can be realized by capacitors in the variable "s_1", "B_1" can be realized by capacitors in the variable "s_2", "C_1" can be realized by conductances and "G" can be realized by a m-port gyrator like before. In the both cases, the realizations of "A_1", "B_1" and "C_1" may require transformers.

We can prove that, under certain conditions, det $[D_m]$ (determinant of the matrix "D_m") gives a 2-variable VSHP. To show this, let "A" be partitioned as

$$A = \begin{bmatrix} A_{11} & \vdots & A_{12} \\ \cdots\cdots & \vdots & \cdots\cdots \\ 0 & \vdots & A_{22} \end{bmatrix} \tag{68}$$

where "A_{11}" is "a" (k x k) upper-triangular matrix, A_{12} is a (k x (m-k)) rectangular matrix and A_{22} is a ((m-k) x (m-k)) upper-triangular matrix. Similarly, for B, C and G, one can write

$$B = \begin{bmatrix} B_{11} & \vdots & B_{12} \\ \cdots\cdots & \vdots & \cdots\cdots \\ 0 & \vdots & B_{22} \end{bmatrix} \tag{69}$$

$$C = \begin{bmatrix} C_{11} & \vdots & C_{12} \\ \cdots\cdots & \vdots & \cdots\cdots \\ 0 & \vdots & C_{22} \end{bmatrix} \tag{70}$$

$$G = \begin{bmatrix} G_{11} & \vdots & G_{12} \\ \cdots\cdots & \vdots & \cdots\cdots \\ G_{12}^T & \vdots & G_{22} \end{bmatrix} \quad (71)$$

Now, the product AA^T can be written as

$$AA^T = \begin{bmatrix} A_{11} & A_{12} \\ 0 & A_{22} \end{bmatrix} \begin{bmatrix} A_{11}^T & 0 \\ A_{12}^T & A_{22}^T \end{bmatrix} = \begin{bmatrix} A_{11}A_{11}^T + A_{12}A_{12}^T & \vdots & A_{12}A_{22}^T \\ \cdots\cdots\cdots\cdots & \vdots & \cdots\cdots \\ A_{22}A_{12}^T & \vdots & A_{22}A_{22}^T \end{bmatrix} \quad (72)$$

The determinant of "AA^T" contains "s_1^m" as a term, which is not permitted in a VSHP. Therefore, "A_{11}" can be set equal to zero. Thus, the power of s_1 is reduced to (m-k). Similarly, "B_{11}" can also be set equal to zero, but it is not necessary to set "C_{11}" equal to zero.

Now A and B are positive-semi-definite matrices of rank (m-k) and C is a positive-definite matrix of rank m. In this case,

$$AA^T = \begin{bmatrix} A_{12}A_{12}^T & A_{12}A_{22}^T \\ A_{22}A_{12}^T & A_{22}A_{22}^T \end{bmatrix} \quad (73)$$

and the value of the determinant AA^T is zero. Similarly, the value of the determinant BB^T is zero. Therefore, "D_m" becomes

$$D_m = \begin{bmatrix} A_{12}A_{12}^T & A_{12}A_{22}^T \\ A_{22}A_{12}^T & A_{22}A_{22}^T \end{bmatrix} s_1 + \begin{bmatrix} B_{12}B_{12}^T & B_{12}B_{22}^T \\ B_{22}B_{12}^T & B_{22}B_{22}^T \end{bmatrix} s_2 +$$

$$\begin{bmatrix} C_{11}C_{11}^T + C_{12}A_{12}^T & C_{12}A_{22}^T \\ C_{22}C_{12}^T & C_{22}A_{22}^T \end{bmatrix} + \begin{bmatrix} G_{11} & G_{12} \\ G_{12}^T & G_{22} \end{bmatrix} = \begin{bmatrix} D_{11} & D_{12} \\ D_{21} & D_{22} \end{bmatrix} \tag{74}$$

where

$$D_{11} = A_{12}A_{12}^T s_1 + B_{12}B_{12}^T s_1 + C_{11}C_{11}^T s_1 + C_{12}C_{12}^T s_1 + G_{11} \tag{75a}$$

$$D_{12} = A_{12}A_{22}^T s_1 + B_{12}B_{22}^T s_2 + C_{12}C_{22}^T s_1 + G_{12} \tag{75b}$$

$$D_{21} = A_{22}A_{12}^T s_1 + B_{22}B_{12}^T s_1 + C_{22}C_{12}^T s_1 - G_{12}^T \tag{75c}$$

$$D_{22} = A_{22}A_{22}^T s_1 + B_{22}B_{22}^T s_2 + C_{22}C_{22}^T s_1 + G_{22} \tag{75d}$$

The determinant of "D_m" can be evaluated in many ways. Two of these techniques are described in the following subsections.

V.2. Method A

The $\det[D_m]$ can be written as

$$\det D_m = |D_{22}|\, |D_{11} - D_{12}\, D_{22}^{-1}\, D_{21}| \tag{76}$$

Let D_{22} be of order 2 x 2. Consider the s_1-portion. It will be

$$A_{22}A_{22}^T = \begin{bmatrix} a_{m-1,m-1} & a_{m-1,m} \\ 0 & a_{m,m} \end{bmatrix} \begin{bmatrix} a_{m-1,m-1} & 0 \\ a_{m-1,m} & a_{m,m} \end{bmatrix}$$

$$= \begin{bmatrix} a^2_{m-1,m-1} + a^2_{m-1,m} & a_{m-1,m}\, a_{m,m} \\ a_{m-1,m}\, a_{m,m} & a^2_{m,m} \end{bmatrix} \tag{77}$$

The coefficient $a_{m-1,m-1}$ must be set equal to zero to remove the s_1^2-term, a condition of all VSHPs. A similar situation arises for s_2^2-term and hence $b_{m-1,m-1} = 0$. The matrix "D_{22}" can now be written as

$$D_{22} = \begin{bmatrix} a^2_{m-1,m} & a_{m-1,m}a_{m,m} \\ a_{m-1,m}a_{m,m} & a^2_{m,m} \end{bmatrix} s_1 + \begin{bmatrix} b^2_{m-1,m} & b_{m-1,m}b_{m,m} \\ b_{m-1,m}b_{m,m} & b^2_{m,m} \end{bmatrix} s_2$$
$$+ \begin{bmatrix} c^2_{m-1,m-1} + c^2_{m-1,m} & c_{m-1,m}c_{m,m} \\ c_{m-1,m}c_{m,m} & c^2_{m,m} \end{bmatrix} + \begin{bmatrix} 0 & g_{m-1,m} \\ -g_{m-1,m} & 0 \end{bmatrix}$$
$$= \begin{bmatrix} d_{11} & d_{12} \\ d_{21} & d_{22} \end{bmatrix} \tag{78}$$

where

$$d_{11} = a^2_{m-1,m}\, s_1 + b^2_{m-1,m}\, s_2 + c^2_{m-1,m-1} + c^2_{m-1,m} \tag{79a}$$

$$d_{12} = a_{m-1,m}\, a_{m,m}\, s_1 + b_{m-1,m}\, b_{m,m}\, s_2 + c_{m-1,m}\, c_{m,m} + g_{m-1,m} \tag{79b}$$

$$d_{21} = a_{m-1,m}\, a_{m,m}\, s_1 + b_{m-1,m}\, b_{m,m}\, s_2 + c_{m-1,m}\, c_{m,m} + g_{m-1,m} \tag{79c}$$

$$d_{22} = a^2_{m,m}\, s_1 + b^2_{m,m}\, s_2 + c^2_{m,m} \tag{79d}$$

It can be easily verified that the factors of the s_1^2-term and the s_2^2-term are zero. Therefore, det[D_{22}] can be written an

$$\det D_{22} = \alpha\, s_1 s_2 + \beta\, s_1 + \gamma\, s_2 + \delta \tag{80}$$

where the factors of the 2-variable polynomial can be obtained as follows:

$$\alpha = \left(a_{m,m}\, b_{m-1,m} - a_{m-1,m}\, b_{m,m}\right)^2 \tag{81a}$$

$$\beta = \left(a_{m,m}\, c_{m-1,m} - a_{m-1,m}\, c_{m,m}\right)^2 + a^2_{m,m}\, c^2_{m-1,m-1} \tag{81b}$$

$$\gamma = \left(b_{m,m}\, c_{m-1,m} - b_{m-1,m}\, c_{m,m}\right)^2 + b^2_{m,m}\, c^2_{m-1,m-1} \tag{81c}$$

$$\delta = c^2_{m-1,m-1}\, c^2_{m,m} + g^2_{m-1,m} \tag{81d}$$

All the coefficients will be positive, provided we have $a_{m-1,m-1} = 0$ and $b_{m-1,m-1} = 0$ and the rank of the matrix

$$\begin{bmatrix} a_{m-1,m} & b_{m-1,m} \\ a_{m,m} & b_{m,m} \end{bmatrix} \tag{82}$$

is two, hence, the condition of VSHP is preserved. It is further noted that the rank of the matrix

$$\begin{bmatrix} 0 & g_{m-1,m} \\ -g_{m-1,m} & 0 \end{bmatrix} \tag{83}$$

can be either zero or two.

To ease the computation of det D_m, A and B can be partitioned as follows:

$$A = \begin{bmatrix} A_{11} & 0 \\ 0 & A_{22} \end{bmatrix} \tag{84a}$$

and

$$B = \begin{bmatrix} B_{11} & 0 \\ 0 & B_{22} \end{bmatrix} \tag{84b}$$

where

$$A_{11} = \begin{bmatrix} a_{11} & a_{12} & a_{13} & \cdots & a_{1\,m-2} \\ 0 & a_{22} & a_{23} & \cdots & a_{2\,m-2} \\ 0 & 0 & a_{33} & \cdots & a_{3\,m-3} \\ \cdot & \cdot & \cdot & \cdots & \cdot \\ \cdot & \cdot & \cdot & \cdots & \cdot \\ 0 & 0 & 0 & \cdots & a_{m-2\,m-2} \end{bmatrix} \tag{85a}$$

$$A_{22} = \begin{bmatrix} 0 & a_{m-1,m} \\ & \\ 0 & a_{m,m} \end{bmatrix} \tag{85b}$$

$$B_{11} = \begin{bmatrix} b_{11} & b_{12} & b_{13} \cdots\cdots & b_{1,m-2} \\ 0 & b_{22} & b_{23} \cdots\cdots & b_{2,m-2} \\ 0 & 0 & b_{33} \cdots\cdots & b_{3,m-2} \\ \cdots & \cdots & \cdots & \cdots \\ 0 & 0 & 0 & b_{m-2,m-2} \end{bmatrix} \tag{85c}$$

$$B_{22} = \begin{bmatrix} 0 & b_{m-1,m} \\ & \\ 0 & b_{m,m} \end{bmatrix} \tag{85d}$$

That is, by setting $A_{12} = 0$ and $B_{12} = 0$, the computation of $\det[D_m]$ becomes easier. The process can be continued for the subsequent (m-2) x (m-2) sub-matrices "A_{11}" and "B_{11}". This leads to the result

$$\det D_m = \prod_{i=1}^{m/2} \left(\alpha_{11_i} s_1 s_2 + \alpha_{10_i} s_1 + \alpha_{01_i} s_2 + \alpha_{00_i} \right) \tag{86}$$

which is a product of VSHP's and is, therefore, a VSHP.

V. 3 Method B

Presented here is an alternative method to partition matrix "A". The matrix "A" can be partitioned as

$$A = \begin{bmatrix} a_{11} & a_{12} & . & \\ 0 & a_{22} & . & 0 \\ .... & .... & . & \\ & 0 & . & A_{22} \end{bmatrix} \tag{87}$$

where "A_{22}" is an (m-2) x (m-2) upper triangular matrix. This gives

$$AA^T = \begin{bmatrix} a_{11}^2 + a_{12}^2 & a_{12}a_{22} & . & \\ a_{12}a_{22} & a_{22}^2 & . & 0 \\ .... & .... & . & \\ & 0 & . & A_{22}A_{22}^T \end{bmatrix} \tag{88}$$

In (88), if the s_1^2-term is to be absent, "a_{11}" should be zero.

Similarly, the matrix "B" can be partitioned as

$$B = \begin{bmatrix} b_{11} & b_{12} & . & \\ 0 & b_{22} & . & 0 \\ .... & .... & . & \\ & 0 & . & B_{22} \end{bmatrix} \tag{89}$$

where "B_{22}" is an (m-2) x (m-2) upper triangular matrix. This gives

$$BB^T = \begin{bmatrix} b_{11}^2 + b_{12}^2 & b_{12}b_{22} & . & \\ b_{12}b_{22} & b_{22}^2 & . & 0 \\ .... & .... & . & \\ 0 & & . & B_{22}B_{22}^T \end{bmatrix} \quad (90)$$

In (90), if the s_2^2-term is to be absent, "b_{11}" should be zero.

The matrix "D_m" can now be written as

$$D_m = \begin{bmatrix} D_{11} & 0 \\ 0 & D_{m-2} \end{bmatrix} \quad (91)$$

where

$$D_{11} = \begin{bmatrix} a_{12}^2 s_1 + b_{12} s_2 + c_{11}^2 + c_{12}^2, & \vdots\ a_{12}a_{22}s_1 + b_{12}b_{22}s_2 + c_{12}c_{22} + g_{12} \\ & \vdots \\ a_{12}a_{22}s_1 + b_{12}b_{22}s_2 + c_{12}c_{22} - g_{12}, & \vdots\ a_{22}^2 s_1 + b_{22}^2 s_2 + c_{22} \end{bmatrix} \quad (92a)$$

and

$$D_{m-2} = A_{22}A_{22}^T s_1 + B_{22}B_{22}^T s_2 + C_{22}C_{22}^T + G_{22} \quad (92b)$$

The determinant of "D_{11}" yields

$$\det D_{11} = \left(a_{22}b_{12} - a_{12}b_{22}\right)^2 s_1 s_2 + \left\{\left(a_{12}c_{22} - a_{22}c_{12}\right)^2 + a_{22}^2\, c_{11}^2\right\} s_1 +$$

$$\left\{\left(b_{12}c_{22} - b_{22}c_{12}\right)^2 + b_{22}^2\, c_{11}^2\right\} s_2 + \left(c_{11}^2 c_{22}^2 + g_{12}^2\right) \tag{93}$$

Equation (93) constitutes a VSHP, provided that

$$\begin{vmatrix} a_{12} & b_{12} \\ a_{22} & b_{22} \end{vmatrix} \neq 0 \tag{94}$$

The process can be continued for the subsequent (m-2) x (m-2) sub-matrices "A_{22}" and "B_{22}" This leads to the same type of result found in (86). In fact, eqns. (80) and (93) along with their respective conditions (82) and (94) provide the generation of VSHPs having unity degree in each of the variables s_1 and s_2.

V. 4. EXAMPLE

In this example, a VSHP with the order equal to two, and one with respect to "s_1" and "s_2" is generated.

$$D_{21} = \begin{bmatrix} a_{11} & 0 & 0 \\ 0 & 0 & a_{23} \\ 0 & 0 & a_{33} \end{bmatrix} \begin{bmatrix} a_{11} & 0 & 0 \\ 0 & 0 & 0 \\ 0 & a_{23} & a_{33} \end{bmatrix} s_1 + \begin{bmatrix} b_{11} & 0 & 0 \\ 0 & 0 & b_{23} \\ 0 & 0 & b_{33} \end{bmatrix} \begin{bmatrix} b_{11} & 0 & 0 \\ 0 & 0 & 0 \\ 0 & b_{23} & b_{33} \end{bmatrix} s_2$$

$$+ \begin{bmatrix} c_{11} & c_{12} & c_{13} \\ 0 & c_{22} & c_{23} \\ 0 & 0 & c_{33} \end{bmatrix} \begin{bmatrix} c_{11} & 0 & 0 \\ c_{12} & c_{22} & 0 \\ c_{13} & c_{23} & c_{33} \end{bmatrix} + \begin{bmatrix} 0 & g_{12} & g_{13} \\ -g_{12} & 0 & g_{23} \\ -g_{13} & -g_{23} & 0 \end{bmatrix} \tag{95}$$

$$\begin{aligned} D(s_1, s_2) = \det D_{21} &= \left[a_{11}^2 (a_{23} b_{33} - a_{33} b_{23})^2\right] s_1^2 s_2 + \\ & a_{11}^2 \left[a_{33}^2 c_{22}^2 + \left(a_{23} c_{33} - a_{33} c_{23}\right)^2\right] s_1^2 \\ & + a_{11}^2 \left[b_{33}^2 c_{22}^2 + \left(b_{23} c_{33} - b_{33} c_{23}\right)^2\right] s_1 s_2 \\ & + \{a_{11}^2 \left(c_{22}^2 c_{33}^2 + g_{23}^2\right) + c_{11}^2 \left[a_{33}^2 c_{22}^2 + (a_{23} c_{33} - a_{33} c_{23})^2\right] + \\ & (a_{33} g_{12} - a_{23} g_{13})^2 + \left(a_{23} c_{12} c_{33} + a_{33} c_{13} c_{22}\right)^2 \} s_1 + \\ & \{ c_{11}^2 \left[b_{33}^2 c_{22}^2 + \left(b_{23} g_{13} - b_{33} g_{12}\right)^2\right] \\ & + \left(b_{23} c_{12} c_{33} + b_{33} c_{13} c_{22} - b_{33} c_{13} c_{23}\right)^2 \} s_2 + \\ & \{c_{11}^2 (c_{22}^2 c_{33}^2 + g_{23}^2) + (c_{22} g_{13} - c_{12} g_{23})^2 + (c_{13} g_{23} - c_{23} g_{13} + c_{33} g_{12})^2\} \end{aligned} \tag{96}$$

Equation (96) constitutes a VSHP, provided $(a_{23}b_{33} - a_{33}b_{23}) \neq 0$. If a polynomial with the orders of unity in s_1 and two in "s_2" is desired, one can make the required transformation of interchanging of the variables s_1 and s_2 in (96). A higher order of VSHP can be generated by cascading VSHPs of lower order like the ones in (93) and (96).

V. 5 APPLICATIONS TO 2-D FILTER DESIGN

In this design method, a 2-variable VSHP is generated using the method discussed earlier and then assigned to the denominator of an analog transfer function of the form

$$H_a(s_1, s_2) = \frac{A(s_1, s_2)}{B(s_1, s_2)} = \frac{\sum\limits_{i=0}^{N}\sum\limits_{=0}^{N} a_{ij} s_1^i s_2^j}{\sum\limits_{i=0}^{N}\sum\limits_{j=0}^{N} b_{ij} s_1^i s_2^j} \tag{97}$$

We now apply the double bilinear transformations to obtain the discrete version of the filter:

$$H_d(z_1,z_2) = H(s_1,s_2)\Big|_{s_i=\frac{2}{T}\frac{1-z_i^{-1}}{1+z_i^{-1}}} \quad i = 1, 2$$

$$= \frac{\sum_{i=0}^{N}\sum_{j=0}^{N} a'_{ij} z_1^{-i} z_2^{-j}}{\sum_{i=0}^{N}\sum_{j=0}^{N} b'_{ij} z_1^{-i} z_2^{-j}} \tag{98}$$

The new parameters of the 2-D digital filter are used as the variables in the optimization process in such a way so that the desired objective function is minimized. The objective function is defined as

$$\begin{aligned} E_{\ell_2}(j\omega_{1m}, j\omega_{2n}, \psi) = & \sum_{m,n\,\varepsilon\, I_{ps}}\sum E^2_{mag}(j\omega_{1m}, j\omega_{2n}, \psi) \\ & + \sum_{m,n\,\varepsilon\, I_p}\sum E^2_{\tau_1}(j\omega_{1m}, j\omega_{2n}, \psi) \\ & + \sum_{m,n\,\varepsilon\, I_p}\sum E^2_{\tau 2}(j\omega_{1m}, j\omega_{2n}, \psi) \end{aligned} \tag{99}$$

where ψ is the coefficient vector to be calculated.

$$E_{mag} = |H_I| - |H_D|,$$

$|H_I|$ and $|H_D|$ are ideal and designed magnitude responses, respectively.

$$E_{\tau_i} = \tau_I T - \tau_{d_i}(\omega_{1m}, \omega_{2n}, \psi), \quad i = 1, 2,$$

where τ_I is the ideal group delay response and its value is chosen to be equal to the order of the filter [27],

τ_{d_i} for i = 1, 2 are group delay responses of the designed filter with respect to ω_i, i = 1, 2 respectively,

I_{p_s} = the set of all discrete frequency pairs along the ω_1 and ω_2 axes in the pass-band and the stop-band of the filter,

I_p = the set of all discrete frequency pairs along the ω_1 and ω_2 axes in the pass-band of the filter.

In this example, we design a fourth order digital filter with the following specifications:

$$\left|H_I\left(e^{j\omega_{1m}}, e^{j\omega_{2n}}\right)\right| = \begin{cases} 1, & 0 \leq \sqrt{\omega_{1m}^2 + \omega_{2n}^2} \leq 1.0 \text{ rad/sec.} \\ 0, & 1.5 \leq \sqrt{\omega_{1m}^2 + \omega_{2n}^2} \leq \pi \text{ rad/sec.} \end{cases} \tag{100}$$

and a constant group delay characteristic$(\tau_I = 4)$, while ω_s is the sampling frequency of 2π radians/second. The optimization technique reported in the reference[28] is then used to minimize the objective function of eqn.(99). Table 4 shows the values of the coefficients of the designed filter. Figures 3(a), (b) and (c) respectively show the magnitude plot and the group delay responses of the designed filter.

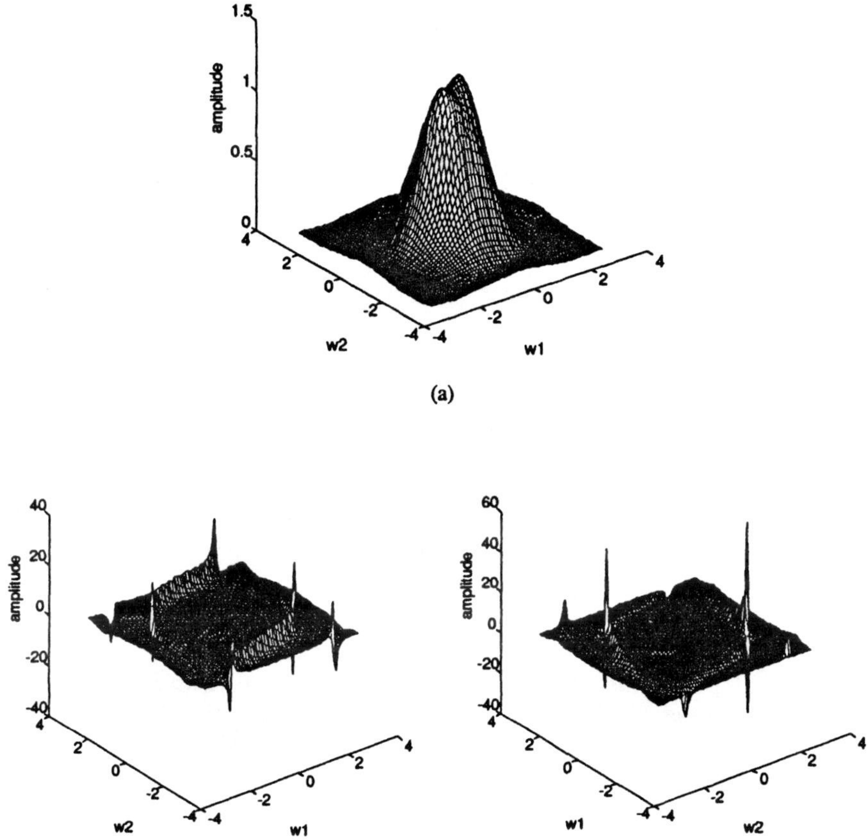

(a)

(b) (c)

Fig (3) (a) amplitude response (b) group delay with respect to ω_1 (c) group delay with respect to ω_2

TABLE 4 (eqn.(2))

Numerator Coefficients	Denominator Coefficients
a_{00} = 0.24250460E + 01	b_{00} = 0.16435898E + 03
a_{01} = 0.40828800E + 01	b_{01} = -0.10108646E + 03
a_{02} = 0.66948986E + 01	b_{02} = 0.29635269E + 02
a_{10} = 0.52808161E + 01	b_{10} = -0.64718079E + 02
a_{11} = 0.83481598E + 01	b_{11} = -0.13981224E + 02
a_{12} = 0.75414143E + 01	b_{12} = 0.15327198E + 02
a_{20} = 0.57536539E + 01	b_{20} = -0.36191940E + 01
a_{21} = 0.71639252E + 01	b_{21} = 0.32147812E + 02
a_{22} = 0.37494564E + 01	b_{22} = -0.10551155E + 02

ACKNOWLEDGMENT

The author wishes to express his gratitude to Mr. Hamidreza Safiri for spending a tireless effort in diligently proofreading of the manuscript and providing examples for the text.

REFERENCES

[1] E.I. Jury, *"Stability of Multi-dimensional Scalar and Matrix Polynomials"*, Proc. of IEEE, Vol. 66, pp. 1018-1047, 1978.

[2] P.K. Rajan, M.N.S. Swamy, *"Design of Separable Denominator 2-Dimensional Digital Filter Possessing Real Circular Symmetric Frequency Responses"*, Proc. of IEE, Pt.G, Vol. 129, pp. 235-240, 1982.

[3] C. Charalambous, *"Design of 2-Dimensional Circular Symmetric Digital Filters"*, Proc. of IEE, Pt.G, Vol. 129, pp. 47-54, 1982.

[4] J. M. Costa, A. N. Venetsanopoulos, *"Design of Circular Symmetric 2-D Recursive Filters"*, IEEE Trans. on Acoustics, Speech and Signal Processing, Vol. 24, pp. 145-147, 1976.

[5] M. Ahmadi, M.T. Boraie, V. Ramachandran, C.S. Gargour, *"2-D Recursive Digital Filters With Constant Group-Delay Characteristics Using Separable Denominator Transfer Function and a New Stability Test"*, IEEE Trans. on Acoustics, Speech and Signal Processing, Vol. 33, pp. 1306-1308, 1985.

[6] J. F. Abramatic, F. Germain, E. Roencher, *"Design of Two-Dimensional Separable Denominator Recursive Filters"*, IEEE Trans. on Acoustics, Speech and Signal Processing, Vol. 27, pp. 445-453, 1979.

[7] M. Ohki, M. Kawamata, T. Higuchi, *"Efficient Design of 2-Dimensional Separable Denominator Digital Filters Using Symmetries"*, IEEE Trans. on Circuits and Systems, Vol. 37, pp. 114-120, 1990.

[8] M. Ahmadi, H.J.J. Lee, M. Shridhar, V. Ramachandran, *"An Efficient Algorithm for the Design of Circular Symmetric Linear Phase Recursive Digital Filters With Separable Denominator Transfer Function"*, Journal of the Franklin Institute, Vol. 327, pp. 359-367, 1990.

[9] M. Ahmadi, H.J.J. Lee, M. Shridhar, V. Ramachandran, *"Design of 2-D Recursive Digital Filters With Non-Circular Symmetric Cut-off Boundary and Constant Group-Delay Responses"*, Proc. of IEE, Pt. G, Vol. 136, pp. 255-259, 1989.

[10] A. Mazinani, M. Ahmadi, M. Shridhar, R.S. Lashkari, *"A Novel Approach to the Design of 2-D Recursive Digital Filters"*, Journal of the Franklin Institute, Vol. 329, pp. 127-133, 1992.

[11] P Karivaratharajan, M.N.S. Swamy, *"Quadrantal Symmetry Associated With Two-Dimensional Digital Transfer Function"*, IEEE Trans. on Circuits and Systems, Vol. 25, pp. 340-343, 1978.

[12] A. Antoniou, M. Ahmadi, C. Charalambous, *"Design of Factorable Lowpass 2-Dimensional Digital Filters Satisfying Prescribed Specifications"*, IEE Proc., Vol. 128, Part G, No. 2, pp. 53-60, 1981.

[13] A. Antoniou, **Digital Filters; Analysis, Design and Applications**, 2nd edition McGraw Hill, 1993.

[14] R. King, M. Ahmadi, R. Gorgui-Naguib, A. Kwabwe, M. Azimi-Sadjadi, **Digital Filtering in One and Two-Dimensions, Design and Applications**, Plenum Publishing Co., 1989.

[15] M. Ahmadi, V. Ramachandran, *"New Method for Generating Two-Variable VSHPs and Its Applications in the Design of Two-Dimensional Recursive Digital Filters With Prescribed Magnitude and Constant Group-Delay Responses"*, Proc. IEE, Vol. 131, No. 4, pp. 151-155, 1984.

[16] M. Ahmadi, M. Shridhar, H. Lee and V. Ramachandran, *"A Method for The Design of 1-D Recursive Digital Filters Satisfying a Given Magnitude and Constant Group-Delay Response"*, J. Franklin Institute, Vol. 326, No. 3, pp. 381-393, 1989.

[17] J.A. Nelson and R. Mead, *"A Simplex Method For Function Minimization"*, Computer Journal, No. 7, pp. 308-313, 1965.

[18] J. M. Costa, A. N. Venetsanopoulos, *"Design of Circular Symmetric Two-Dimensional Recursive Filters"*, IEEE Trans., Acoust., Speech, Signal Processing, ASSP. 22, pp. 432-443, 1974.

[19] M. Ahmadi, A.G. Constantinides, R.A. King, *"Design Technique for a Class of Stable 2-Dimensional Recursive Digital Filters"*, Proc. of IEEE Inter. Conf. on Acoust., Speech and Signal Processing, Philadelphia, USA, pp. 145-147, 1976.

[20] R.A. King, A.H. Kayran, *"A New Transformation Technique For The Design of 2-Dimensional Stable Recursive Digital Filters"*, Proc. of IEEE, Int. Symp. on Circuits and Systems, Chicago, USA, pp. 196-199, 1981.

[21] P.A. Ramamoorthy, L.T. Bruton, *"Design of Stable Two-Dimensional Analogue and Digital Filters With Applications in Image Processing"*, Int. J. Circuit Theory Appl, 7, pp. 224-245, 1979.

[22] D. Goodman, *"Some Difficulties With the Double Bi-Linear Transformation in 2-D Recursive Filter Design"*, Proc. IEEE, Vol. 66, No. 7, pp. 796-797, 1978.

[23] P.K. Rajan, H.C. Reddy, M.N.S. Swamy, V. Ramachandran, *"Generation of Two-Dimensional Digital Function Without Non-Essential Singularities of The Second Kind"*, IEEE Trans. on Acoust., Speech and Signal Processing, Vol. ASSP-28, No. 2, pp. 216-223, 1980.

[24] V. Ramachandran, M. Ahmadi, *"Design of 2-D Stable Analog and Digital Recursive Filters Using Properties of the Derivatives of Even or Odd Parts of Hurwitz Polynomials"*, J. Franklin Institute, Vol. 315, No. 4, pp. 259-267, 1983.

[25] M. Ahmadi, V. Ramachandran, *"New Method of Generating Two-Variable VSHP and Its Application in the Design of Two-Dimensional Recursive Digital Filters With Prescribed Magnitude and Constant Group-Delay Responses"*, Proc. IEE, Vol. 131, Pt.G, No. 4, pp. 151-155, 1984.

[26] F.E. Hohn, **Elementary Matrix Algebra**, McMillan Co., 1964.

[27] A.T. Chottera, G.A. Jullien, *"Design of Two-Dimensional Recursive Digital Filters Using Linear Programming"*, IEEE Trans. on Circuits and Systems, Vol. CAS-29, No. 12, pp. 817-826, 1982.

[28] R. Fletcher, M.J.D. Powell, *"A Rapidly Convergent Descent Method for Minimization"*, Computer Journal, Vol. 6, No. 2, pp. 163-168, 1963.

A Periodic Fixed-Architecture Approach to Multirate Digital Control Design

Wassim M. Haddad

Vikram Kapila

School of Aerospace Engineering
Georgia Institute of Technology
Atlanta, GA 30332-0150

I. INTRODUCTION

Many applications of feedback control involve continuous-time systems subject to digital (discrete-time) control. Furthermore, in practical applications, the control-system actuators and sensors have differing bandwidths. For example, in flexible structure control it is not unusual to attenuate the low-frequency, high amplitude modes by means of low-bandwidth actuators that are relatively heavy and hence able to exert high force/torque to control the higher frequency modes. Obviously, the high-bandwidth actuators would require sensors that are sampled at high rates while low bandwidth actuators require only sensors sampled at low data rates. As a consequence, the use of various sensor data rates leads to a multirate control problem. To properly use such data, a multirate controller must carefully account for the timing sequence of incoming data. The purpose of this chapter is to develop a general approach to full- and reduced-order steady-state multirate dynamic compensation.

Multirate control problems have been of interest for many years with increased emphasis in recent years [1–16]. A common feature of these papers is the realization that the multirate sampling process leads to periodically time-varying dynamics. Hence with a suitable reinterpretation, results on multirate control can also be applied to single rate or multirate problems involving systems with periodically time-varying dynamics. The principal challenge of these problems is to arrive at a tractable control design formulation in spite of the extreme complexity of such systems. In order to account for the periodic-time-varying dynamics of multirate systems a periodically time-varying control law architecture was proposed in [7–11] which appears promising in this regard. An alternative approach which has been proposed for the multirate control problem is the use of an expanded state-space formulation [3]. However, this approach results in very high order systems and is often numerically intractable. Finally, a cost translation and a lifting approach to the multirate LQG problem has been proposed in [13–15] which does not lead to an increase in the state dimension. Specifically, it is shown in [15] how to translate a multirate sampled-data LQG problem into an equivalent, modified, single rate, shift invariant problem via a lifted isomorphism. However, this approach results in an equivalent system involving more inputs and outputs than the original system and hence results in increased controller implementation complexity. The interested reader is referred to [7, 8, 17–20] for further discussions on multirate and periodic control.

For generality in our development, we consider both full- and reduced-order dynamic compensators as well as static output-feedback controllers. In the discrete-time case this problem was considered in [21] while single rate sampled-data aspects were addressed in [22]. The approach of the present chapter is the fixed-structure Riccati equation technique developed in [21, 23]. Essentially, this approach addresses controller complexity by explicitly imposing implementation constraints on the controller structure and optimizing over that class of controllers. Specifically, in addressing the problem of reduced-order dynamic compensation it is shown in [21] that optimal reduced-order, steady-state dynamic compensators can be characterized by means of an algebraic system of Riccati/Lyapunov equations coupled by a projection matrix which arises as a direct consequence of optimality and which represents a breakdown of the separation between the operations of state estimation and state estimate feedback; that is, the certainty equivalence principle is no longer valid. The proof is based on expressing the closed-loop quadratic cost functional as a function of the design parameters, i.e., the compensator gains, and the utilization of Lagrange multiplier arguments for optimizing the quadratic performance criterion over the parameter space. Thus, this approach provides a constrained optimal control methodology in which we do not seek to optimize a performance measure per se, but rather a performance measure within a class of *a priori* fixed-structure controllers.

In the present chapter, analog-to-digital conversions are employed within a multirate setting to obtain periodically time-varying dynamics. The compensator is thus assigned a corresponding discrete-time periodic structure to account for the multirate measurements. It is shown that the optimal reduced-order multirate dynamic compensator is characterized by a periodically time-varying system of four equations consisting of two modified Riccati equations and two modified Lyapunov equations corresponding to each intermediate point of the periodicity interval. Because of the time-varying nature of the problem, the necessary conditions for optimality now involve multiple projections corresponding to each intermediate point of the periodic interval and whose rank along the periodic interval is equal to the order of the compensator. Similar extensions to reduced-order multirate estimation are addressed in [24].

The contents of this chapter are as follows. In Section II, the statements of the fixed-structure multirate static and dynamic output-feedback control problems are presented. Section III addresses the analysis and synthesis of the multirate static output-feedback control problem. Specifically, Lemma 1 shows that under the assumption of cylostationary disturbances the closed-loop covariance equation reaches a steady-state periodic trajectory under periodic dynamics. Theorem 2 presents necessary conditions for optimality which characterize solutions to the multirate sampled-data static output-feedback control problem. Section IV generalizes the results of Section III to reduced-order dynamic compensation. In Section V, using the identities of Van Loan [25], we derive formulae for integrals of matrix exponentials arising in the continuous-time/sampled-data conversion. Section VI presents a Newton homotopy prediction/correction algorithm for the full-order multirate design equations. In Section VII, we apply our results to a rigid body spacecraft with a flexible appendage and a lightly damped flexible beam structure. Finally, Section VIII gives some conclusions and discusses future extensions.

NOMENCLATURE

$I_r, 0_{r\times s}, 0_r$	–	$r \times r$ identity matrix, $r \times s$ zero matrix, $0_{r\times r}$
$()^T, ()^{-1}$	–	transpose, inverse
$\mathrm{tr}(), \rho()$	–	trace, spectral radius
$\mathcal{E}$	–	expected value
$\mathcal{R}, \mathcal{R}^{r\times s}, \mathcal{R}^r$	–	real numbers, $r \times s$ real matrices, $\mathcal{R}^{r\times 1}$
k, α	–	discrete-time indices
n, m, l_k, n_c	–	positive integers; $1 \leq n_c \leq n$
x, y, x_c, u	–	$n-, l_k-, n_c-, m-$ dimensional vectors
$A, B, C(k)$	–	$n \times n, n \times m, l_k \times n$ matrices
$A_c(k), B_c(k), C_c(k), D_c(k)$	–	$n_c \times n_c, n_c \times l_k, m \times n_c, m \times l_k$ matrices
R_1, R_2	–	$n \times n, m \times m$ state and control weightings; $R_1 \geq 0, R_2 > 0$
R_{12}	–	$n \times m$ cross weighting; $R_1 - R_{12}R_2^{-1}R_{12}^T \geq 0$
$\tilde{n}$	–	$n + n_c$

II. THE FIXED-ARCHITECTURE MULTIRATE STATIC AND DYNAMIC DIGITAL CONTROL PROBLEMS

In this section we state the fixed-structure static and dynamic, sampled-data, multirate output-feedback control problems. In the problem formulation the sample intervals h_k and dynamic compensator order n_c are fixed and the optimization is performed over the compensator parameters ($A_c(\cdot)$, $B_c(\cdot)$, $C_c(\cdot)$, $D_c(\cdot)$). For design trade-off studies h_k and n_c can be varied and the problem can be solved for each pair of values of interest.

Fixed-Structure Multirate Static Output-Feedback Control Problem. Consider the nth–order continuous-time system

$$\dot{x}(t) = Ax(t) + Bu(t) + w_1(t), \quad t \in [0, \infty), \tag{1}$$

with multirate sampled-data measurements

$$y(t_k) = C(t_k)x(t_k) + w_2(t_k), \quad k = 1, 2, \ldots . \tag{2}$$

Then design a static output-feedback multirate sampled-data control law

$$u(t_k) = D_c(t_k)y(t_k), \tag{3}$$

which, with D/A zero-order-hold controls

$$u(t) = u(t_k), \quad t \in [t_k, t_{k+1}), \tag{4}$$

minimizes the quadratic performance criterion

$$\mathcal{J}_s(D_c(\cdot)) \triangleq \lim_{t\to\infty} \frac{1}{t}\mathcal{E} \int_0^t [x^T(s)R_1x(s) + 2x^T(s)R_{12}u(s) + u^T(s)R_2u(s)]\, ds. \tag{5}$$

Fixed-Structure Multirate Dynamic Output-Feedback Control Problem. Given the nth-order continuous-time system (1) with multirate sampled-data measurements (2) design an n_c^{th}- order $(1 \le n_c \le n)$ multirate sampled-data dynamic compensator

$$x_c(t_{k+1}) = A_c(t_k)x_c(t_k) + B_c(t_k)y(t_k), \tag{6}$$

$$u(t_k) = C_c(t_k)x_c(t_k) + D_c(t_k)y(t_k), \tag{7}$$

which, with D/A zero-order-hold controls (4), minimizes the quadratic performance criterion (5) with $\mathcal{J}_s(D_c(\cdot))$ denoted by $\mathcal{J}_c(A_c(\cdot), B_c(\cdot), C_c(\cdot), D_c(\cdot))$.

The key feature of both problems is the time-varying nature of the output equation (2) which represents sensor measurements available at different rates. Figure 1 provides a typical multirate timing diagram for a three-sensor model. For generality, we do not assume that the sample intervals $h_k \triangleq t_{k+1} - t_k$ are uniform (note the sample times for sensor # 3 in

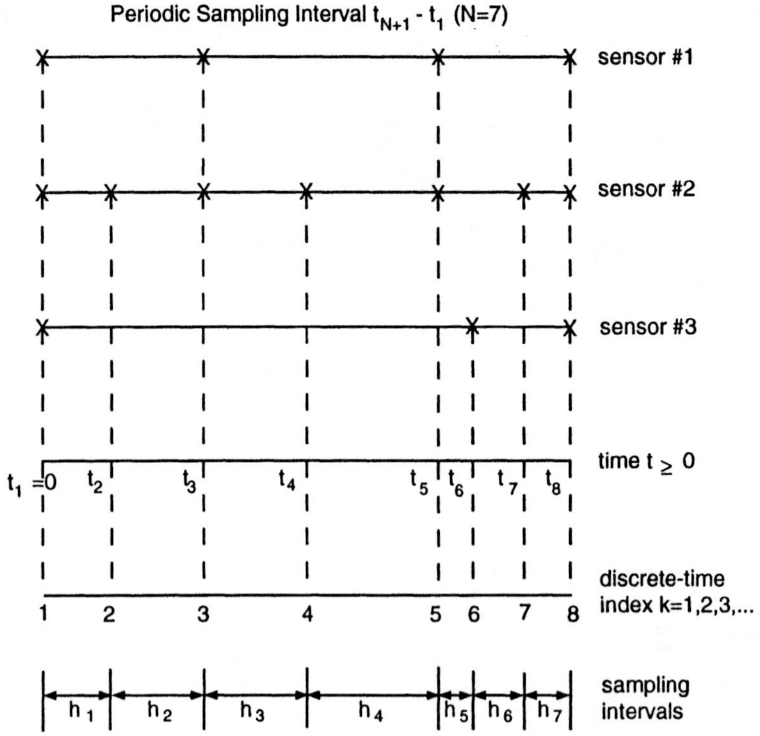

Figure 1: Multirate Timing Diagram for Sampled-Data Control

Figure 1). However, we do assume that the overall timing sequence of intervals $[t_k, t_{k+N}]$, $k = 1, 2, \ldots$ is periodic over $[0, \infty)$, where N represents the periodic interval. Note that $h_{k+N} = h_k$, $k = 1, 2, \ldots$. Since different sensor measurements are available at different times t_k, the dimension l_k of the measurements $y(t_k)$ may also vary periodically. Finally, in subsequent analysis the static output-feedback law (3) and dynamic compensator (6)–(7) are assigned periodic gains corresponding to the periodic timing sequence of the multirate measurements.

In the above problem formulation, $w_1(t)$ denotes a continuous-time stationary white noise process with nonnegative-definite intensity $V_1 \in \mathcal{R}^{n \times n}$, while $w_2(t_k)$ denotes a variable-dimension discrete-time white noise process with positive-definite covariance $V_2(t_k) \in \mathcal{R}^{l_k \times l_k}$. We assume $w_2(t_k)$ is cyclostationary, that is, $V_2(t_{k+N}) = V_2(t_k)$, $k = 1, 2, \ldots$.

In what follows we shall simplify the notation considerably by replacing the real-time sample instant t_k by the discrete-time index k. With this minor abuse of notation we replace $x(t_k)$ by $x(k)$, $x_c(t_k)$ by $x_c(k)$, $y(t_k)$ by $y(k)$, $u(t_k)$ by $u(k)$, $w_2(t_k)$ by $w_2(t_k)$, $A_c(t_k)$ by $A_c(k)$ (and similarly for $B_c(\cdot), C_c(\cdot)$, and $D_c(\cdot)$), $C(t_k)$ by $C(k)$, and $V_2(t_k)$ by $V_2(k)$. The context should clarify whether the argument is "k" or "t_k". With this notation our periodicity assumption on the compensator implies

$$A_c(k+N) = A_c(k), \quad k = 1, 2, \ldots,$$

and similarly for $B_c(\cdot), C_c(\cdot)$, and $D_c(\cdot)$. Also, by assumption, $C(k+N) = C(k)$, for $k = 1, 2, \ldots$.

Next, we model the propagation of the plant over one time step. For notational convenience define

$$H(k) \triangleq \int_0^{h_k} e^{As} ds.$$

Theorem 1. For the fixed-order, multirate sampled-data control problem, the plant dynamics (1) and quadratic performance criterion (5) have the equivalent discrete-time representation

$$x(k+1) = A(k)x(k) + B(k)u(k) + w_1'(k), \tag{8}$$

$$y(k) = C(k)x(k) + w_2(k), \tag{9}$$

$$\mathcal{J} = \delta_\infty + \lim_{K\to\infty} \frac{1}{K} \mathcal{E} \sum_{k=1}^{K} \left[x^T(k) R_1(k) x(k) + 2x^T(k) R_{12}(k) u(k) + u^T(k) R_2(k) u(k) \right], \tag{10}$$

where

$$A(k) \triangleq e^{Ah_k}, \quad B(k) \triangleq H(k)B, \quad w_1'(k) \triangleq \int_0^{h_k} e^{A(h_k - s)} w_1(k+s) ds,$$

$$\delta_\infty \triangleq \lim_{K\to\infty} \frac{1}{K} \mathrm{tr} \sum_{k=1}^{K} \frac{1}{h_k} \int_0^{h_k} \int_0^{s} e^{Ar} V_1 e^{A^T r} R_1 \, dr \, ds,$$

$$R_1(k) \triangleq \frac{1}{h_k} \int_0^{h_k} e^{A^T s} R_1 e^{As} ds,$$

$$R_{12}(k) \triangleq \frac{1}{h_k} \int_0^{h_k} e^{A^T s} R_1 H(s) B ds + \frac{1}{h_k} H^T(k) R_{12},$$

$$R_2(k) \triangleq R_2 + \frac{1}{h_k}\int_0^{h_k} [B^T H^T(s) R_1 H(s) B + R_{12}^T H(s) B + B^T H^T(s) R_{12}]\, ds,$$

and $w_1'(k)$ is a zero-mean, discrete-time white noise process with

$$\mathcal{E}\{w_1'(k) w_1'^T(k)\} = V_1(k),$$

where

$$V_1(k) \triangleq \int_0^{h_k} e^{As} V_1 e^{A^T s} ds.$$

Note that by the sampling periodicity assumption, $A(k+N) = A(k)$, $k = 1, 2 \ldots$.

The proof of this theorem is a straightforward calculation involving integrals of white noise signals, and hence is omitted. See Refs. [22, 26] for related details.

The above formulation assumes that a discrete-time multirate measurement model is available. One can assume, alternatively, that analog measurements corrupted by continuous-time white noise are available instead, that is,

$$y(t) = Cx(t) + w_2(t).$$

In this case one can develop an equivalent discrete-time model that employs an averaging-type A/D device [22, 26–28]

$$\hat{y}(k) = \frac{1}{h_k}\int_{t_k}^{t_{k+1}} y(t) dt.$$

It can be shown that the resulting averaged measurements depend upon delayed samples of the state. In this case the equivalent discrete-time model can be captured by a suitably augmented system. For details see [22, 26].

Remark 1. The equivalent discrete-time quadratic performance criterion (10) involves a constant offset δ_∞[1] which is a function of sampling rates and effectively imposes a lower bound on sampled-data performance due to the discretization process.

---

[1] As will be shown by Lemma 1, due to the periodicity of h_k, δ_∞ is a constant.

III. THE FIXED-ARCHITECTURE MULTIRATE DIGITAL STATIC OUTPUT-FEEDBACK PROBLEM

In this section we obtain necessary conditions that characterize solutions to the multirate sampled-data static output-feedback control problem. First, we form the closed-loop system for (8), (9), and (3) to obtain

$$x(k+1) = \hat{A}(k)x(k) + \hat{w}(k), \tag{11}$$

where

$$\hat{A}(k) \triangleq A(k) + B(k)D_c(k)C(k).$$

The closed-loop disturbance

$$\hat{w}(k) \triangleq w_1'(k) + B(k)D_c(k)w_2(k), \qquad k = 1, 2, \ldots,$$

has nonnegative-definite covariance

$$\hat{V}(k) \triangleq V_1(k) + B(k)D_c(k)V_2(k)D_c^T(k)B^T(k),$$

where we assume that the noise correlation $V_{12}(k) \triangleq \mathcal{E}[w_1'(k)w_2^T(k)] = 0$, that is, the continuous-time plant noise and the discrete-time measurement noise are uncorrelated. The cost functional (10) can now be expressed in terms of the closed-loop second-moment matrix. The following result is immediate.

Proposition 1. For given $D_c(\cdot)$ the second-moment matrix

$$Q(k) \triangleq \mathcal{E}[x(k)x^T(k)], \tag{12}$$

satisfies

$$Q(k+1) = \hat{A}(k)Q(k)\hat{A}^T(k) + \hat{V}(k). \tag{13}$$

Furthermore,

$$\begin{aligned}\mathcal{J}_s(D_c(\cdot)) = \delta_\infty + \lim_{K\to\infty} \frac{1}{K}\mathrm{tr}\sum_{k=1}^{K}[Q(k)\hat{R}(k) \\ + D_c^T(k)R_2(k)D_c(k)V_2(k)],\end{aligned} \tag{14}$$

where

$$\hat{R}(k) \triangleq R_1(k) + R_{12}(k)D_c(k)C(k) + C^T(k)D_c^T(k)R_{12}^T(k) + C^T(k)D_c^T(k)R_2(k)D_c(k)C(k).$$

Remark 2. Equation (13) is a periodic Lyapunov equation which has been extensively studied in [19, 29, 30].

We now show that the covariance Lyapunov equation (13) reaches a steady-state periodic trajectory as $K \to \infty$. For the next result we introduce the parameterization, $k = \alpha + \beta N$, where the index α satisfies $1 \leq \alpha \leq N$, and $\beta = 1, 2, \ldots$. We now restrict our attention to output-feedback controllers having the property that the closed-loop transition matrix over one period

$$\hat{\Phi}_p(\alpha) \triangleq \hat{A}(\alpha + N - 1)\hat{A}(\alpha + N - 2) \cdots \hat{A}(\alpha), \tag{15}$$

is stable for $\alpha = 1, \ldots, N$. Note that since $\hat{A}(\cdot)$ is periodic, the eigenvalues of $\hat{\Phi}_p(\alpha)$ are actually independent of α. Hence it suffices to require that $\hat{\Phi}_p(1) = \hat{A}(N)\hat{A}(N-1) \cdots \hat{A}(1)$ is stable.

Lemma 1. Suppose $\hat{\Phi}_p(1)$ is stable. Then for given $D_c(k)$ the covariance Lyapunov equation (13) reaches a steady state periodic trajectory as $k \to \infty$, that is,

$$\lim_{k \to \infty} [Q(k), Q(k+1), \ldots, Q(k+N-1)] = [Q(\alpha), Q(\alpha+1), \ldots, Q(\alpha+N-1)]. \tag{16}$$

In this case the covariance $Q(k)$ defined by (12) satisfies

$$Q(\alpha + 1) = \hat{A}(\alpha)Q(\alpha)\hat{A}^T(\alpha) + \hat{V}(\alpha), \quad \alpha = 1, \ldots, N, \tag{17}$$

where

$$Q(N + 1) = Q(1). \tag{18}$$

Furthermore, the quadratic performance criterion (14) is given by

$$\mathcal{J}_s(D_c(\cdot)\,) = \delta + \frac{1}{N} \operatorname{tr} \sum_{\alpha=1}^{N} [Q(\alpha)\hat{R}(\alpha) + D_c^T(\alpha)R_2(\alpha)D_c(\alpha)V_2(\alpha)], \tag{19}$$

where

$$\delta \triangleq \frac{1}{N}\text{tr}\sum_{\alpha=1}^{N}\frac{1}{h_\alpha}\int_0^{h_\alpha}\int_0^{s} e^{Ar}V_1 e^{A^T r}R_1 dr\, ds.$$

Proof. See Appendix A. □

For the statement of the main result of this section define the set

$$\mathcal{S}_s \triangleq \{D_c(\cdot) : \hat{\Phi}_p(\alpha) \text{ is stable, for } \alpha = 1,\ldots,N\}. \tag{20}$$

In addition to ensuring that the covariance Lyapunov equation (13) reaches a steady state periodic trajectory as $k \to \infty$, the set $\mathcal{S}_s$ constitutes sufficient conditions under which the Lagrange multiplier technique is applicable to the fixed-order multirate sampled-data static output-feedback control problem. The asymptotic stability of the transition matrix $\hat{\Phi}_p(\alpha)$ serves as a normality condition which further implies that the dual $P(\alpha)$ of $Q(\alpha)$ is nonnegative-definite.

For notational convenience in stating the multirate sampled-data static output-feedback result, define the notation

$$\begin{aligned}
R_{2a}(\alpha) &\triangleq B^T(\alpha)P(\alpha+1)B(\alpha) + \frac{1}{N}R_2(\alpha),\\
V_{2a}(\alpha) &\triangleq C(\alpha)Q(\alpha)C^T(\alpha) + V_2(\alpha),\\
P_a(\alpha) &\triangleq B^T(\alpha)P(\alpha+1)A(\alpha) + \frac{1}{N}R_{12}^T(\alpha),\\
Q_a(\alpha) &\triangleq A(\alpha)Q(\alpha)C^T(\alpha),
\end{aligned}$$

for arbitrary $Q(\alpha)$ and $P(\alpha) \in \mathcal{R}^{n\times n}$ and $\alpha = 1,\ldots,N$.

Theorem 2. Suppose $D_c(\cdot) \in \mathcal{S}_s$ solves the multirate sampled-data static output-feedback control problem. Then there exist $n\times n$ nonnegative-definite matrices $Q(\alpha)$ and $P(\alpha)$ such that, for $\alpha = 1,\ldots,N$, $D_c(\alpha)$ is given by

$$D_c(\alpha) = -R_{2a}^{-1}(\alpha)P_a(\alpha)Q(\alpha)C^T(\alpha)V_{2a}^{-1}(\alpha), \tag{21}$$

and such that $Q(\alpha)$ and $P(\alpha)$ satisfy

$$Q(\alpha+1) = A(\alpha)Q(\alpha)A^T(\alpha) + V_1(\alpha) - Q_a(\alpha)V_{2a}^{-1}(\alpha)Q_a^T(\alpha)$$

$$\begin{aligned}
&+[Q_a(\alpha) + B(\alpha)D_c(\alpha)V_{2a}(\alpha)]V_{2a}^{-1}(\alpha) \\
&\cdot[Q_a(\alpha) + B(\alpha)D_c(\alpha)V_{2a}(\alpha)]^T, \qquad (22)\\
P(\alpha) &= A^T(\alpha)P(\alpha+1)A(\alpha) + \frac{1}{N}R_1(\alpha) - P_a^T(\alpha)R_{2a}^{-1}(\alpha)P_a(\alpha) \\
&+[P_a(\alpha) + R_{2a}(\alpha)D_c(\alpha)C(\alpha)]^T R_{2a}^{-1}(\alpha) \\
&\cdot[P_a(\alpha) + R_{2a}(\alpha)D_c(\alpha)C(\alpha)]. \qquad (23)
\end{aligned}$$

Furthermore, the minimal cost is given by

$$\begin{aligned}
\mathcal{J}_s(D_c(\cdot)) &= \delta + \frac{1}{N}\mathrm{tr}\sum_{\alpha=1}^{N} Q(\alpha)\left[R_1(\alpha) - 2R_{12}(\alpha)R_{2a}^{-1}(\alpha)P_a(\alpha)Q(\alpha)\right. \\
&\cdot C^T(\alpha)V_{2a}^{-1}(\alpha)C(\alpha) + P_a^T(\alpha)R_{2a}^{-1}(\alpha)R_2(\alpha)R_{2a}^{-1}(\alpha) \\
&\left.\cdot P_a(\alpha)Q(\alpha)C^T(\alpha)V_{2a}^{-1}(\alpha)C(\alpha)\right]. \qquad (24)
\end{aligned}$$

Proof. To optimize (19) subject to constraint (17) over the open set $\mathcal{S}_s$ form the Lagrangian

$$\begin{aligned}
&\mathcal{L}(D_c(\alpha), Q(\alpha), P(\alpha+1), \lambda) \triangleq \\
&\mathrm{tr}\sum_{\alpha=1}^{N}\{\lambda\frac{1}{N}[Q(\alpha)\hat{R}(\alpha) + D_c^T(\alpha)R_2(\alpha)D_c(\alpha)V_2(\alpha)] \\
&+ [(\hat{A}(\alpha)Q(\alpha)\hat{A}^T(\alpha) + \hat{V}(\alpha) - Q(\alpha+1))P(\alpha+1)]\}, \qquad (25)
\end{aligned}$$

where the Lagrange multipliers $\lambda \geq 0$ and $P(\alpha+1) \in \mathcal{R}^{n\times n}$, $\alpha = 1,\ldots,N$, are not all zero. We thus obtain

$$\frac{\partial\mathcal{L}}{\partial Q(\alpha)} = \hat{A}^T(\alpha)P(\alpha+1)\hat{A}(\alpha) + \lambda\frac{1}{N}\hat{R}(\alpha) - P(\alpha), \quad \alpha = 1,\ldots,N. \qquad (26)$$

Setting $\frac{\partial\mathcal{L}}{\partial Q(\alpha)} = 0$ yields

$$P(\alpha) = \hat{A}^T(\alpha)P(\alpha+1)\hat{A}(\alpha) + \lambda\frac{1}{N}\hat{R}(\alpha), \quad \alpha = 1,\ldots,N. \qquad (27)$$

Next, propagating (27) from α to $\alpha+N$ yields

$$\begin{aligned}
P(\alpha) &= \hat{A}^T(\alpha)\cdots\hat{A}^T(\alpha+N-1)P(\alpha)\hat{A}(\alpha+N-1)\cdots\hat{A}(\alpha) \\
&+\lambda\frac{1}{N}[\hat{A}^T(\alpha)\cdots\hat{A}^T(\alpha+N-2)\hat{R}(\alpha+N-1)\hat{A}(\alpha+N-1) \\
&\cdots\hat{A}(\alpha) + \hat{A}^T(\alpha)\cdots\hat{A}^T(\alpha+N-3)\hat{R}(\alpha+N-2)\hat{A}(\alpha+N-3) \\
&\cdots\hat{A}(\alpha) + \cdots + \hat{R}(\alpha)]. \qquad (28)
\end{aligned}$$

Note that since $\hat{A}(\alpha+N-1)\cdots\hat{A}(\alpha)$ is assumed to be stable, $\lambda = 0$ implies $P(\alpha) = 0, \alpha = 1,\ldots,N$. Hence, it can be assumed without loss of generality that $\lambda = 1$. Furthermore, $P(\alpha)$, $\alpha = 1,\ldots,N$, is nonnegative-definite. Thus, with $\lambda = 1$, the stationary conditions are given by

$$\frac{\partial \mathcal{L}}{\partial Q(\alpha)} = \hat{A}^T(\alpha)P(\alpha+1)\hat{A}(\alpha) + \frac{1}{N}\hat{R}(\alpha) - P(\alpha) = 0, \tag{29}$$

$$\frac{\partial \mathcal{L}}{\partial D_c(\alpha)} = R_{2a}(\alpha)D_c(\alpha)V_{2a}(\alpha) + P_a(\alpha)Q(\alpha)C^T(\alpha) = 0, \tag{30}$$

for $\alpha = 1,\ldots,N$. Now, (30) implies (21). Next, with $D_c(\alpha)$ given by (21), equations (22) and (23) are equivalent to (17) and (29) respectively. □

Remark 3. In the full-state feedback case we take $C(\alpha) = I$, $V_2(\alpha) = 0$, and $R_{12}(\alpha) = 0$ for $\alpha = 1,\ldots,N$. In this case (21) becomes

$$D_c(\alpha) = -R_{2a}^{-1}(\alpha)B^T(\alpha)P(\alpha+1)A(\alpha), \tag{31}$$

and (23) specializes to

$$\begin{aligned} P(\alpha) &= A^T(\alpha)P(\alpha+1)A(\alpha) + \frac{1}{N}R_1(\alpha) \\ &\quad - A^T(\alpha)P(\alpha+1)B(\alpha)R_{2a}^{-1}(\alpha)B^T(\alpha)P(\alpha+1)A(\alpha), \end{aligned} \tag{32}$$

while (22) is superfluous and can be omitted. Finally, we note that if we assume a single rate architecture the plant dynamics are constant and (32) collapses to the standard discrete-time regulator Riccati equation.

IV. THE FIXED-ARCHITECTURE MULTIRATE DIGITAL DYNAMIC OUTPUT-FEEDBACK PROBLEM

In this section we consider the fixed-order multirate sampled-data dynamic compensation problem. As in Section III, we first form the closed-loop system for (8), (9), (6), and (7), to obtain

$$\tilde{x}(k+1) = \tilde{A}(k)\tilde{x}(k) + \tilde{w}(k), \tag{33}$$

where

$$\tilde{x}(k) \triangleq \begin{bmatrix} x(k) \\ x_c(k) \end{bmatrix},$$

and

$$\begin{aligned} \tilde{A}(k) &\triangleq \begin{bmatrix} A(k)+B(k)D_c(k)C(k) & B(k)C_c(k) \\ B_c(k)C(k) & A_c(k) \end{bmatrix}, \\ \tilde{A}(k+N) &= \tilde{A}(k), \quad k=1,2,\ldots. \end{aligned}$$

The closed-loop disturbance

$$\tilde{w}(k) \triangleq \begin{bmatrix} w_1'(k)+B(k)D_c(k)w_2(k) \\ B_c(k)w_2(k) \end{bmatrix}, \quad k=1,2,\ldots,$$

has nonnegative-definite covariance

$$\tilde{V}(k) \triangleq \begin{bmatrix} V_1(k) & B(k)D_c(k)V_2(k)B_c^T(k) \\ +B(k)D_c(k)V_2(k)D_c^T(k)B^T(k) & \\ & \\ B_c(k)V_2(k)D_c^T(k)B^T(k) & B_c(k)V_2(k)B_c^T(k) \end{bmatrix},$$

where once again we assume that the continuous-time plant noise and the discrete-time measurement noise are uncorrelated, i.e., $V_{12}(k) \triangleq \mathcal{E}\,[w_1'(k)\, w_2^T(k)] = 0$. As for the static output-feedback case, the cost functional (10) can now be expressed in terms of the closed-loop second-moment matrix.

Proposition 2. For given $(A_c(\cdot), B_c(\cdot), C_c(\cdot), D_c(\cdot))$ the second-moment matrix

$$\tilde{Q}(k) \triangleq \mathcal{E}[\tilde{x}(k)\tilde{x}^T(k)], \tag{34}$$

satisfies

$$\tilde{Q}(k+1) = \tilde{A}(k)\tilde{Q}(k)\tilde{A}^T(k) + \tilde{V}(k). \tag{35}$$

Furthermore,

$$\begin{aligned} \mathcal{J}_c(A_c(\cdot), B_c(\cdot), C_c(\cdot), D_c(\cdot)\,) = \delta_\infty + \lim_{K\to\infty}\frac{1}{K}\operatorname{tr}\sum_{k=1}^{K}[\tilde{Q}(k)\tilde{R}(k) \\ +D_c^T(k)R_2(k)D_c(k)V_2(k)], \end{aligned} \tag{36}$$

where the performance weighting matrix $\tilde{R}(k)$ for the closed-loop system is given by

$$\tilde{R}(k) \triangleq \begin{bmatrix} R_1(k) + R_{12}(k)D_c(k)C(k) & R_{12}(k)C_c(k) \\ +C^T(k)D_c^T(k)R_{12}^T(k) & +C^T(k)D_c^T(k)R_2(k)C_c(k) \\ +C^T(k)D_c^T(k)R_2(k)D_c(k)C(k) & \\ & \\ C_c^T(k)R_{12}^T(k) & C_c^T(k)R_2(k)C_c(k) \\ +C_c^T(k)R_2(k)D_c(k)C(k) & \end{bmatrix}.$$

Next, it follows from Lemma 1 with Q and $\hat{R}$ replaced by $\tilde{Q}$ and $\tilde{R}$, respectively, that the covariance Lyapunov equation (35) reaches a steady-state periodic trajectory as $K \to \infty$ under the assumption that the transition matrix over one period for the closed-loop system (33) given by

$$\tilde{\Phi}_p(\alpha) \triangleq \tilde{A}(\alpha + N - 1)\tilde{A}(\alpha + N - 2) \cdots \tilde{A}(\alpha), \tag{37}$$

is stable for $\alpha = 1, \ldots, N$. Hence, the following result is immediate.

Lemma 2. Suppose $\tilde{\Phi}_p(1)$ is stable. Then for given $(A_c(k), B_c(k), C_c(k), D_c(k))$ the covariance Lyapunov equation (35) reaches a steady state periodic trajectory as $k \to \infty$, that is,

$$\lim_{k\to\infty} [\tilde{Q}(k), \tilde{Q}(k+1), \ldots, \tilde{Q}(k+N-1)] = [\tilde{Q}(\alpha), \tilde{Q}(\alpha+1), \ldots, \tilde{Q}(\alpha+N-1)]. \tag{38}$$

In this case the covariance $\tilde{Q}(k)$ defined by (34) satisfies

$$\tilde{Q}(\alpha+1) = \tilde{A}(\alpha)\tilde{Q}(\alpha)\tilde{A}^T(\alpha) + \tilde{V}(\alpha), \quad \alpha = 1, \ldots, N, \tag{39}$$

where

$$\tilde{Q}(N+1) = \tilde{Q}(1). \tag{40}$$

Furthermore, the quadratic performance criterion (36) is given by

$$\mathcal{J}_c(A_c(\cdot), B_c(\cdot), C_c(\cdot), D_c(\cdot)\ ) = \delta + \frac{1}{N} \operatorname{tr} \sum_{\alpha=1}^{N} [\tilde{Q}(\alpha)\tilde{R}(\alpha) + D_c^T(\alpha)R_2(\alpha)D_c(\alpha)V_2(\alpha)]. \tag{41}$$

Proof. The proof is identical to the proof of Lemma 1 with Q and $\hat{R}$ replaced by $\tilde{Q}$ and $\tilde{R}$, respectively. □

For the next result, define the compensator transition matrix over one period by

$$\Phi_{cp}(\alpha) \triangleq A_c(\alpha+N-1)A_c(\alpha+N-2)\cdots A_c(\alpha). \tag{42}$$

Note that since $A_c(\alpha)$ is required to be periodic, the eigenvalues of $\Phi_{cp}(\alpha)$ are actually independent of α.

In the following we obtain necessary conditions that characterize solutions to the fixed-order multirate sampled-data dynamic compensation problem. Derivation of these conditions requires additional technical assumptions. Specifically, we further restrict $(A_c(\cdot), B_c(\cdot), C_c(\cdot), D_c(\cdot))$ to the set

$$\begin{aligned}\mathcal{S}_c \triangleq \{(A_c(\alpha), B_c(\alpha), C_c(\alpha), D_c(\alpha)) : \tilde{\Phi}_p(\alpha) \text{ is stable and} \\ (\Phi_{cp}(\alpha), B_{cp}(\alpha), C_{cp}(\alpha)) \text{ is controllable and} \\ \text{observable}, \ \alpha = 1, \ldots, N\},\end{aligned} \tag{43}$$

where

$$\begin{aligned}B_{cp}(\alpha) \triangleq\ & [A_c(\alpha+N-1)A_c(\alpha+N-2)\cdots A_c(\alpha+1)B_c(\alpha), \\ & A_c(\alpha+N-1)A_c(\alpha+N-2)\cdots A_c(\alpha+2)B_c(\alpha+1), \\ & \ldots, B_c(\alpha+N-1)],\end{aligned} \tag{44}$$

$$C_{cp}(\alpha) \triangleq \begin{bmatrix} C_c(\alpha+N-1)A_c(\alpha+N-2)\cdots A_c(\alpha) \\ C_c(\alpha+N-2)A_c(\alpha+N-3)\cdots A_c(\alpha) \\ \vdots \\ C_c(\alpha) \end{bmatrix}. \tag{45}$$

The set $\mathcal{S}_c$ constitutes sufficient conditions under which the Lagrange multiplier technique is applicable to the fixed-order multirate sampled-data control problem. This is similar to concepts involving *moving equilibria* for periodic Lyapunov/Riccati equations discussed in [17, 19]. Specifically, the formulae for the lifted isomorphism (44) and (45) are equivalent to assuming the stability of $\tilde{A}(\cdot)$ along with the reachability and observability of $(A_c(\cdot),\ B_c(\cdot),\ C_c(\cdot))$ [8, 19]. The asymptotic stability of the transition matrix $\tilde{\Phi}_p(\alpha)$ serves as a normality condition which further implies that the dual $\tilde{P}(\alpha)$ of $\tilde{Q}(\alpha)$ is nonnegative-definite. Furthermore, the assumption that $(\Phi_{cp}(\alpha), B_{cp}(\alpha), C_{cp}(\alpha))$ is controllable and observable is a nondegeneracy condition which implies that the lower right $n_c \times n_c$ subblocks of $\tilde{Q}(\alpha)$ and $\tilde{P}(\alpha)$ are positive definite thus yielding explicit gain expressions for $A_c(\alpha), B_c(\alpha), C_c(\alpha)$, and $D_c(\alpha)$.

In order to state the main result we require some additional notation and a lemma concerning pairs of nonnegative-definite matrices.

Lemma 3. Let $\hat{Q}, \hat{P}$ be $n\times n$ nonnegative-definite matrices and assume rank $\hat{Q}\hat{P} = n_c$. Then there exist $n_c \times n$ matrices G, Γ and an $n_c \times n_c$ invertible matrix M, unique except for a change of basis in $\mathcal{R}^{n_c}$, such that

$$\hat{Q}\hat{P} = G^T M \Gamma, \tag{46}$$

$$\Gamma G^T = I_{n_c}. \tag{47}$$

Furthermore, the $n \times n$ matrices

$$\tau \triangleq G^T\Gamma, \quad \tau_\perp \triangleq I_n - \tau, \tag{48}$$

are idempotent and have rank n_c and $n - n_c$, respectively.

Proof. See Ref. [31]. □

The following result gives necessary conditions that characterize solutions to the fixed-order multirate sampled-data control problem. For convenience in stating this result, recall the definitions of $R_{2a}(\cdot)$, $V_{2a}(\cdot)$, $P_a(\cdot)$, and $Q_a(\cdot)$ and define the additional notation

$$M(\alpha) \triangleq \begin{bmatrix} I_n \\ D_c(\alpha)C(\alpha) \end{bmatrix}, \quad \hat{M}(\alpha) \triangleq \begin{bmatrix} I_n \\ -R_{2a}^{-1}(\alpha)P_a(\alpha) \end{bmatrix},$$

$$\hat{R}(\alpha) \triangleq \begin{bmatrix} R_1(\alpha) & R_{12}(\alpha) \\ R_{12}^T(\alpha) & R_2(\alpha) \end{bmatrix},$$

for arbitrary $P(\alpha) \in \mathcal{R}^{n\times n}$ and $\alpha = 1,\ldots,N$.

Theorem 3. Suppose $(A_c(\cdot), B_c(\cdot), C_c(\cdot), D_c(\cdot)) \in \mathcal{S}_c$ solves the fixed-order multirate sampled-data dynamic output-feedback control problem.

Then there exist $n \times n$ nonnegative-definite matrices $Q(\alpha)$, $P(\alpha)$, $\hat{Q}(\alpha)$, and $\hat{P}(\alpha)$ such that, for $\alpha = 1, \ldots, N$, $A_c(\alpha), B_c(\alpha), C_c(\alpha)$, and $D_c(\alpha)$ are given by

$$\begin{aligned} A_c(\alpha) &= \Gamma(\alpha+1)[A(\alpha) - B(\alpha)R_{2a}^{-1}(\alpha)P_a(\alpha) - Q_a(\alpha)V_{2a}^{-1}(\alpha)C(\alpha) \\ &\quad -B(\alpha)D_c(\alpha)C(\alpha)]G^T(\alpha), && (49) \\ B_c(\alpha) &= \Gamma(\alpha+1)[Q_a(\alpha)V_{2a}^{-1}(\alpha) + B(\alpha)D_c(\alpha)], && (50) \\ C_c(\alpha) &= -[R_{2a}^{-1}(\alpha)P_a(\alpha) + D_c(\alpha)C(\alpha)]G^T(\alpha), && (51) \\ D_c(\alpha) &= -R_{2a}^{-1}(\alpha)P_a(\alpha)Q(\alpha)C^T(\alpha)V_{2a}^{-1}(\alpha), && (52) \end{aligned}$$

and such that $Q(\alpha), P(\alpha), \hat{Q}(\alpha)$, and $\hat{P}(\alpha)$ satisfy

$$\begin{aligned} Q(\alpha+1) &= A(\alpha)Q(\alpha)A^T(\alpha) + V_1(\alpha) - Q_a(\alpha)V_{2a}^{-1}(\alpha)Q_a^T(\alpha) \\ &\quad +\tau_\perp(\alpha+1)[\{A(\alpha) - B(\alpha)R_{2a}^{-1}(\alpha)P_a(\alpha)\}\hat{Q}(\alpha) \\ &\quad \cdot\{A(\alpha) - B(\alpha)R_{2a}^{-1}(\alpha)P_a(\alpha)\}^T \\ &\quad +\{Q_a(\alpha) + B(\alpha)D_c(\alpha)V_{2a}(\alpha)\}V_{2a}^{-1}(\alpha) \\ &\quad \cdot\{Q_a(\alpha) + B(\alpha)D_c(\alpha)V_{2a}(\alpha)\}^T]\tau_\perp^T(\alpha+1), && (53) \\ P(\alpha) &= A^T(\alpha)P(\alpha+1)A(\alpha) + \frac{1}{N}R_1(\alpha) - P_a^T(\alpha)R_{2a}^{-1}(\alpha)P_a(\alpha) \\ &\quad +\tau_\perp^T(\alpha)[\{A(\alpha) - Q_a(\alpha)V_{2a}^{-1}(\alpha)C(\alpha)\}^T\hat{P}(\alpha+1) \\ &\quad \cdot\{A(\alpha) - Q_a(\alpha)V_{2a}^{-1}(\alpha)C(\alpha)\} \\ &\quad +\{P_a(\alpha) + R_{2a}(\alpha)D_c(\alpha)C(\alpha)\}^T R_{2a}^{-1}(\alpha) \\ &\quad \cdot\{P_a(\alpha) + R_{2a}(\alpha)D_c(\alpha)C(\alpha)\}]\tau_\perp(\alpha), && (54) \\ \hat{Q}(\alpha+1) &= \tau(\alpha+1)[\{A(\alpha) - B(\alpha)R_{2a}^{-1}(\alpha)P_a(\alpha)\}\hat{Q}(\alpha) \\ &\quad \cdot\{A(\alpha) - B(\alpha)R_{2a}^{-1}(\alpha)P_a(\alpha)\}^T \\ &\quad +\{Q_a(\alpha) + B(\alpha)D_c(\alpha)V_{2a}(\alpha)\}V_{2a}^{-1}(\alpha) \\ &\quad \cdot\{Q_a(\alpha) + B(\alpha)D_c(\alpha)V_{2a}(\alpha)\}^T]\tau^T(\alpha+1), && (55) \\ \hat{P}(\alpha) &= \tau^T(\alpha)[\{A(\alpha) - Q_a(\alpha)V_{2a}^{-1}(\alpha)C(\alpha)\}^T\hat{P}(\alpha+1) \\ &\quad \cdot\{A(\alpha) - Q_a(\alpha)V_{2a}^{-1}(\alpha)C(\alpha)\} \\ &\quad +\{P_a(\alpha) + R_{2a}(\alpha)D_c(\alpha)C(\alpha)\}^T R_{2a}^{-1}(\alpha) \\ &\quad \cdot\{P_a(\alpha) + R_{2a}(\alpha)D_c(\alpha)C(\alpha)\}]\tau(\alpha), && (56) \\ \operatorname{rank}\hat{Q}(\alpha) &= \operatorname{rank}\hat{P}(\alpha) = \operatorname{rank}\hat{Q}(\alpha)\hat{P}(\alpha) = n_c. && (57) \end{aligned}$$

Furthermore, the minimal cost is given by

$$\begin{aligned}\mathcal{J}_c(A_c(\cdot), B_c(\cdot), C_c(\cdot), D_c(\cdot)) &= \delta + \frac{1}{N}\text{tr} \sum_{\alpha=1}^{N}[\{M(\alpha)Q(\alpha)M^T(\alpha) \\ &\quad +\hat{M}(\alpha)\hat{Q}(\alpha)\hat{M}^T(\alpha)\}\hat{\hat{R}}(\alpha)]. \qquad (58)\end{aligned}$$

Proof. See Appendix B. □

Theorem 3 provides necessary conditions for the fixed-order multirate sampled-data control problem. These necessary conditions consist of a system of two modified periodic difference Lyapunov equations and two modified periodic difference Riccati equations coupled by projection matrices $\tau(\alpha)$, $\alpha = 1, \ldots, N$. As expected, these equations are periodically time-varying over the period $1 \leq \alpha \leq N$ in accordance with the multirate nature of the measurements. As discussed in [21] the fixed-order constraint on the compensator gives rise to the projection τ which characterizes the optimal reduced-order compensator gains. In the multirate case however, it is interesting to note that the time-varying nature of the problem gives rise to multiple projections corresponding to each of the intermediate points of the periodicity interval and whose rank along the periodic interval is equal to the order of the compensator.

Remark 4. As in the linear time-invariant case [21] to obtain the full-order multirate LQG controller, set $n_c = n$. In this case, the projections $\tau(\alpha)$, and $\Gamma(\alpha)$ and $G(\alpha)$, for $\alpha = 1, \ldots, N$, become the identity. Consequently, equations (55) and (56) play no role and hence can be omitted. In order to draw connections with existing full-order multirate LQG results set $D_c(\alpha) = 0$ and $R_{12}(\alpha) = 0$, $\alpha = 1, \ldots, N$, so that

$$\begin{aligned}A_c(\alpha) &= A(\alpha) - B(\alpha)R_{2a}^{-1}(\alpha)B^T(\alpha)P(\alpha+1)A(\alpha) \\ &\quad - A(\alpha)Q(\alpha)C^T(\alpha)V_{2a}^{-1}(\alpha)C(\alpha), \qquad (59)\\ B_c(\alpha) &= A(\alpha)Q(\alpha)C^T(\alpha)V_{2a}^{-1}(\alpha), \qquad (60)\\ C_c(\alpha) &= -R_{2a}^{-1}(\alpha)B^T(\alpha)P(\alpha+1)A(\alpha), \qquad (61)\end{aligned}$$

where $Q(\alpha)$ and $P(\alpha)$ satisfy

$$Q(\alpha+1) = A(\alpha)Q(\alpha)A^T(\alpha) + V_1(\alpha)$$

$$-A(\alpha)Q(\alpha)C^T(\alpha)V_{2a}^{-1}(\alpha)C(\alpha)Q(\alpha)A^T(\alpha), \tag{62}$$

$$\begin{aligned} P(\alpha) &= A^T(\alpha)P(\alpha+1)A(\alpha) + \frac{1}{N}R_1(\alpha) \\ &\quad -A^T(\alpha)P(\alpha+1)B(\alpha)R_{2a}^{-1}(\alpha)B^T(\alpha)P(\alpha+1)A(\alpha). \end{aligned} \tag{63}$$

Thus the full-order multirate sampled-data controller is characterized by two decoupled periodic difference Riccati equations (observer and regulator Riccati equations) over the period $\alpha = 1, \ldots, N$. This corresponds to the results obtained in [7, 8]. Next, assuming a single rate architecture yields time-invariant plant dynamics while (62) and (63) specialize to the discrete-time observer and regulator Riccati equations. Alternatively, retaining the reduced-order constraint and assuming single rate sampling, Theorem 3 yields the sampled-data optimal projection equations for reduced-order dynamic compensation given in [22].

V. NUMERICAL EVALUATION OF INTEGRALS INVOLVING MATRIX EXPONENTIALS

To evaluate the integrals involving matrix exponentials appearing in Theorem 1, we utilize the approach of Ref [25]. The idea is to eliminate the need for integration by computing the matrix exponential of appropriate block matrices. Numerical matrix exponentiation is discussed in [32].

Proposition 3. For $\alpha = 1, \ldots, N$, consider the following partitioned matrix exponentials

$$\begin{bmatrix} E_1 & E_2 & E_3 & E_4 \\ 0_n & E_5 & E_6 & E_7 \\ 0_n & 0_n & E_8 & E_9 \\ 0_{m\times n} & 0_{m\times n} & 0_{m\times n} & I_m \end{bmatrix} \triangleq \exp \begin{bmatrix} -A^T & I_n & 0_n & 0_{n\times m} \\ 0_n & -A^T & R_1 & 0_{n\times m} \\ 0_n & 0_n & A & B \\ 0_{m\times n} & 0_{m\times n} & 0_{m\times n} & 0_m \end{bmatrix} h_\alpha,$$

$$\begin{bmatrix} E_{10} & E_{11} & E_{12} & E_{13} \\ 0_n & E_{14} & E_{15} & E_{16} \\ 0_n & 0_n & E_{17} & E_{18} \\ 0_{m\times n} & 0_{m\times n} & 0_{m\times n} & I_m \end{bmatrix} \triangleq \exp \begin{bmatrix} -A^T & I_n & 0_n & 0_{n\times m} \\ 0_n & -A^T & R_1 & R_{12} \\ 0_n & 0_n & A & B \\ 0_{m\times n} & 0_{m\times n} & 0_{m\times n} & 0_m \end{bmatrix} h_\alpha,$$

$$\begin{bmatrix} E_{19} & E_{20} & E_{21} \\ 0_n & E_{22} & E_{23} \\ 0_n & 0_n & E_{24} \end{bmatrix} \triangleq \exp \begin{bmatrix} -A & I_n & 0_n \\ 0_n & -A & V_1 \\ 0_n & 0_n & A^T \end{bmatrix} h_\alpha,$$

of orders $(3n+m)\times(3n+m)$, $(3n+m)\times(3n+m)$, and $3n\times 3n$, respectively. Then

$$\begin{aligned} A(\alpha) &= E_{24}^T, \quad B(\alpha) = E_{18}, \quad V_1(\alpha) = E_{24}^T E_{23}, \\ R_1(\alpha) &= \frac{1}{h_\alpha} E_{17}^T E_{15}, \quad R_{12}(\alpha) = \frac{1}{h_\alpha} E_{17}^T E_{16}, \\ R_2(\alpha) &= R_2 + \frac{1}{h_\alpha}[B^T E_{17}^T E_{13} + E_{13}^T E_{17} B - B^T E_{17}^T E_4], \\ \delta &= \frac{1}{N}\sum_{\alpha=1}^{N} \frac{1}{h_\alpha} \operatorname{tr} R_1 E_{24}^T E_{21}. \end{aligned}$$

The proof of the above proposition involves straightforward manipulations of matrix exponentials and hence is omitted.

VI. HOMOTOPY ALGORITHM FOR MULTIRATE DYNAMIC COMPENSATION

In this section we present a new class of numerical algorithms using homotopic continuation methods based on a predictor/corrector scheme for solving the design equations (62) and (63) for the full-order multirate control problem. Homotopic continuation methods operate by first replacing the original problem by a simpler problem with a known solution. The desired solution is then reached by integrating along a path (homotopy path) that connects the starting problem to the original problem. The advantage of such algorithms is that they are global in nature. In particular, homotopy methods facilitate the finding of (multiple) solutions to a problem, and the convergence of the homotopy algorithm is generally not dependent upon having initial conditions which are in some sense close to the actual solution. These ideas have been illustrated for the H_2 reduced-order control problem in [33] and H_∞ constrained problem in [34]. A complete description of the homotopy algorithm for the reduced-order H_2 problem is given in [35].

In the following we use the notation $Q_\alpha \triangleq Q(\alpha)$. To solve (62) for $\alpha = 1, \ldots, N$, consider the equivalent discrete-time algebraic Riccati equation (See [17])

$$Q_\alpha = \bar{\Phi}_{\alpha+N,\alpha} Q_\alpha \bar{\Phi}^T_{\alpha+N,\alpha} + W_{\alpha+N,\alpha}, \tag{64}$$

where

$$\begin{aligned} \bar{\Phi}_{\alpha+N,i} &\triangleq A(\alpha+N-1)A(\alpha+N-2)\cdots A(i), \quad \alpha+N>i, \\ \bar{\Phi}_{\alpha+N,\alpha+N} &= I_{n^2}, \end{aligned} \tag{65}$$

and $W_{\alpha+N,\alpha}$ is the reachability Gramian defined by

$$W_{\alpha+N,\alpha} \triangleq \sum_{i=\alpha}^{\alpha+N-1} \left[\bar{\Phi}_{\alpha+N,i+1}[V_{1_i} - Q_{a_i} V_{2a_i}^{-1} Q_{a_i}^T] \bar{\Phi}^T_{\alpha+N,i+1}\right]. \tag{66}$$

To define the homotopy map we assume that the plant matrices (A_α, C_α) and the disturbance intensities $(V_{1_\alpha}, V_{2_\alpha})$ are functions of the homotopy parameter $\lambda \in [0,1]$. In particular, let

$$A_\alpha(\lambda) = A_{\alpha_0} + \lambda(A_{\alpha_f} - A_{\alpha_0}), \tag{67}$$

$$C_\alpha(\lambda) = C_{\alpha_0} + \lambda(C_{\alpha_f} - C_{\alpha_0}), \tag{68}$$

where the subscripts '0' and 'f' denote initial and final values and

$$\begin{bmatrix} V_{1_\alpha}(\lambda) & 0 \\ 0 & V_{2_\alpha}(\lambda) \end{bmatrix} = L_R(\lambda) L_R^T(\lambda), \tag{69}$$

where

$$L_R(\lambda) = L_{R,0} + \lambda(L_{R,f} - L_{R,0}), \tag{70}$$

and $L_{R,0}$ and $L_{R,f}$ satisfy

$$L_{R,0}L_{R,0}^T = \begin{bmatrix} V_{1,0_\alpha} & 0 \\ 0 & V_{2,0_\alpha} \end{bmatrix}, \tag{71}$$

$$L_{R,f}L_{R,f}^T = \begin{bmatrix} V_{1,f_\alpha} & 0 \\ 0 & V_{2,f_\alpha} \end{bmatrix}. \tag{72}$$

The homotopy map for (64) is defined by the equation

$$\begin{aligned} Q_\alpha(\lambda) = {} & A_{\alpha+N-1}\cdots A_\alpha(\lambda)Q_\alpha(\lambda)A_\alpha^T(\lambda)\cdots A_{\alpha+N-1}^T \\ & +A_{\alpha+N-1}\cdots A_{\alpha+1}[V_{1_\alpha}(\lambda) - Q_{a_\alpha}(\lambda)V_{2a_\alpha}^{-1}(\lambda)Q_{a_\alpha}^T(\lambda)]A_{\alpha+1}^T \\ & \cdots A_{\alpha+N-1}^T + \sum_{j=\alpha+1}^{\alpha+N-1} \bar{\Phi}_{\alpha+N,j+1}[V_{1_j} - Q_{a_j}V_{2a_j}^{-1}Q_{a_j}^T] \\ & \cdot\bar{\Phi}_{\alpha+N,j+1}^T, \end{aligned} \tag{73}$$

where

$$Q_{a_\alpha} \triangleq A_\alpha(\lambda)Q_\alpha(\lambda)C_\alpha^T(\lambda), \qquad V_{2a_\alpha}(\lambda) \triangleq V_{2_\alpha}(\lambda) + C_\alpha(\lambda)Q_\alpha(\lambda)C_\alpha^T(\lambda).$$

The homotopy algorithm presented in this section uses a predictor/corrector numerical scheme. The predictor step requires the derivative $Q'_\alpha(\lambda)$, where $Q'_\alpha \triangleq dQ_\alpha/d\lambda$, while the correction step is based on using the Newton correction, denoted here as ΔQ_α. Below, we derive the matrix equations that can be used to solve for the derivative and correction. For notational simplicity we omit the argument λ in the derived equations.

Differentiating (73) with respect to λ gives the discrete-time matrix Lyapunov equation

$$Q'_\alpha = \mathcal{A}_Q Q'_\alpha \mathcal{A}_Q^T + \mathcal{V}_Q, \tag{74}$$

where

$$\mathcal{A}_Q \triangleq \bar{\Phi}_{\alpha+N,\alpha+1}[A_\alpha - Q_{a_\alpha}V_{2a_\alpha}^{-1}C_\alpha],$$

and

$$\begin{aligned} \mathcal{V}_Q \triangleq {} & \bar{\Phi}_{\alpha+N,\alpha+1}[\mathcal{A}_{1q}Q_\alpha\mathcal{A}_{2q}^T + \mathcal{A}_{2q}Q_\alpha\mathcal{A}_{1q}^T + V'_{1_\alpha} \\ & +Q_{a_\alpha}V_{2a_\alpha}^{-1}V'_{2_\alpha}V_{2a_\alpha}^{-1}Q_{a_\alpha}^T]\bar{\Phi}_{\alpha+N,\alpha+1}^T, \\ \mathcal{A}_{1q} \triangleq {} & A'_\alpha - Q_{a_\alpha}V_{2a_\alpha}^{-1}C'_\alpha, \qquad \mathcal{A}_{2q} \triangleq A_\alpha - Q_{a_\alpha}V_{2a_\alpha}^{-1}C_\alpha. \end{aligned}$$

The correction equation is developed with λ at some fixed value, say λ^*. The development of the correction is based on the following discussion. Below, we use the notation

$$f'(\theta) \triangleq \frac{df}{d\theta}. \tag{75}$$

Let $f : \mathcal{R}^n \to \mathcal{R}^n$ be C^1 continuous and consider the equation

$$f(\theta) = 0. \tag{76}$$

If $\theta^{(i)}$ is the current approximation to the solution of (76), then the Newton correction $\Delta\theta$ is defined by

$$\theta^{(i+1)} - \theta^{(i)} \triangleq \Delta\theta = -f'(\theta^{(i)})^{-1}e, \tag{77}$$

where

$$e \triangleq f(\theta(i)). \tag{78}$$

Now let $\theta^{(i)}$ be an approximation to θ satisfying (76). Then with $e = f(\theta(i))$ construct the following homotopy to solve $f(\theta) = 0$

$$(1-\beta)e = f(\theta(\beta)), \qquad \beta \in [0,1]. \tag{79}$$

Note that at $\beta = 0$, (79) has a solution $\theta(0) = \theta^{(i)}$ while $\theta(1)$ satisfies $f(\theta) = 0$. Then differentiating (79) with respect to β gives

$$\left.\frac{d\theta}{d\beta}\right|_{\beta=0} = -f'(\theta^{(i)})^{-1}e. \tag{80}$$

Remark 5. Note that the Newton correction $\Delta\theta$ in (77) and the derivative $\left.\frac{d\theta}{d\beta}\right|_{\beta=0}$ in (80) are identical. Hence, the Newton correction $\Delta\theta$ can be found by constructing a homotopy of the form (79) and solving for the resulting derivative $\left.\frac{d\theta}{d\beta}\right|_{\beta=0}$. As seen below, this insight is particularly useful when deriving Newton corrections for equations that have a matrix structure.

Now we use the insights of Remark 5 to derive the equation that needs to be solved for the Newton correction ΔQ_α. We begin by recalling that λ

is assumed to have some fixed value, say λ^*. Also, it is assumed that Q^*_α is the current approximation to $Q_\alpha(\lambda^*)$ and that E_Q is the error in equation (64) with $\lambda = \lambda^*$ and $Q_\alpha(\lambda)$ replaced by Q^*_α.

Next we form the homotopy map for (64) as follows

$$(1-\beta)E_Q = \bar{\Phi}_{\alpha+N,\alpha}Q_\alpha(\beta)\bar{\Phi}^T_{\alpha+N,\alpha} + W_{\alpha+N,\alpha}(\beta) - Q_\alpha(\beta), \quad (81)$$

where

$$W_{\alpha+N,\alpha}(\beta) \triangleq A_{\alpha+N-1}\cdots A_{\alpha+1}[V_{1_\alpha} - Q_{a_\alpha}(\beta)V^{-1}_{2a_\alpha}(\beta)Q^T_{a_\alpha}(\beta)]A^T_{\alpha+1} \cdots A^T_{\alpha+N-1} + \sum_{i=\alpha+1}^{\alpha+N-1}\left[\bar{\Phi}_{\alpha+N,i+1}[V_{1_i} - Q_{a_i}V^{-1}_{2a_i}Q^T_{a_i}]\bar{\Phi}^T_{\alpha+N,i+1}\right],$$

and

$$V_{2a_\alpha}(\beta) \triangleq C_\alpha Q_\alpha(\beta)C^T_\alpha + V_{2_\alpha}, \quad Q_{a_\alpha}(\beta) \triangleq A_\alpha Q_\alpha(\beta)C^T_\alpha.$$

Differentiating (81) with respect to β and using Remark 5 to make the replacement

$$\Delta Q_\alpha = \left.\frac{dQ}{d\beta}\right|_{\beta=0}, \quad (82)$$

gives the Newton correction equation

$$\Delta Q_\alpha = \mathcal{A}_Q \Delta Q_\alpha \mathcal{A}^T_Q + E_Q. \quad (83)$$

Note that (83) is a discrete-time algebraic Lyapunov equation.

Algorithm 1. To solve the design equation (62), carry out the following steps:

Step 1. Initialize loop $= 0$, $\lambda = 0$, $\Delta\lambda \in (0,1]$, $Q(\alpha) = Q_0(\alpha)$, $\alpha = 1,\ldots N$.

Step 2. Let loop=loop+1. If loop=1, then go to step 4.

Step 3. Advance the homotopy parameter λ and predict $Q_\alpha(\lambda)$ as follows.

3a. Let $\lambda_0 = \lambda$.

3b. Let $\lambda = \lambda_0 + \Delta\lambda$.

3c. Compute $Q'_\alpha(\lambda)$ using (74).

3d. Predict $Q_\alpha(\lambda)$ using $Q_\alpha(\lambda) = Q_\alpha(\lambda_0) + (\lambda - \lambda_0)Q'_\alpha(\lambda)$.

3e. Compute the error E_Q in equation (64). If E_Q satisfies some preassigned tolerance then continue else reduce $\Delta\lambda$ and go to step 3b.

Step 4. Correct the current approximation $Q_\alpha(\lambda)$ as follows.

4a. Compute the error E_Q in equation (64).

4b. Solve (83) to obtain a Newton homotopy correction.

4c. Let $Q_\alpha \leftarrow Q_\alpha + \Delta Q_\alpha$.

4d. Propagate (62) over the period.

4e. Recompute the error E_Q in equation (64). If E_Q satisfies some preassigned tolerance then continue else reduce ΔQ_α and go to Step 4c.

Step 5. If $\lambda = 1$ then stop. Else go to Step 2.

Equivalently, to solve (63) for $\alpha = 1, \ldots, N$, consider the dual discrete-time algebraic Riccati equation

$$P_\alpha = \bar{\Phi}^T_{\alpha+N,\alpha} P_\alpha \bar{\Phi}_{\alpha+N,\alpha} + \hat{W}_{\alpha,\alpha+N}, \tag{84}$$

where $P_\alpha \triangleq P(\alpha)$, and $\hat{W}_{\alpha,\alpha+N}$ is the observability Gramian defined by

$$\hat{W}_{\alpha,\alpha+N} \triangleq \sum_{i=\alpha}^{\alpha+N-1} \left[\bar{\Phi}^T_{i,\alpha} [\frac{1}{N} R_{1_i} - P^T_{a_i} R^{-1}_{2a_i} P_{a_i}] \bar{\Phi}_{i,\alpha} \right]. \tag{85}$$

Next, in a similar fashion we get the dual prediction and Newton correction equations

$$P'_\alpha = \mathcal{A}^T_P P'_\alpha \mathcal{A}_P + \mathcal{R}_P, \tag{86}$$

$$\Delta P_\alpha = \mathcal{A}^T_P \Delta P_\alpha \mathcal{A}_P + E_P, \tag{87}$$

respectively, where

$$\begin{aligned}
\mathcal{A}_P &\triangleq [A_{\alpha+N-1} - B_{\alpha+N-1}R_{2a_{\alpha+N-1}}^{-1}P_{a_{\alpha+N-1}}]\bar{\Phi}_{\alpha+N-1,\alpha},\\
\mathcal{R}_P &\triangleq \bar{\Phi}_{\alpha+N-1,\alpha}^T[\mathcal{A}_{1p}^T P_\alpha \mathcal{A}_{2p} + \mathcal{A}_{2p}^T P_\alpha \mathcal{A}_{1p} + \frac{1}{N}R'_{1_{\alpha+N-1}}\\
&\quad -\frac{1}{N}R'_{12_{\alpha+N-1}}R_{2a_{\alpha+N-1}}^{-1}P_{a_{\alpha+N-1}}\\
&\quad -(\frac{1}{N}R'_{12_{\alpha+N-1}}R_{2a_{\alpha+N-1}}^{-1}P_{a_{\alpha+N-1}})^T\\
&\quad +\frac{1}{N}P_{a_{\alpha+N-1}}^T R_{2a_{\alpha+N-1}}^{-1}R'_{2_{\alpha+N-1}}R_{2a_{\alpha+N-1}}^{-1}P_{a_{\alpha+N-1}}]\bar{\Phi}_{\alpha+N-1,\alpha},\\
\mathcal{A}_{1p} &\triangleq A'_{\alpha+N-1} - B'_{\alpha+N-1}R_{2a_{\alpha+N-1}}^{-1}P_{a_{\alpha+N-1}},\\
\mathcal{A}_{2p} &\triangleq A_{\alpha+N-1} - B_{\alpha+N-1}R_{2a_{\alpha+N-1}}^{-1}P_{a_{\alpha+N-1}},
\end{aligned}$$

and E_P is the error in the equation (84) with the current approximation for P_α for $\alpha = 1, \ldots, N$. To solve design equation (63), we can now apply the steps in Algorithm 1.

VII. ILLUSTRATIVE NUMERICAL EXAMPLES

For illustrative purposes we consider two numerical examples. Our first example involves a rigid body with a flexible appendage (Figure 2) and is reminiscent of a single-axis spacecraft involving unstable dynamics and sensor fusion of slow, accurate spacecraft attitude sensors (such as horizon sensors or star trackers) with fast, less accurate rate gyroscopes. The motivation for slow/fast sensor configuration is that rate information can be used to improve the attitude control between attitude measurements. Hence define

$$\begin{aligned}
A &= \begin{bmatrix} 0 & 1 & 0 & 0 \\ 0 & 0 & 0 & 0 \\ 0 & 0 & 0 & 1 \\ 0 & 0 & -1 & -0.01 \end{bmatrix}, \qquad B = \begin{bmatrix} 0 \\ 1 \\ 0 \\ 0 \end{bmatrix},\\
C &= \begin{bmatrix} 0 & 1 & 0 & 1 \\ 1 & 0 & 1 & 0 \end{bmatrix}, \qquad D = \begin{bmatrix} 0.1 & 0 & 0.1 & 0 \\ 0 & 1 & 0 & 1 \end{bmatrix}^T,\\
V_1 &= DD^T, \qquad V_2 = I_2,\\
E &= \begin{bmatrix} 1 & 0 & 0 & 0 \\ 0 & 0 & 1 & 0 \end{bmatrix}, \qquad R_1 = E^T E, \qquad R_2 = 1.
\end{aligned}$$

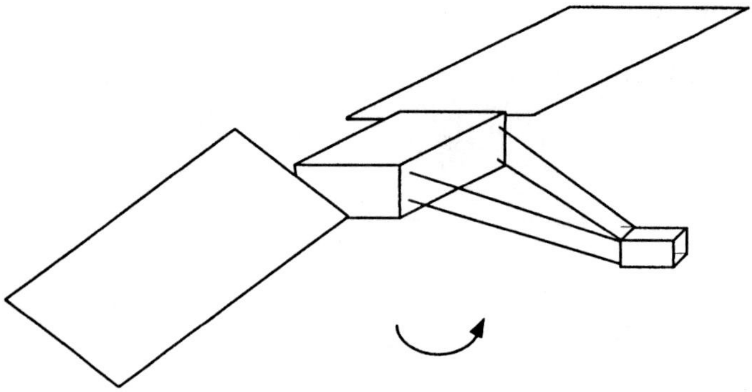

Figure 2: Rigid Body with Flexible Appendage

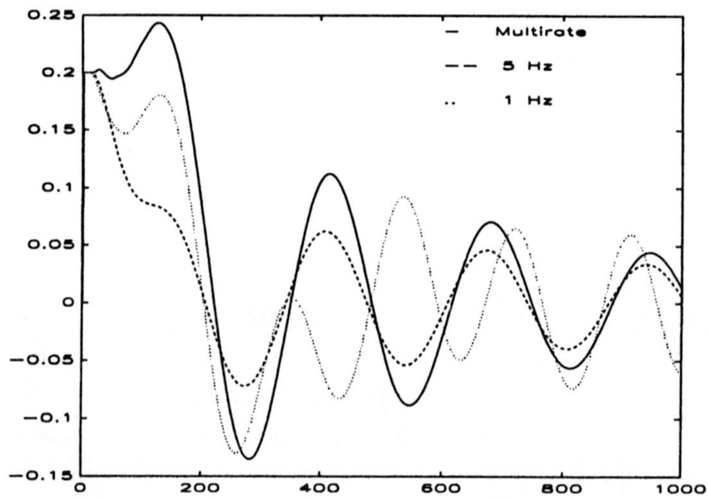

Figure 3: Rigid Body Position vs. Sample Number

Note that the dynamic model involves one rigid body mode along with one flexible mode at a frequency of 1 rad/s with 0.5% damping. The matrix C captures the fact that the rigid body angular position and tip velocity of the flexible appendage are measured. Also, note that the rigid body position measurement is corrupted by the flexible mode (i.e., observation spillover). To reflect a plausible mission we assume that the rigid body angular position is measured by an attitude sensor sampling at 1 Hz while the tip appendage velocity is measured by a rate gyro sensor sampling at 5 Hz. The matrix R_1 expresses the desire to regulate the rigid body and tip appendage positions, and the matrix V_1 was chosen to capture the type of noise correlation that arises when the dynamics are transformed into a modal basis [36].

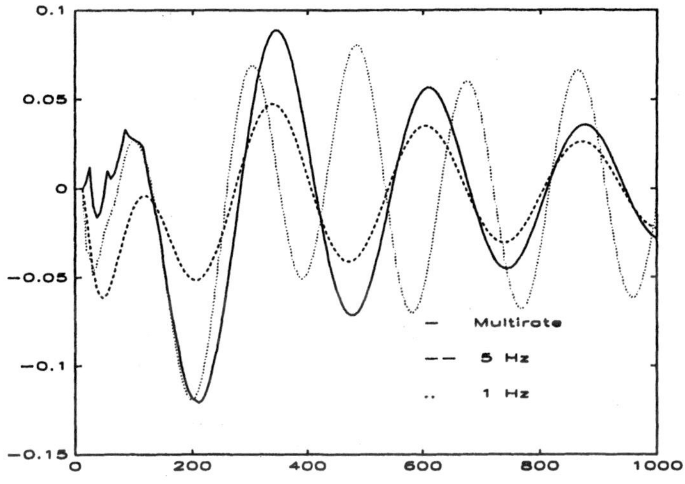

Figure 4: Rigid Body Velocity vs. Sample Number

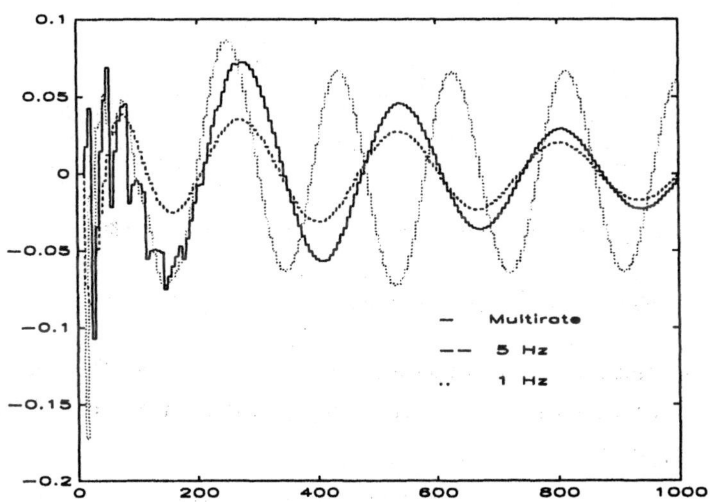

Figure 5: Control Torque vs. Sample Number

For $n_c = 4$ discrete-time single rate and multirate controllers were obtained from (59)–(63) using Theorem 1 for continuous-time to discrete-time conversions. Different measurement schemes were considered and the resulting designs are compared in Figures 3–5. The results are summarized as follows. Figures 3 and 4 show controlled rigid body position and velocity responses, respectively. Finally, Figure 5 shows control torque versus sample number. Figures 3 and 4 demonstrate the fact that the multirate design involving one sensor operating at 5 Hz and the other sensor operating at 1 Hz has responses very close to the single rate design involving two fast sensors. Finally, the three designs were compared using the performance criterion (58). The results are summarized in Table I.

Table I: Summary of Design: Example 8.1

Measurement Scheme	Optimal Cost
Two 1 Hz sensors	65.6922
Two 5 Hz sensors	53.9930
Multirate scheme (1 Hz and 5 Hz sensors)	54.8061

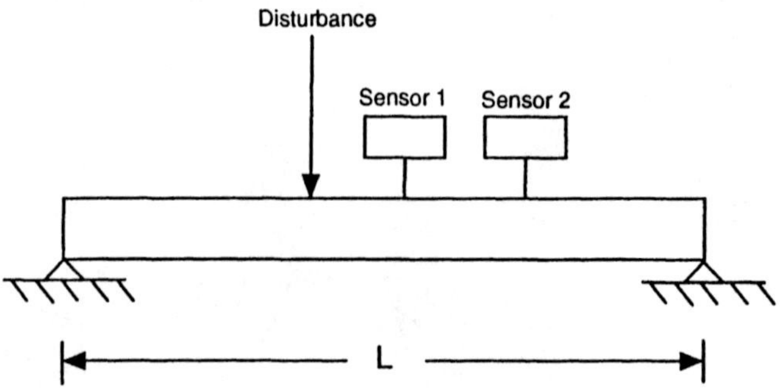

Figure 6: Simply Supported Euler-Bernoulli Beam

As a second example consider a simply supported Euler-Bernoulli beam (Figure 6) whose partial differential equation for the transverse deflection $w(x,t)$ is given by

$$m(x)\frac{\partial^2 w(x,t)}{\partial t^2} = -\frac{\partial^2}{\partial x^2}\left[EI(x)\frac{\partial^2 w(x,t)}{\partial x^2}\right] + f(x,t), \tag{88}$$

$$w(x,t)|_{x=0,L} = 0, \quad EI(x)\frac{\partial^2 w(x,t)}{\partial x^2}\bigg|_{x=0,L} = 0,$$

where $m(x)$ is the mass per unit length of the beam, $EI(x)$ is the flexural rigidity with E denoting Young's modulus of elasticity and $I(x)$ denoting the moment of inertia about an axis normal to the plane of vibration and passing through the center of the cross-sectional area. Finally, $f(x,t)$ is a distributed disturbance acting on the beam. Assuming uniform beam properties, the modal decomposition of this system has the form

$$\begin{aligned} w(x,t) &= \sum_{r=1}^{\infty} W_r(x)q_r(t), \\ \int_0^L mW_r^2(x)dx &= 1, \\ W_r(x) &= \sqrt{\frac{2}{ml}}\sin\frac{r\pi x}{L}, \end{aligned}$$

where, assuming uniform proportional damping, the modal coordinates q_r satisfy

$$\ddot{q}_r(t) + 2\zeta\omega_r\dot{q}_r(t) + \omega_r^2 q_r(t) = \int_0^L f(x,t)W_r(x)dx, \quad r = 1,2,\ldots. \tag{89}$$

For simplicity assume $L = \pi$ and $m = EI = 2/\pi$ so that $\sqrt{\frac{2}{mL}} = 1$. We assume two sensors located at $x = 0.55\pi$ and $x = 0.65\pi$ sampling at 60 Hz and 30 Hz respectively. Furthermore, we assume that a point force is applied by an actuator located at $x = 0.45\pi$, while a white noise disturbance of unit intensity acts on the beam at $x = 0.45\pi$. Finally, modeling the first five modes and defining the plant states as $x = [q_1, \dot{q}_1, \ldots, q_5, \dot{q}_5]^T$, the resulting state-space model and problem data is

$$\begin{aligned} A &= \underset{i=1,\ldots,5}{\text{block-diag}}\begin{bmatrix} 0 & 1 \\ -\omega_i^2 & -2\zeta\omega_i \end{bmatrix}, \quad \omega_i = i^2, \quad i = 1,\ldots,5, \quad \zeta = 0.005, \\ B &= \begin{bmatrix} 0 & 0.9877 & 0 & 0.3090 & 0 & -0.8900 & 0 & -0.5878 & 0 & 0.7071 \end{bmatrix}^T, \\ C &= \begin{bmatrix} 0.9877 & 0 & -0.3090 & 0 & -0.8910 & 0 & 0.5878 & 0 & 0.7071 & 0 \\ 0.8910 & 0 & -0.8090 & 0 & -0.1564 & 0 & 0.9511 & 0 & -0.7071 & 0 \end{bmatrix}, \\ E &= \begin{bmatrix} 0.9877 & 0 & -0.3090 & 0 & -0.8910 & 0 & 0.5878 & 0 & 0.7071 & 0 \end{bmatrix}, \\ R_1 &= E^T E, \quad R_2 = 0.1, \quad V_1 = BB^T, \quad V_2 = \begin{bmatrix} 0.01 & 0 \\ 0 & 0.01 \end{bmatrix}. \end{aligned}$$

For $n_c = 10$ discrete-time single rate and multirate controllers were obtained from (59)–(63) using Theorem 1 for continuous-time to discrete-time conversions. Different measurement schemes were considered and the resulting controllers were compared using the performance criterion (58). The results are summarized in Table II.

Table II: Summary of Design: Example 8.2

Measurement Scheme	Optimal Cost
One 30 Hz sensor @ $x = 0.65\pi$	0.4549
Two 30 Hz sensors @ $x = 0.55\pi$ and $x = 0.65\pi$	0.3753
One 60 Hz sensor @ $x = 0.55\pi$	0.3555
Two 60 Hz sensors @ $x = 0.55\pi$ and $x = 0.65\pi$	0.3404
Multirate scheme (30 Hz and 60 Hz sensors)	0.3446

It is interesting to note that the multirate architecture gives the least cost for the cases considered with the exception to the two 60 Hz sensor scheme which is to be expected. In this case, the improvement in the cost of two 60 Hz sensor scheme over the multirate scheme is minimal. However, the multirate scheme provides sensor complexity reduction over the twc 60 Hz sensor scheme.

VIII. CONCLUSION

This chapter developed a periodic fixed-structure control framework (temporal) for multirate systems. An equivalent discrete-time representation was obtained for the given continuous-time system. Optimality conditions were derived for the problems of optimal multirate sampled-data static output-feedback as well as multirate fixed-order sampled-data dynamic compensation. Furthermore, a novel homotopy continuation algorithm was developed to obtain numerical solutions to the full-order design equations. Future work will use these results to develop numerical algorithms for reduced-order, multirate dynamic compensator design as well as extensions to decentralized (spatial) multirate controller architectures.

APPENDIX A. PROOF OF LEMMA 1

It follows from (17) that

$$\text{vec } Q(k+1) = (\hat{A}(k) \otimes \hat{A}(k))\text{vec } Q(k) + \text{vec } \hat{V}(k), \tag{90}$$

where $\otimes$ denotes Kronecker product and "vec" is the column stacking operator [37]. Next, define the notation $q(k) \triangleq \text{vec } Q(k)$, $\mathcal{A}(k) \triangleq \hat{A}(k) \otimes \hat{A}(k)$, and $v(k) \triangleq \text{vec } \hat{V}(k)$, so that

$$q(k+1) = \mathcal{A}(k)q(k) + v(k). \tag{91}$$

It now follows with $k = \alpha + \beta N$, that

$$q(k+\beta N) = \Phi(\alpha+\beta N, 1)q(1) + \sum_{i=1}^{\alpha+\beta N-1} \Phi(\alpha+\beta N, i+1)v(i), \tag{92}$$

where

$$\begin{aligned}
\Phi(\alpha+\beta N, i+1) &\triangleq \mathcal{A}(\alpha+\beta N-1)\mathcal{A}(\alpha+\beta N-2)\cdots\mathcal{A}(i+1), \\
&\qquad \alpha+\beta N > i+1, \\
\Phi(\alpha+\beta N, i+1) &= I_{n^2}, \quad \alpha+\beta N = i+1.
\end{aligned} \tag{93}$$

Next, note that

$$\begin{aligned}
&\sum_{i=1}^{\alpha+\beta N-1} \Phi(\alpha+\beta N, i+1)v(i) = \\
&\quad \sum_{i=1}^{N+(\alpha-1)} \Phi(\alpha+\beta N, i+1)v(i) + \sum_{i=N+\alpha}^{2N+(\alpha-1)} \Phi(\alpha+\beta N, i+1)v(i) \\
&\quad + \sum_{i=2N+\alpha}^{3N+(\alpha-1)} \Phi(\alpha+\beta N, i+1)v(i) \\
&\quad + \cdots + \sum_{i=(\beta-1)N+\alpha}^{\beta N+(\alpha-1)} \Phi(\alpha+\beta N, i+1)v(i).
\end{aligned} \tag{94}$$

Using the identities

$$\begin{aligned}
\Phi(\alpha+\beta N, 1) &= \Phi^{\beta}(\alpha+N, 1), \\
\Phi(\alpha+\beta N, \alpha+\gamma N) &= \Phi^{\beta-\gamma}(\alpha+N, \alpha),
\end{aligned}$$

it now follows that (94) is equivalent to

$$\sum_{i=1}^{\alpha+\beta N-1} \Phi(\alpha+\beta N, i+1)v(i) =$$

$$\Phi^{\beta-1}(\alpha+N,\alpha) \sum_{i=1}^{N+(\alpha-1)} \Phi(\alpha+N, i+1)v(i)$$

$$+\Phi^{\beta-2}(\alpha+N,\alpha) \sum_{i=\alpha}^{N+(\alpha-1)} \Phi(\alpha+N, i+1)v(i)$$

$$+\Phi^{\beta-3}(\alpha+N,\alpha) \sum_{i=\alpha}^{N+(\alpha-1)} \Phi(\alpha+N, i+1)v(i) + \cdots$$

$$+ \sum_{i=\alpha}^{N+(\alpha-1)} \Phi(\alpha+N, i+1)v(i), \tag{95}$$

which implies that

$$\sum_{i=1}^{\alpha+\beta N-1} \Phi(\alpha+\beta N, i+1)v(i) =$$

$$\Phi_\alpha^{\beta-1} \sum_{i=1}^{\alpha-1} \Phi(\alpha+N, i+1)v(i)$$

$$+[I+\Phi_\alpha+\Phi_\alpha^2+\cdots+\Phi_\alpha^{\beta-1}] \sum_{i=\alpha}^{N+(\alpha-1)} \Phi(\alpha+N, i+1)v(i), \tag{96}$$

where

$$\Phi_\alpha \triangleq \Phi(\alpha+N,\alpha).$$

Since $\rho(\Phi_\alpha) < 1$ it follows from (92) and (96) that

$$\lim_{k\to\infty} q(\alpha+\beta N) = (I-\Phi_\alpha)^{-1} \sum_{i=\alpha}^{N+(\alpha-1)} \Phi(\alpha+N, i+1)v(i), \tag{97}$$

which shows that the second moment converges to a steady-state periodic trajectory.

To prove (19), rewrite (14) as

$$\mathcal{J}_c(D_c(\cdot)) = \delta_\infty + \lim_{K\to\infty} \frac{1}{K} \operatorname{tr} \sum_{k=1}^{K} [Q(k)\hat{R}(k) + D_c^T(k)R_2(k)D_c(k)V_2(k)]. \tag{98}$$

Due to the periodicity of the closed-loop second-moment matrix $Q(\cdot)$ and the noise covariance matrix $V_2(\cdot)$, we obtain

$$\mathcal{J}_c(D_c(\cdot)) = \delta + \frac{1}{N}\mathrm{tr}\sum_{\alpha=1}^{N}[Q(\alpha)\hat{R}(\alpha) + D_c^T(\alpha)R_2(\alpha)D_c(\alpha)V_2(\alpha)]. \quad (99)$$

□

APPENDIX B. PROOF OF THEOREM 3

To optimize (41) subject to constraint (39) over the open set $\mathcal{S}_c$ form the Lagrangian

$$\begin{aligned}
&\mathcal{L}(A_c(\alpha), B_c(\alpha), C_c(\alpha), D_c(\alpha), \tilde{Q}(\alpha), \tilde{P}(\alpha+1), \lambda) \triangleq \\
&\quad \mathrm{tr}\sum_{\alpha=1}^{N}\{\lambda\frac{1}{N}[\tilde{Q}(\alpha)\tilde{R}(\alpha) + D_c^T(\alpha)R_2(\alpha)D_c(\alpha)V_2(\alpha)] \\
&\quad +[(\tilde{A}(\alpha)\tilde{Q}(\alpha)\tilde{A}^T(\alpha) + \tilde{V}(\alpha) - \tilde{Q}(\alpha+1))\tilde{P}(\alpha+1)]\}, \quad (100)
\end{aligned}$$

where the Lagrange multipliers $\lambda \geq 0$ and $\tilde{P}(\alpha+1) \in \mathcal{R}^{(n+n_c)\times(n+n_c)}$, $\alpha = 1, \ldots, N$, are not all zero. We thus obtain

$$\frac{\partial \mathcal{L}}{\partial \tilde{Q}(\alpha)} = \tilde{A}^T(\alpha)\tilde{P}(\alpha+1)\tilde{A}(\alpha) + \lambda\frac{1}{N}\tilde{R}(\alpha) - \tilde{P}(\alpha), \quad \alpha = 1, \ldots, N. \quad (101)$$

Setting $\frac{\partial \mathcal{L}}{\partial \tilde{Q}(\alpha)} = 0$ yields

$$\tilde{P}(\alpha) = \tilde{A}^T(\alpha)\tilde{P}(\alpha+1)\tilde{A}(\alpha) + \lambda\frac{1}{N}\tilde{R}(\alpha), \quad \alpha = 1, \ldots, N. \quad (102)$$

Next, propagating (102) from α to $\alpha + N$ yields

$$\begin{aligned}
\tilde{P}(\alpha) = {} & \tilde{A}^T(\alpha)\cdots\tilde{A}^T(\alpha+N-1)\tilde{P}(\alpha)\tilde{A}(\alpha+N-1)\cdots\tilde{A}(\alpha) \\
& +\lambda\frac{1}{N}[\tilde{A}^T(\alpha)\cdots\tilde{A}^T(\alpha+N-2)\tilde{R}(\alpha+N-1) \\
& \cdot\tilde{A}(\alpha+N-2)\cdots\tilde{A}(\alpha) + \tilde{A}^T(\alpha)\cdots\tilde{A}^T(\alpha+N-3) \\
& \cdot\tilde{R}(\alpha+N-2)\tilde{A}(\alpha+N-3)\cdots\tilde{A}(\alpha) + \cdots + \tilde{R}(\alpha)]. \quad (103)
\end{aligned}$$

Note that since $\tilde{A}(\alpha+N-1)\cdots\tilde{A}(\alpha)$ is assumed to be stable, $\lambda = 0$ implies $\tilde{P}(\alpha) = 0, \alpha = 1, \ldots, N$. Hence, it can be assumed without loss of generality that $\lambda = 1$. Furthermore, $\tilde{P}(\alpha), \alpha = 1, \ldots, N$, is nonnegative-definite.

Now partition the $(n + n_c) \times (n + n_c)$ matrices $\tilde{Q}(\alpha)$ and $\tilde{P}(\alpha)$ into $n \times n, n \times n_c$, and $n_c \times n_c$ subblocks as

$$\tilde{Q}(\alpha) = \begin{bmatrix} Q_1(\alpha) & Q_{12}(\alpha) \\ Q_{12}^T(\alpha) & Q_2(\alpha) \end{bmatrix}, \quad \tilde{P}(\alpha) = \begin{bmatrix} P_1(\alpha) & P_{12}(\alpha) \\ P_{12}^T(\alpha) & P_2(\alpha) \end{bmatrix}.$$

Thus, with $\lambda = 1$, the stationary conditions are given by

$$\frac{\partial \mathcal{L}}{\partial \tilde{P}(\alpha+1)} = \tilde{A}(\alpha)\tilde{Q}(\alpha)\tilde{A}^T(\alpha) + \tilde{V}(\alpha) - \tilde{Q}(\alpha+1) = 0, \tag{104}$$

$$\begin{aligned} \frac{\partial \mathcal{L}}{\partial A_c(\alpha)} = {} & P_{12}^T(\alpha+1)A(\alpha)Q_{12}(\alpha) \\ & + P_{12}^T(\alpha+1)B(\alpha)D_c(\alpha)C(\alpha)Q_{12}(\alpha) \\ & + P_{12}^T(\alpha+1)B(\alpha)C_c(\alpha)Q_2(\alpha) \\ & + P_2(\alpha+1)B_c(\alpha)C(\alpha)Q_{12}(\alpha) \\ & + P_2(\alpha+1)A_c(\alpha)Q_2(\alpha) = 0, \end{aligned} \tag{105}$$

$$\begin{aligned} \frac{\partial \mathcal{L}}{\partial B_c(\alpha)} = {} & P_{12}^T(\alpha+1)A(\alpha)Q_1(\alpha)C^T(\alpha) \\ & + P_{12}^T(\alpha+1)B(\alpha)D_c(\alpha)C(\alpha)Q_1(\alpha)C^T(\alpha) \\ & + P_{12}^T(\alpha+1)B(\alpha)C_c(\alpha)Q_{12}^T(\alpha)C^T(\alpha) \\ & + P_2(\alpha+1)B_c(\alpha)C(\alpha)Q_1(\alpha)C^T(\alpha) \\ & + P_2(\alpha+1)A_c(\alpha)Q_{12}^T(\alpha)C^T(\alpha) \\ & + P_{12}^T(\alpha+1)B(\alpha)D_c(\alpha)V_2(\alpha) \\ & + P_{12}^T(\alpha+1)V_{12}(\alpha) \\ & + P_2(\alpha+1)B_c(\alpha)V_2(\alpha) = 0, \end{aligned} \tag{106}$$

$$\begin{aligned} \frac{\partial \mathcal{L}}{\partial C_c(\alpha)} = {} & \frac{1}{N}[R_2(\alpha)D_c(\alpha)C(\alpha)Q_{12}(\alpha) + R_{12}^T(\alpha)Q_{12}(\alpha) \\ & + R_2(\alpha)C_c(\alpha)Q_2(\alpha)] \\ & + B^T(\alpha)P_1(\alpha+1)A(\alpha)Q_{12}(\alpha) \\ & + B^T(\alpha)P_1(\alpha+1)B(\alpha)D_c(\alpha)C(\alpha)Q_{12}(\alpha) \\ & + B^T(\alpha)P_{12}(\alpha+1)B_c(\alpha)C(\alpha)Q_{12}(\alpha) \\ & + B^T(\alpha)P_{12}(\alpha+1)A_c(\alpha)Q_2(\alpha) \\ & + B^T(\alpha)P_1(\alpha+1)B(\alpha)C_c(\alpha)Q_2(\alpha) = 0, \end{aligned} \tag{107}$$

$$
\begin{aligned}
\frac{\partial \mathcal{L}}{\partial D_c(\alpha)} = & \frac{1}{N}[R_{12}^T(\alpha)Q_1(\alpha)C^T(\alpha) \\
& +R_2(\alpha)C_c(\alpha)Q_{12}^T(\alpha)C^T(\alpha) \\
& +R_2(\alpha)D_c(\alpha)C(\alpha)Q_1(\alpha)C^T(\alpha) \\
& +R_2(\alpha)D_c(\alpha)V_2(\alpha)] \\
& +B^T(\alpha)P_1(\alpha+1)A(\alpha)Q_1(\alpha)C^T(\alpha) \\
& +B^T(\alpha)P_{12}(\alpha+1)B_c(\alpha)C(\alpha)Q_1(\alpha)C^T(\alpha) \\
& +B^T(\alpha)P_1(\alpha+1)B(\alpha)C_c(\alpha)Q_{12}^T(\alpha)C^T(\alpha) \\
& +B^T(\alpha)P_{12}(\alpha+1)A_c(\alpha)Q_{12}^T(\alpha)C^T(\alpha) \\
& +B^T(\alpha)P_1(\alpha+1)V_{12}(\alpha) \\
& +B^T(\alpha)P_{12}(\alpha+1)B_c(\alpha)V_2(\alpha) \\
& +B^T(\alpha)P_1(\alpha+1)B(\alpha)D_c(\alpha)C(\alpha)Q_1(\alpha)C^T(\alpha) \\
& +B^T(\alpha)P_1(\alpha+1)B(\alpha)D_c(\alpha)V_2(\alpha) = 0, \qquad (108)
\end{aligned}
$$

for $\alpha = 1, \ldots, N$. Expanding (39) and (102) yields

$$
\begin{aligned}
0 = & A(\alpha)Q_1(\alpha)A^T(\alpha) + A(\alpha)Q_1(\alpha)C^T(\alpha)D_c^T(\alpha)B^T(\alpha) \\
& +B(\alpha)D_c(\alpha)C(\alpha)Q_1(\alpha)A^T(\alpha) \\
& +B(\alpha)D_c(\alpha)C(\alpha)Q_1(\alpha)C^T(\alpha)D_c^T(\alpha)B^T(\alpha) \\
& +B(\alpha)C_c(\alpha)Q_{12}^T(\alpha)A^T(\alpha) \\
& +B(\alpha)C_c(\alpha)Q_{12}^T(\alpha)C^T(\alpha)D_c^T(\alpha)B^T(\alpha) \\
& +A(\alpha)Q_{12}(\alpha){C_c}^T(\alpha)B^T(\alpha) \\
& +B(\alpha)D_C(\alpha)C(\alpha)Q_{12}(\alpha)C_c^T(\alpha)B^T(\alpha) \\
& +B(\alpha)C_c(\alpha)Q_2(\alpha)C_c^T(\alpha)B^T(\alpha) \\
& +B(\alpha)D_c(\alpha)V_2(\alpha)D_c^T(\alpha)B^T(\alpha) + B(\alpha)D_c(\alpha)V_{12}^T(\alpha) \\
& +V_{12}^T(\alpha)D_c^T(\alpha)B^T(\alpha) + V_1(\alpha) - Q_1(\alpha+1), \qquad (109) \\
0 = & A(\alpha)Q_1(\alpha)C^T(\alpha)B_c^T(\alpha) \\
& +B(\alpha)D_c(\alpha)C(\alpha)Q_1(\alpha)C^T(\alpha)B_c^T(\alpha) \\
& +B(\alpha)C_c(\alpha)Q_{12}^T(\alpha)C^T(\alpha)B_c^T(\alpha) + A(\alpha)Q_{12}(\alpha)A_c^T(\alpha) \\
& +B(\alpha)D_c(\alpha)C(\alpha)Q_{12}(\alpha)A_c^T(\alpha) + B(\alpha)C_c(\alpha)Q_2(\alpha)A_c^T(\alpha) \\
& +B(\alpha)D_c(\alpha)V_2(\alpha)B_c^T(\alpha) + V_{12}(\alpha)B_c^T(\alpha) - Q_{12}(\alpha+1), \qquad (110)
\end{aligned}
$$

$$\begin{aligned}
0 \;=\;& B_c(\alpha)C(\alpha)Q_1(\alpha)C^T(\alpha)B_c^T(\alpha) + A_c(\alpha)Q_{12}^T(\alpha)C^T(\alpha)B_c^T(\alpha) \\
&+B_c(\alpha)C(\alpha)Q_{12}(\alpha)A_c^T(\alpha) + A_c(\alpha)Q_2(\alpha)A_c^T(\alpha) \\
&+B_c(\alpha)V_2(\alpha)B_c^T(\alpha) - Q_2(\alpha+1), \qquad (111)\\
0 \;=\;& A^T(\alpha)P_1(\alpha+1)A(\alpha) + A^T(\alpha)P_1(\alpha+1)B(\alpha)D_c(\alpha)C(\alpha) \\
&+C^T(\alpha)D_c^T(\alpha)B^T(\alpha)P_1(\alpha+1)A(\alpha) \\
&+C^T(\alpha)D_c^T(\alpha)B^T(\alpha)P_1(\alpha+1)B(\alpha)D_c(\alpha)C(\alpha) \\
&+C^T(\alpha)B_c^T(\alpha)P_{12}^T(\alpha+1)A(\alpha) \\
&+C^T(\alpha)B_c^T(\alpha)P_{12}^T(\alpha+1)B(\alpha)D_c(\alpha)C(\alpha) \\
&+A^T(\alpha)P_{12}(\alpha+1)B_c(\alpha)C(\alpha) \\
&+C^T(\alpha)D_c^T(\alpha)B^T(\alpha)P_{12}(\alpha+1)B_c(\alpha)C(\alpha) \\
&+C^T(\alpha)B_c^T(\alpha)P_2(\alpha+1)B_c(\alpha)C(\alpha) \\
&+\frac{1}{N}[R_1(\alpha) + R_{12}(\alpha)D_c(\alpha)C(\alpha) \\
&+C^T(\alpha)D_C^T(\alpha)R_2(\alpha)D_c(\alpha)C(\alpha) \\
&+C^T(\alpha)D_c^T(\alpha)R_{12}^T(\alpha)] - P_1(\alpha), \qquad (112)\\
0 \;=\;& A^T(\alpha)P_1(\alpha+1)B(\alpha)C_c(\alpha) \\
&+C^T(\alpha)D_c^T(\alpha)B^T(\alpha)P_1(\alpha+1)B(\alpha)C_c(\alpha) \\
&+C^T(\alpha)B_c^T(\alpha)P_{12}^T(\alpha+1)B(\alpha)C_c(\alpha) \\
&+A^T(\alpha)P_{12}(\alpha+1)A_c(\alpha) \\
&+C^T(\alpha)D_c^T(\alpha)B^T(\alpha)P_{12}(\alpha+1)A_c(\alpha) \\
&+C^T(\alpha)B_C^T(\alpha)P_2(\alpha+1)A_c(\alpha) \\
&+\frac{1}{N}[R_{12}(\alpha)C_c(\alpha) + C^T(\alpha)D_c^T(\alpha)R_2(\alpha)C_c(\alpha)] - P_{12}(\alpha), \qquad (113)\\
0 \;=\;& C_c^T(\alpha)B^T(\alpha)P_1(\alpha+1)B(\alpha)C_c(\alpha) + A_c^T(\alpha)P_{12}^T(\alpha+1)B(\alpha)C_c(\alpha) \\
&+C_c^T(\alpha)B^T(\alpha)P_{12}(\alpha+1)A_c(\alpha) + A_c^T(\alpha)P_2(\alpha+1)A_c(\alpha) \\
&+\frac{1}{N}C_c^T(\alpha)R_2(\alpha)C_c(\alpha) - P_2(\alpha). \qquad (114)
\end{aligned}$$

Lemma 4. $Q_2(\alpha)$ and $P_2(\alpha)$ are positive-definite for $\alpha = 1, \ldots, N$.

Proof. By a minor extension of the results from [38], (111) can be written as

$$Q_2(\alpha+1) \;=\; \left(A_c(\alpha) + B_c(\alpha)C(\alpha)Q_{12}(\alpha)Q_2{}^+(\alpha)\right) Q_2(\alpha)$$

$$\cdot \big(A_c(\alpha) + B_c(\alpha)C(\alpha)Q_{12}(\alpha)Q_2{}^+(\alpha)\big)^T$$
$$+B_c(\alpha)V_{2a}(\alpha)B_c^T(\alpha), \quad \alpha = 1, \ldots, N, \tag{115}$$

where $Q_2{}^+(\alpha)$ is the Moore-Penrose or Drazin generalized inverse of $Q_2(\alpha)$. Next, propagating (115) from α to $\alpha + N$ yields

$$\begin{aligned} Q_2(\alpha) &= A_{cs}(\alpha+N-1)\cdots A_{cs}(\alpha)Q_2(\alpha)A_{cs}^T(\alpha)\cdots A_{cs}^T(\alpha+N-1) \\ &\quad +A_{cs}(\alpha+N-1)\cdots A_{cs}(\alpha+1)B_c(\alpha)V_{2a}(\alpha) \\ &\quad \cdot B_c^T(\alpha)A_{cs}^T(\alpha+1)\cdots A_{cs}^T(\alpha+N-1) \\ &\quad +A_{cs}(\alpha+N-1)\cdots A_{cs}(\alpha+2)B_c(\alpha+1) \\ &\quad \cdot V_{2a}(\alpha+1)B_c^T(\alpha+1)A_{cs}^T(\alpha+2)\cdot\cdots A_{cs}^T(\alpha+N-1)+ \\ &\quad \cdots + B_c(\alpha+N-1)V_{2a}(\alpha+N-1)B_c^T(\alpha+N-1), \end{aligned} \tag{116}$$

where $A_{cs}(\cdot) \triangleq A_c(\cdot) + B_c(\cdot)C(\cdot)Q_{12}(\cdot)Q_2{}^+(\cdot)$. Next, note that the controllability of $(\Phi_{csp}(\alpha), B_{cs}(\alpha))$ implies that $(\Phi_{csp}(\alpha), B_{cs}(\alpha)V_{2s}^{1/2}(\alpha))$ is also controllable, where

$$\begin{aligned} \Phi_{csp}(\alpha) &\triangleq A_{cs}(\alpha+N-1)A_{cs}(\alpha+N-2)A_{cs}(\alpha+N-3)\cdots A_{cs}(\alpha), \\ B_{cs}(\alpha) &\triangleq [A_{cs}(\alpha+N-1)\cdots A_{cs}(\alpha+1)B_c(\alpha), \\ &\qquad A_{cs}(\alpha+N-1)\cdots A_{cs}(\alpha+2)B_c(\alpha+1), \ldots, B_c(\alpha+N-1)], \\ V_{2s}(\alpha) &\triangleq \text{block-diagonal}\,[V_{2a}(\alpha), V_{2a}(\alpha+1), \ldots, V_{2a}(\alpha+N-1)]. \end{aligned}$$

Next, using the above notation, (116) becomes

$$Q_2(\alpha) = \Phi_{csp}(\alpha)Q_2(\alpha)\Phi_{csp}^T(\alpha) + B_{cs}(\alpha)V_{2s}(\alpha)B_{cs}^T(\alpha). \tag{117}$$

Now, since $\Phi_{csp}(\cdot)$ is stable and $B_{cs}(\cdot)V_{2s}(\cdot)B_{cs}^T(\cdot)$ is nonnegative-definite, Lemma 12.2, p.282, of [39], implies that $Q_2(\cdot)$ is positive-definite. Using similar arguments we can show that $P_2(\cdot)$ is positive-definite. □

Since $Q_2(\alpha), P_2(\alpha), V_2(\alpha), R_2(\alpha), V_{2a}(\alpha)$, and $R_{2a}(\alpha)$, for $\alpha = 1, \ldots, N$, are invertible (105)–(108) can be written as

$$\begin{aligned} A_c(\alpha) &= -P_2^{-1}(\alpha+1)P_{12}^T(\alpha+1)A(\alpha)Q_{12}(\alpha)Q_2^{-1}(\alpha) \\ &\quad -B_c(\alpha)C(\alpha)Q_{12}(\alpha)Q_2^{-1}(\alpha) \\ &\quad -P_2^{-1}(\alpha+1)P_{12}^T(\alpha+1)B(\alpha)C_c(\alpha) \end{aligned}$$

$$
\begin{aligned}
& -P_2^{-1}(\alpha+1)P_{12}^T(\alpha+1)B(\alpha)D_c(\alpha) \\
& \cdot C(\alpha)Q_{12}(\alpha)Q_2^{-1}(\alpha), \qquad (118) \\
B_c(\alpha) &= -P_2^{-1}(\alpha+1)P_{12}^T(\alpha+1)Q_a(\alpha)V_{2a}^{-1}(\alpha) \\
& -P_2^{-1}(\alpha+1)P_{12}^T(\alpha+1)B(\alpha)D_c(\alpha), \qquad (119) \\
C_c(\alpha) &= -R_{2a}^{-1}(\alpha)P_a(\alpha)Q_{12}(\alpha)Q_2^{-1}(\alpha) \\
& -D_c(\alpha)C(\alpha)Q_{12}(\alpha)Q_2^{-1}(\alpha), \qquad (120) \\
D_c(\alpha) &= -[R_{2a}^{-1}(\alpha)\{B^T(\alpha)P(\alpha+1)A(\alpha)Q(\alpha)C^T(\alpha) \\
& +\frac{1}{N}R_{12}^T(\alpha)Q(\alpha)C^T(\alpha) \\
& +B^T(\alpha)P(\alpha+1)V_{12}(\alpha)\}V_{2a}^{-1}(\alpha)], \qquad (121)
\end{aligned}
$$

for $\alpha = 1, \ldots, N$. Now, for $\alpha = 1, \ldots, N$, define $n_c \times n, n_c \times n_c$, and $n_c \times n$ matrices

$$G(\alpha) \triangleq Q_2^{-1}(\alpha)Q_{12}^T(\alpha), \quad M(\alpha) \triangleq Q_2(\alpha)P_2(\alpha), \quad \Gamma(\alpha) \triangleq -P_2^{-1}(\alpha)P_{12}^T(\alpha),$$

and the $n \times n$ matrices

$$
\begin{aligned}
Q(\alpha) &\triangleq Q_1(\alpha) - Q_{12}(\alpha)Q_2^{-1}(\alpha)Q_{12}^T(\alpha), \\
P(\alpha) &\triangleq P_1(\alpha) - P_{12}(\alpha)P_2^{-1}(\alpha)P_{12}^T(\alpha), \\
\hat{Q}(\alpha) &\triangleq Q_{12}(\alpha)Q_2^{-1}(\alpha)Q_{12}^T(\alpha), \\
\hat{P}(\alpha) &\triangleq P_{12}(\alpha)P_2^{-1}(\alpha)P_{12}^T(\alpha).
\end{aligned}
$$

Now, for $\alpha = 1, \ldots, N$, computing $\Gamma(\alpha+1)(110)-(111) = 0$ and $G(\alpha)(113)+(114) = 0$ yields

$$-P_2^{-1}(\alpha+1)P_{12}^T(\alpha+1)Q_{12}(\alpha+1)Q_2^{-1}(\alpha+1) = I_{n_c}, \qquad (122)$$

$$-P_2^{-1}(\alpha)P_{12}^T(\alpha)Q_{12}(\alpha)Q_2^{-1}(\alpha) = I_{n_c}. \qquad (123)$$

Note that $Q(\alpha), P(\alpha), \hat{P}(\alpha)$, and $\hat{Q}(\alpha)$, for $\alpha = 1, \ldots, N$, are nonnegative-definite. Next, note that with the above definitions (122) and (123) are equivalent to (47) and (46) holds. Hence, $\tau(\alpha) = G^T(\alpha)\Gamma(\alpha)$ is idempotent, i.e., $\tau^2(\alpha) = \tau(\alpha)$, for $\alpha = 1, \ldots, N$. Sylvester's inequality yields (57). Note also that

$$\hat{Q}(\alpha) = \tau(\alpha)\hat{Q}(\alpha), \quad \hat{P}(\alpha) = \hat{P}(\alpha)\tau(\alpha).$$

The components of $\tilde{Q}(\alpha)$ and $\tilde{P}(\alpha)$ can now be written in terms of $Q(\alpha)$, $P(\alpha)$, $\hat{Q}(\alpha)$, $\hat{P}(\alpha)$, $G(\alpha)$, and $\Gamma(\alpha)$ as

$$
\begin{aligned}
Q_1(\alpha) &= Q(\alpha) + \hat{Q}(\alpha), & P_1(\alpha) &= P(\alpha) + \hat{P}(\alpha),\\
Q_{12}(\alpha) &= \hat{Q}(\alpha)\Gamma^T(\alpha), & P_{12}(\alpha) &= -\hat{P}(\alpha)G^T(\alpha),\\
Q_2(\alpha) &= \Gamma(\alpha)\hat{Q}(\alpha)\Gamma^T(\alpha), & P_2(\alpha) &= G(\alpha)\hat{P}(\alpha)G^T(\alpha).
\end{aligned}
$$

The expressions (49)–(52) follow from (105)–(108) by using above identities. Substituting these expressions for $A_c(\alpha), B_c(\alpha), C_c(\alpha)$, and $D_c(\alpha)$ in (109)–(114) it follows that (111)$= \Gamma(\alpha+1)$(110) and (114)$= -G(\alpha)$(113). Hence, it follows that (111) and (114) are superfluous and can be omitted. Thus, (109)–(114) reduce to

$$
\begin{aligned}
0 &= A(\alpha)Q(\alpha)A^T(\alpha) + \left[A(\alpha) - B(\alpha)R_{2a}^{-1}(\alpha)P_a(\alpha)\right]\hat{Q}(\alpha)\\
&\quad \cdot \left[A(\alpha) - B(\alpha)R_{2a}^{-1}(\alpha)P_a(\alpha)\right]^T\\
&\quad + \left[Q_a(\alpha) + B(\alpha)D_c(\alpha)V_{2a}(\alpha)\right]V_{2a}^{-1}(\alpha)\\
&\quad \cdot \left[Q_a(\alpha) + B(\alpha)D_c(\alpha)V_{2a}(\alpha)\right]^T\\
&\quad + V_1(\alpha) - Q_a(\alpha)V_{2a}^{-1}(\alpha)Q_a^T(\alpha) - Q(\alpha+1) - \hat{Q}(\alpha+1), \qquad (124)\\
0 &= [\{A(\alpha) - B(\alpha)R_{2a}^{-1}(\alpha)P_a(\alpha)\}\hat{Q}(\alpha)\\
&\quad \cdot\{A(\alpha) - B(\alpha)R_{2a}^{-1}(\alpha)P_a(\alpha)\}^T\\
&\quad +\{Q_a(\alpha) + B(\alpha)D_c(\alpha)V_{2a}(\alpha)\}V_{2a}^{-1}(\alpha)\\
&\quad \cdot\{Q_a(\alpha) + B(\alpha)D_c(\alpha)V_{2a}(\alpha)\}^T - \hat{Q}(\alpha+1)]\Gamma^T(\alpha+1), \qquad (125)\\
0 &= A^T(\alpha)P(\alpha+1)A(\alpha) + \left[A(\alpha) - Q_a(\alpha)V_{2a}^{-1}(\alpha)C(\alpha)\right]^T\\
&\quad \cdot\hat{P}(\alpha+1)\left[A(\alpha) - Q_a(\alpha)V_{2a}^{-1}(\alpha)C(\alpha)\right]\\
&\quad + \left[P_a(\alpha) + R_{2a}(\alpha)D_c(\alpha)C(\alpha)\right]^T\\
&\quad \cdot R_{2a}^{-1}(\alpha)\left[P_a(\alpha) + R_{2a}(\alpha)D_c(\alpha)C(\alpha)\right]\\
&\quad + \frac{1}{N}R_1(\alpha) - P_a^T(\alpha)R_{2a}^{-1}(\alpha)P_a(\alpha) - P(\alpha) - \hat{P}(\alpha), \qquad (126)\\
0 &= [\{A(\alpha) - Q_a(\alpha)V_{2a}^{-1}(\alpha)C(\alpha)\}^T\hat{P}(\alpha+1)\\
&\quad \cdot\{A(\alpha) - Q_a(\alpha)V_{2a}^{-1}(\alpha)C(\alpha)\}\\
&\quad +\{P_a(\alpha) + R_{2a}(\alpha)D_c(\alpha)C(\alpha)\}^T R_{2a}^{-1}(\alpha)\\
&\quad \cdot\{P_a(\alpha) + R_{2a}(\alpha)D_c(\alpha)C(\alpha)\} - \hat{P}(\alpha)]G^T(\alpha). \qquad (127)
\end{aligned}
$$

Next, computing

$$
\begin{aligned}
(124) + G^T(\alpha+1)\Gamma(\alpha+1)(125)G(\alpha+1) \\
-(125)G(\alpha+1) - [(125)G(\alpha+1)]^T &= 0,
\end{aligned}
$$

and

$$
(126) + \Gamma^T(\alpha)G(\alpha)(127)\Gamma(\alpha) - (127)\Gamma(\alpha) - [(127)\Gamma(\alpha)]^T = 0,
$$

yields (53) and (54) respectively. Finally (55) and (56) are obtained by computing $G^T(\alpha+1)\Gamma(\alpha+1)(125)G(\alpha+1) = 0$ and $\Gamma^T(\alpha)G(\alpha)(127)\Gamma(\alpha) = 0$, respectively. □

ACKNOWLEDGEMENT

The authors wish to thank Prof. D.S. Bernstein for several helpful discussions, and Drs. E.G. Collins, Jr. and L.D. Davis for several helpful suggestions concerning the numerical algorithm and illustrative examples of Sections VI and VII. This research was supported in part by the National Science Foundation under Grants ECS-9109558 and ECS-9496249.

REFERENCES

1. M. Al-Rahamani and G.F. Franklin, "A New Optimal Multirate Control of Linear Periodic and Time-Invariant Systems," *IEEE Trans. Autom. Contr.*, **35**, pp. 406–415, (1990).
2. R. Aracil, A.J. Avella, and V. Felio, "Multirate Sampling Technique in Digital Control Systems Simulation," *IEEE Trans. Sys. Man Cyb.*, **SMC-14**, pp. 776–780, (1984).
3. M. Araki and K. Yamamoto, "Multivariable Multirate Sampled-Data Systems: State-Space Description, Transfer Characteristics and Nyquist Criterion," *IEEE Trans. Autom. Contr.*, **AC-31**, pp. 145–154, (1986).
4. M.C. Berg, N. Amit, and D. Powell, "Multirate Digital Control System Design," *IEEE Trans. Autom. Contr.*, **33**, pp. 1139–1150, (1988).
5. M.C. Berg and G.-S. Yang, "A New Algorithm for Multirate Digital Control Law Synthesis," *Proc. IEEE Conf. Dec. Contr.*, Austin, TX, pp. 1685–1690, (1988).
6. J.R. Broussard and N. Haylo, "Optimal Multirate Output Feedback," *Proc. IEEE Conf. Dec. Contr.*, Las Vegas, NV, pp. 926–929, (1984).
7. P. Colaneri, R. Scattolini, and N. Schiavoni, "The LQG Problem for Multirate Sampled-Data Systems," *Proc. IEEE Conf. Dec. Contr.*, Tampa, FL, pp. 469–474, (1989).
8. P. Colaneri, R. Scattolini, and N. Schiavoni, "LQG Optimal Control of Multirate Sampled-Data Systems," *IEEE Trans. Autom. Contr.*, **37**, pp. 675–682, (1992).
9. D.P. Glasson, *Research in Multirate Estimation and Control, Rep. TR–1356–1*, The Analytic Science Corp., Reading, MA, (1980).
10. D.P. Glasson, "A New Technique for Multirate Digital Control," *AIAA J. Guid., Contr., Dyn.*, **5**, pp. 379–382, (1982).
11. D.P. Glasson, "Development and Applications of Multirate Digital Control," *IEEE Contr. Sys. Mag.*, **3**, pp. 2–8, (1983).

12. G.S. Mason and M.C. Berg, "Reduced-Order Multirate Compensator Synthesis," *AIAA J. Guid., Contr., Dyn.*, **15**, pp. 700–706, (1992).

13. D.G. Meyer, "A New Class of Shift-Varying Operators, Their Shift-Invariant Equivalents, and Multirate Digital Systems," *IEEE Trans. Autom. Contr.*, **35**, pp. 429–433, (1990).

14. D.G. Meyer, "A Theorem on Translating the General Multi-Rate LQG Problem to a Standard LQG Problem via Lifts," *Proc. Amer. Contr. Conf.*, Boston, MA, pp. 179–183, (1991).

15. D.G. Meyer, "Cost Translation and a Lifting Approach to the Multirate LQG Problem," *IEEE Trans. Autom. Contr.*, **37**, pp. 1411–1415, (1992).

16. D.P. Stanford, "Stability for a Multi-Rate Digital Control Design and Sample Rate Selection," *AIAA J. Guid., Contr., Dyn.*, **5**, pp. 379–382, (1982).

17. S. Bittanti, P. Colaneri, and G. DeNicolao, "The Difference Periodic Riccati Equation for the Periodic Prediction Problem," *IEEE Trans. Autom. Contr.*, **33**, pp. 706–712, (1988).

18. S. Bittanti, P. Colaneri, and G. DeNicolao, "An Algebraic Riccati Equation for the Discrete-Time Periodic Prediction Problem," *Sys. Contr. Lett.*, **14**, pp. 71–78, (1990).

19. P. Bolzern and P. Colaneri, "The Periodic Lyapunov Equation," *SIAM J. Matr. Anal. Appl.*, **9**, pp. 499–512, (1988).

20. W.M. Haddad, V. Kapila, and E.G., Collins, Jr., "Optimality Conditions for Reduced-Order Modeling, Estimation, and Control for Discrete-Time Linear Periodic Plants," *J. Math. Sys. Est. Contr.*, to appear.

21. D.S. Bernstein, L.D. Davis, and D.C. Hyland, "The Optimal Projection Equations for Reduced-Order Discrete-Time Modeling, Estimation and Control," *AIAA J. Guid., Contr., Dyn.*, **9**, pp. 288–293, (1986).

22. D.S. Bernstein, L.D. Davis, and S.W. Greeley, "The Optimal Projection Equations for Fixed-Order, Sampled-Data Dynamic Compensa-

tion with Computational Delay," *IEEE Trans. Autom. Contr.*, **AC-31**, pp. 859–862, (1986).

23. D.S. Bernstein and D.C. Hyland, "Optimal Projection Approach to Robust Fixed-Structure Control Design," in *Mechanics and Control of Large Flexible Structures*, J.L. Junkins, Ed., AIAA Inc., pp. 237–293, (1990).

24. W.M. Haddad, D.S. Bernstein, and V. Kapila, "Reduced-Order Multirate Estimation," *AIAA J. Guid., Contr., Dyn.*, **17**, pp. 712–721, (1994).

25. C.F. Van Loan, "Computing Integrals Involving the Matrix Exponential," *IEEE Trans. Autom. Contr.*, **AC-23**, pp. 395–404, (1978).

26. W.M. Haddad, D.S. Bernstein, H.-H., Huang, and Y. Halevi, "Fixed-Order Sampled-Data Estimation," *Int. J. Contr.*, **55**, pp. 129–139, (1992).

27. K.J. Åström, *Introduction to Stochastic Control Theory*, Academic Press, New York, (1970).

28. W.M. Haddad, H.-H. Huang, and D.S. Bernstein, "Sampled-Data Observers With Generalized Holds for Unstable Plants," *IEEE Trans. Autom. Contr.*, **39**, pp. 229–234, (1994).

29. S. Bittanti and P. Colaneri, "Lyapunov and Riccati Equations: Periodic Inertia Theorems," *IEEE Trans. Autom. Contr.*, **AC-31**, pp. 659–661, (1986).

30. P. Bolzern and P. Colaneri, "Inertia Theorems for the Periodic Lyapunov Difference Equation and Periodic Riccati Difference Equation," *Lin. Alg. Appl.*, **85**, pp. 249–265, (1987).

31. D.S. Bernstein and W.M. Haddad, "Robust Stability and Performance via Fixed-Order Dynamic Compensation with Guaranteed Cost Bounds," *Math. Contr. Sig. Sys.*, **3**, pp. 139–163, (1990).

32. C. Moler and C.F. Van Loan, "Nineteen Dubious Ways to Compute the Exponential of a Matrix," *SIAM Review*, **20**, pp. 801–836, (1978).

33. S. Richter, "A Homotopy Algorithm for Solving the Optimal Projection Equations for Fixed-Order Compensation: Existence, Conver-

gence and Global Optimality," *Proc. Amer. Contr. Conf.*, Minneapolis, MN, pp. 1527–1531. (1987).

34. D.S. Bernstein and W.M. Haddad, "LQG Control with an H_∞ Performance Bound: A Riccati Equation Approach," *IEEE Trans. Autom. Contr.*, **34**, pp. 293–305, (1989).

35. E.G. Collins, Jr., L.D. Davis, and S. Richter, "Design of Reduced-Order, H_2 Optimal Controllers Using a Homotopy Algorithm," *Int. J. Contr.*, **61**, 97–126, (1995).

36. W.M. Haddad and D.S. Bernstein, "Optimal Reduced-Order Observer-Estimators," *AIAA J. Guid., Contr., Dyn.*, **13**, pp. 1126–1135, (1990).

37. J.W. Brewer, "Kronecker Products and Matrix Calculus in System Theory," *IEEE Trans. Autom. Contr.*, **AC-25**, pp. 772–781, (1976).

38. A. Albert, "Conditions for Positive and Nonnegative Definiteness in Terms of Pseudo Inverses," *SIAM J. Contr. Opt.*, **17**, pp. 434–440, (1969).

39. W.H. Wonham, *Linear Multivariable Control*, Springer, (1983).

Optimal Finite Wordlength Digital Control With Skewed Sampling

Robert E. Skelton

Space Systems Control Lab, Purdue University
West Lafayette, IN 47907

Guoming G. Zhu

Cummins Engine Company, Inc., MC 50197
Columbus, IN 47202

Karolos M. Grigoriadis

Department of Mechanical Engineering, University of Houston
Houston, Texas 77204

1 Introduction

The advances in digital hardware and microprocessor technology have made it possible to build more and more complex and effective real-time digital controllers with decreasing size and cost. Digital controllers are used for implementing control laws in many kinds of engineering technologies. The term "microcontrollers" is commonly denoted for single-chip microprocessors used for digital control in areas of application ranging from automotive controls to the controls of "smart" structures. However, the reduction in size and cost of the digital control hardware provides limitations in the computational speed and the available computer memory. The finite wordlength of the digital computer and the computational time delay causes a degradation of the expected performance (compared with the near infinitely precise control law computed off-line). In this chapter we consider design of digital controllers taking into account the finite wordlength and the computational time delay of the control computer, as well as the finite wordlengths of the A/D and D/A converters. We assume that the control computer operates in fixed point arithmetic, which is most likely the choice in small size, low-cost applications. However, it has been demonstrated that algorithms which perform well in fixed point computation will also perform well in floating point computations [11].

In the field of signal processing, Mullis and Roberts [8] and Hwang [5] first revealed the fact that the influence of round-off errors on digital filter performance depends on the realization chosen for the filter implementation. To minimize round-off errors these papers suggest a special coordinate transformation T prior to filter (or controller) synthesis, see also [10, 2]. In this paper, we consider the linear quadratic optimal control problems that arise with fixed-point arithmetic and the finite wordlengths of digital computers, A/D and D/A converters. The optimum solution is to design controllers which *directly* takes into account the round-off errors associated with a finite wordlength implementation, rather than merely performing a coordinate transformation T on the controller *after* it is designed. The problem of optimum LQG controller design in the presence of round-off error was studied by Kadiman and Williamson [6]. This paper worked with upper bounds and numerical results showed improvement over earlier work, but their algorithm does not provide the necessary conditions for an optimal solution. Liu, Skelton and Grigoriadis [7] provided the necessary conditions and a controller design algorithm for the solution of this problem.

This chapter provides the following contributions beyond [7]: i) we allow skewed sampling to accommodate the computational delay of the control computer, ii) we allow finite precision A/D and D/A computations, iii) we optimize the accuracy (wordlength) of the A/D and D/A devices, and iv) we present the solution of a realistic practical problem (control design for a large flexible structure). We allow the wordlength to be used as a control resource to be optimized in the design. That is, we shall modify the LQG cost function to include penalties on the wordlength assigned for computations in the control computer and the A/D and D/A converters (see also [3], [4]).

If we denote "controller complexity" as the sum of all wordlengths required to implement the controller and the A/D and D/A devices, we can point to the main contribution of this paper as a systematic methodology with which to trade ℓ_2 performance with computational resources (complexity). Furthermore, if we assign the (optimized) wordlength of the ith

channel of the A/D converter as a measure of importance of the ith sensor, then the solution of our optimization problem provides important ***design*** information. For example, if the optimal wordlength for the A/D channel i is much greater than channel j, then it is clear that sensor i is much more important to closed loop performance than sensor j. This suggests which sensors should be made extremely reliable and which could perhaps be purchased from off-the-shelf hardware. Furthermore, such wordlength information might be useful for sensor and actuator selection (e.g., a one or two-bit A/D channel might be eliminated entirely). These same arguments apply to actuators (D/A channels).

A control design example for a large flexible structure is provided to illustrate the suggested methodology and to indicate the performance improvement when the finite wordlength effects are taken into account.

2 Round-Off Error and the LQG Problem

Consider a linear time-invariant continuous-time plant with state space representation

$$\left\{ \begin{array}{rcl} \dot{x}_p & = & A_p x_p + B_p u + D_p w_p \\ y_p & = & C_p x_p \\ z & = & M_p x_p + v \ , \end{array} \right. \tag{1}$$

where x_p is the plant state vector, u is the control input, y_p is the regulated output and z the measured output. The external disturbance w_p and the measurement noise v are assumed to be zero-mean white noise processes with intensities W_p and V, respectively. We assume that the measurement z is sampled with uniform sampling rate $1/\Delta$ (where Δ *seconds* is the sampling period) and the control input u is sampled at the same rate as z but with δ *seconds* delay, where $0 \leq \delta < \Delta$ (we call δ the skewing in the control channel). Using the result in [1], the continuous system (1) with skewing in the control channels, can be discretized as follows

$$\left\{ \begin{array}{rcl} x_p(k+1) & = & A_d x_p(k) + B_d Q_u[u_\delta(k)] + D_d w_p(k) \\ y_p(k) & = & C_d x_p(k) \\ z(k) & = & M_d x_p(k) + v(k) \ , \end{array} \right. \tag{2}$$

where $u_\delta(k) = u(kT+\delta)$, $x_p^T(k) = [x_p^T(t)\ u_\delta^T(t)]_{t=k}$ and $Q_u[\cdot]$ is the quantization operator of the quantization process in the D/A converter. We seek a digital controller of order n_c to provide desired performance to the closed loop system. In a finite wordlength implementation of the digital controller, the controller state vector x_c and the measurement vector z will be quantized in each computation in the control computer and the A/D converter, respectively. The computation process in the digital controller can be described as follows

$$\begin{cases} x_c(k+1) &= A_cQ_x[x_c(k)] + B_cQ_z[z(k)] \\ u_\delta(k) &= C_cQ_x[x_c(k)] + D_cQ_z[z(k)] \, , \end{cases} \tag{3}$$

where $Q_x[\cdot]$ and $Q_z[\cdot]$ are the quantization operators of the digital computer and A/D converter, respectively. Assuming an additive property of the round-off error, we can model the quantization process by

$$\begin{cases} Q_u[u_\delta(k)] &= u_\delta(k) + e_u(k) \\ Q_z[z(k)] &= z(k) + e_z(k) \\ Q_x[x_c(k)] &= x_c(k) + e_x(k) \end{cases} \tag{4}$$

where $e_u(k)$ is the round-off error resulting from the D/A conversion, $e_x(k)$ is the error resulting from the quantization in the control computer, and $e_z(k)$ is the quantization error resulting from the A/D conversion.

It was shown in [9] that, under sufficient excitation conditions, the round-off error $e_x(k)$ can be modeled as a zero-mean, white noise process, independent of $w_p(k)$ and $v_p(k)$, with covariance matrix

$$E_x = diag\ [q_1^x, q_2^x, \ldots, q_{n_c}^x] \ , \ q_i^x \triangleq \frac{1}{12} 2^{-2\beta_i^x} \ , \tag{5}$$

where β_i^x denotes the wordlength of the ith state. Similarly, we assume that the D/A and A/D quantization errors $e_u(k)$ and $e_z(k)$ are zero-mean, mutually independent, white noise processes (also independent of $w_p(k)$, $v(k)$ and $e_x(k)$) with covariance matrices E_u and E_z given by

$$E_u = diag\ [q_1^u, q_2^u, \ldots, q_{n_u}^u] \ , \ q_i^u \triangleq \frac{1}{12} 2^{-2\beta_i^u} \ , \tag{6}$$

$$E_z = diag\ [q_1^z, q_2^z, \ldots, q_{n_z}^z] \ , \ q_i^z \triangleq \frac{1}{12} 2^{-2\beta_i^z} \ , \tag{7}$$

where β_i^u and β_i^z are the wordlengths (fractional part) of the D/A and A/D converters, respectively.

We seek the controller to minimize the following cost function

$$\begin{aligned} J \ = \ & \lim_{k\to\infty} \mathcal{E}\{y_p^T(k)Q_py_p(k) + u_\delta^T(k)Ru_\delta(k)\} \\ & + \textstyle\sum_{i=1}^{n_c} \rho_i^x(q_i^x)^{-1} + \sum_{i=1}^{n_u} \rho_i^u(q_i^u)^{-1} + \sum_{i=1}^{n_z} \rho_i^z(q_i^z)^{-1} \end{aligned} \tag{8}$$

where Q_p and R are positive-semidefinite and positive-definite weighting matrices, respectively, and ρ_i^x, ρ_i^u and ρ_i^z are positive scalars to penalize the wordlengths β_i^x, β_i^u and β_i^z, respectively.

Cost function (8) should be interpreted as a generalization of the trade-off traditionally afforded by LQG, between output performance (weighted by Q_p) and "control effort" (measured by the weighted variance of the control *signal* u). In this generalization we consider the *total* control resource V_u to be the sum of the control effort *and* the controller complexity (weighted sum of wordlengths of the control computer and the A/D and D/A converters). Hence, the total cost J can be decomposed as follows

$$\begin{aligned} J &= V_y + V_u \\ V_y &= \lim_{k\to\infty} \mathcal{E}\{y_p^T(k) Q_p y_p(k)\} \\ V_u &= \text{“}control\ effort\text{''} + \text{“}controller\ complexity\text{''} \\ control\ effort &= \lim_{k\to\infty} \mathcal{E}\{u^T(k) R u(k)\} \\ controller\ complexity &= \sum_{i=1}^{n_c} \rho_i^x (q_i^x)^{-1} + \sum_{i=1}^{n_u} \rho_i^u (q_i^u)^{-1} + \sum_{i=1}^{n_z} \rho_i^z (q_i^z)^{-1} . \end{aligned}$$

Using the following notation for the vectors and matrices:

$$x(k) = \begin{bmatrix} x_p(k) \\ x_c(k) \end{bmatrix} ; \quad y(k) = \begin{bmatrix} y_p(k) \\ u_\delta(k) \end{bmatrix} ; \tag{9a}$$

$$A = \begin{bmatrix} A_d & 0 \\ 0 & 0 \end{bmatrix} ; \quad B = \begin{bmatrix} B_d & 0 \\ 0 & I_{n_c} \end{bmatrix} ; \tag{9b}$$

$$C = \begin{bmatrix} C_d & 0 \\ 0 & 0 \end{bmatrix} ; \quad M = \begin{bmatrix} M_d & 0 \\ 0 & I_{n_c} \end{bmatrix} ; \tag{9c}$$

$$H = \begin{bmatrix} 0 & 0 \\ I_{n_u} & 0 \end{bmatrix} ; \quad G = \begin{bmatrix} D_c & C_c \\ B_c & A_c \end{bmatrix} ; \tag{9d}$$

$$D^T = \begin{bmatrix} B_d^T & 0 \\ D_d^T & 0 \\ 0 & 0 \\ 0 & 0 \end{bmatrix} ; \quad J^T = \begin{bmatrix} 0 & 0 \\ 0 & 0 \\ I_{n_z} & 0 \\ 0 & I_{n_c} \end{bmatrix} ; \tag{9e}$$

$$w(k) = \begin{bmatrix} e_u(k) \\ w_p(k) \\ e_z(k) + v(k) \\ e_x(k) \end{bmatrix} , \tag{9f}$$

the closed-loop system, including the finite wordlength effects, is compactly described by

$$\begin{cases} x(k+1) &= (A+BGM)x(k)+(D+BGJ)w(k) \\ y(k) &= (C+HGM)x(k)+HGJw(k) \end{cases} \tag{10}$$

The cost function (8) may be rewritten as follows

$$J = \lim_{k\to\infty} \mathcal{E}\{y^T(k)Qy(k)\} + \rho_x^T\alpha_x + \rho_u^T\alpha_u + \rho_z^T\alpha_z \ , \tag{11}$$

where

$$\rho_x = \begin{bmatrix} \rho_1^x \\ \rho_2^x \\ \vdots \\ \rho_{n_c}^x \end{bmatrix} ; \ \rho_u = \begin{bmatrix} \rho_1^u \\ \rho_2^u \\ \vdots \\ \rho_{n_u}^u \end{bmatrix} ; \ \rho_z = \begin{bmatrix} \rho_1^z \\ \rho_2^z \\ \vdots \\ \rho_{n_z}^z \end{bmatrix} , \tag{12}$$

$$\alpha_x = \begin{bmatrix} (q_1^x)^{-1} \\ (q_2^x)^{-1} \\ \vdots \\ (q_{n_c}^x)^{-1} \end{bmatrix} ; \ \alpha_u = \begin{bmatrix} (q_1^u)^{-1} \\ (q_2^u)^{-1} \\ \vdots \\ (q_{n_u}^u)^{-1} \end{bmatrix} ; \ \alpha_z = \begin{bmatrix} (q_1^z)^{-1} \\ (q_2^z)^{-1} \\ \vdots \\ (q_{n_z}^z)^{-1} \end{bmatrix} , \tag{13}$$

and

$$Q = block\ diag\ [Q_p \ , \ R] \ . \tag{14}$$

Now, since $e_u(k)$, $e_x(k)$, $e_z(k)$, $w_p(k)$, and $v(k)$ are mutually independent, substitute (10) into (11), to obtain

$$\begin{aligned} J \ = \ & trace\{X[C+HGM]^T Q[C+HGM]\} \\ & +trace\{W(HGJ)^T Q(HGJ)\} \\ & +\rho_x^T\alpha_x + \rho_u^T\alpha_u + \rho_z^T\alpha_z \ , \end{aligned} \tag{15}$$

where X is the state covariance matrix satisfying

$$X = [A+BGM]X[A+BGM]^T + (D+BGM)W(D+BGM)^T \ , \tag{16}$$

and W is defined by

$$W \triangleq block\ diag[E_u, W_p, E_z+V, E_x] \ . \tag{17}$$

We can decompose J in equation (15) into two terms $J = J_s + J_e$ where

$$\begin{aligned} J_s \ = \ & trace\{X_s(C+HGM)^T Q(C+HGM)\} \\ & +trace\{(W_s(HGJ)^T Q(HGJ)\} \\ & +\rho_x^T\alpha_x + \rho_u^T\alpha_u + \rho_z^T\alpha_z \ ; \end{aligned} \tag{18a}$$

$$\begin{aligned} J_e \ = \ & trace\{X_e(C+HGM)^T Q(C+HGM)\} \\ & +trace\{(W_e(HGJ)^T Q(HGJ)\} \end{aligned} \tag{18b}$$

where X_s and X_e are defined by

$$\begin{aligned} X_s &= (A+BGM)X_s(A+BGM)^T \\ &\quad +(D+BGJ)W_s(D+BGJ)^T \qquad (19a) \\ X_e &= (A+BGM)X_e(A+BGM)^T \\ &\quad +(D+BGJ)W_e(D+BGJ)^T \qquad (19b) \end{aligned}$$

and

$$W_s \triangleq block\ diag\ [E_u, W_p, E_z + V, 0] \ ; \ W_e \triangleq block\ diag\ [0,0,0,E_x] \ . \qquad (20)$$

Notice that $X = X_s + X_e$. And also it is clear that J_s is the portion of the performance index contributed by the disturbances $e_u(k)$, $e_z(k)$, $w_p(k)$ and $v(k)$, and that J_e is the portion contributed solely by the state round-off error $e_x(k)$.

To reduce the probability of overflow in the controller state variables computation, we must properly scale the controller state vector. We use the ℓ_2 norm scaling approach which results in the following condition

$$[X_s(2,2)]_{ii} = 1 \ ; \ i = 1,2,\ldots,n_c \ , \qquad (21)$$

where $X_s(2,2)$ is the 2-2 block of matrix X_s (the controller block), and $[\cdot]_{ii}$ stands for the ith diagonal element of the matrix. Equation (21) requires that all controller state variables have variances equal to 1 when the closed-loop system is excited only by system disturbance w_p, measurement noise v, A/D quantization error e_z and D/A quantization error e_u. We call (21) the scaling constraint. Choosing the scaling equal to 1 leaves the largest possible part of the wordlength devoted to the fractional part of the computer word.

Therefore, the optimization problem becomes

$$\min_{G,\beta_i^x,\beta_i^u,\beta_i^z} \{J = J_s + J_e\} \qquad (22)$$

subject to (18) , (19) and (21).

3 Contribution of State Round-Off Error to the LQG Performance Index

In this section, we discuss the J_e term of the cost function, defined in (18). This portion of the cost function is coordinate dependent, it is unbounded from above (that is, it can be arbitrarily large), but it has a lower bound, which can be achieved by an optimal coordinate transformation of the controller states. This lower bound result was obtained in [2]. The construction of the optimal coordinate transformation is discussed in this section.

We first observe that the J_e term of the cost function can be written as:

$$\begin{aligned} J_e &= trace\{K_e(D+BGM)^T W_e(D+BGM)\} \\ &\quad + trace\{W_e(HGJ)^T Q(HGJ)\} \ ; \qquad (23a) \\ K_e &= [A+BGM]^T K_e[A+BGM] \\ &\quad +(C+HGM)^T Q(C+HGM) \ . \qquad (23b) \end{aligned}$$

We can easily check that the minimization of J_e reduces to the problem:

$$\min_{T_c} \ J_e \ , \ J_e = trace\{E_x K_e(2,2)\} \qquad (24)$$

subject to (21). We consider the singular value decompositions

$$X_s(2,2) = U_x^T \Sigma_x U_x \ , \ \Sigma_x^{1/2} U_x K_e(2,2) U_x^T \Sigma_x^{1/2} = U_k^T \Sigma_k U_k \ , \qquad (25)$$

where U_x and U_k are orthonormal and Σ_x and Σ_k are diagonal. The matrix Σ_k is given by

$$\Sigma_k \triangleq diag[\dots \ \lambda_i[K_e(2,2)X_s(2,2)] \ \dots] \ . \qquad (26)$$

Suppose we begin our study with the closed-loop coordinate transformation T defined by

$$T = \begin{bmatrix} I & 0 \\ 0 & U_x^T \Sigma_x^{1/2} U_k^T \end{bmatrix} \ . \qquad (27)$$

Then, after this coordinate transformation, as suggested in [6], we have

$$\begin{aligned} \bar{X}_s(2,2) &= (U_x^T \Sigma_x^{1/2} U_k^T)^{-1} X_s(2,2) (U_x^T \Sigma_x^{1/2} U_k^{-T})^{-T} = I \qquad (28a) \\ \bar{K}_e(2,2) &= (U_x^T \Sigma_x^{1/2} U_k^T)^T K_e(2,2) (U_x^T \Sigma_x^{1/2} U_k^T) = \Sigma_k \ . \qquad (28b) \end{aligned}$$

If we take one more controller coordinate transformation T_c, the cost J_e and its constraint equations, (after we substitute (28) into (23)), become

$$J_e = trace[T_c E_x T_c^T \Sigma_k] \ , \qquad (29)$$

where

$$[T_c^{-1} T_c^{-*}]_{ii} = 1 \ , \ i = 1, \ 2, \ \dots, \ n_c \ . \qquad (30)$$

Since, from Lemma 2 in [7], Σ_k in (26) is coordinate independent, we may ignore the K_e and X_s calculations in (21) and (30) and concentrate on T_c in (22). Then, by applying Result 4.5.5 in [11] in equation (30), we have the following theorem.

Theorem 1 *The round-off error term J_e in the LQG performance index (22), constrained by the scaling constraint equation (21), is controller coordinate dependent. It is unbounded from above when the realization coordinate varies arbitrarily. It is bounded from below by the following lower bound*

$$\underline{J}_e = \frac{\tilde{q}}{n_c}(trace\sqrt{\Sigma_k})^2 \; ; \; \tilde{q} = \frac{trace(E_x\sqrt{\Sigma_k})}{trace\sqrt{\Sigma_k}} \; . \tag{31}$$

The lower bound is achieved by the following controller coordinate transformation

$$\underline{T}_c = U_x^T \Sigma_x^{1/2} U_k^T U_t \Pi_t V_t^T \; , \tag{32}$$

where U_x, U_k, U_t, and V_t are unitary matrices, and Σ_x and Π_t are diagonal matrices, subject to the constraints:

$$X_s(2,2) = U_x^T \Sigma_x U_x \; ; \; \Sigma_x^{1/2} U_x K_e(2,2) U_x^T \Sigma_x^{1/2} = U_k^T \Sigma_k U_k \; ; \tag{33}$$

$$\Pi_t^{-2} = \frac{n_c E_x U_t^T \sqrt{\Sigma_k} U_t}{trace(E_x\sqrt{\Sigma_k})} \; ; \; [V_t \Pi_t^{-2} V_t^T]_{ii} = 1 \; , \; i = 1, \ldots, n_c \; . \tag{34}$$

To find the optimal coordinate transformation $\underline{T}_c$ in (32), we must solve (34) to obtain U_t, Π_t and V_t, as suggested in [7, 11]. These matrices are obtained by a sequence of orthogonal transformation.

4 LQG Controller Design in the Presence of Round-Off Errors

The LQG controller design problem, when finite wordlength effects are taken into account, is described by the equations (11), (18), (19), and (21). This is denoted as the LQG_{FWSC} control design problem. However, the scaling constraint (21) can be always satisfied by properly choosing the coordinates of the controller, so the problem breaks up into two parts: i) finding the optimal controller G, and ii) finding its optimal coordinate transformation T_c to satisfy (34). On the strength of Section 3, we can therefore write the optimization problem as

$$\min_{G,\beta_i^x,\beta_i^u,\beta_i^z,T_c} J = \min_{G,\beta_i^x,\beta_i^u,\beta_i^z,T_c} (J_s + J_e) = \min_{G,\beta_i^x,\beta_i^u,\beta_i^z} [\min_{T_c}(J_s + J_e)] \; . \tag{35}$$

Since J_s is constant in terms of the variation of T_c, we have

$$\min_{G,\beta_i^x,\beta_i^u,\beta_i^z,T_c} J = \min_{G,\beta_i^x,\beta_i^u,\beta_i^z} [J_s + \min_{T_c} J_e] = \min_{G,\beta_i^x,\beta_i^u,\beta_i^z} [J_s + \underline{J}_e] \quad . \tag{36}$$

The following theorem states the necessary conditions of the optimization problem (36).

Theorem 2 *Necessary conditions for G to be the solution of the optimal controller design problem* (36) *are:*

$$\begin{aligned}
X_s &= (A+BGM)X_s(A+BGM)^T \\
&\quad +(D+BGJ)W_s(D+BGJ)^T && (37a)\\
K_e &= (A+BGM)^T K_e(A+BGM) \\
&\quad +(C+HGM)^T Q(C+HGM) && (37b)\\
K_s &= (A+BGM)^T K_s(A+BGM) \\
&\quad +(C+HGM)^T Q(C+HGM)+\nabla_x && (37c)\\
K_T &= (A+BGM)^T K_T(A+BGM)+\nabla_k && (37d)\\
0 &= (H^TQH+B^TK_sB)G(MX_sM^T+JW_sJ^T) \\
&\quad +B^T(K_sAX_s+K_eAK_T)M^T \\
&\quad +(B^TK_eB+H^TQH)GMK_TM^T && (37e)\\
q_i^x &= [n_c\rho_i^x/(trace\sqrt{\Sigma_k}[\sqrt{\Sigma_k}]_{ii})]^{1/2} \;;\; i=1,\,2,\,\ldots,\,n_c && (37f)\\
q_i^u &= [\rho_i^u/[D^TK_sD]_{ii}]^{1/2} \;;\; i=1,\,2,\,\ldots,\,n_u && (37g)\\
q_i^z &= [\rho_i^z/[J^TG^T(H^TQH+B^TK_sB)GJ]_{ii}]^{1/2} \;; \\
&\quad i=n_u+n_w+j \;;\; j=1,\,2,\,\ldots,\,n_z \;, && (37h)
\end{aligned}$$

where $\nabla_x = $ *block diag* $[0, \nabla_x(2,2)]$, $\nabla_k = $ *block diag* $[0, \nabla_k(2,2)]$, *and*

$$\nabla_x(2,2) = \frac{1}{2n_c}\{\sum_{i=1}^{n_c}[trace E_x\Sigma_k^{1/2} + q_i^x trace\Sigma_k^{1/2}] \cdot \frac{K_e(2,2)[E^{-1}]^T_{ith-row}E^T_{ith-col}}{[\Sigma_k^{1/2}]_{ii}}\} \;; \tag{38a}$$

$$\nabla_k(2,2) = \frac{1}{2n_c}\{\sum_{i=1}^{n_c}[trace E_x\Sigma_k^{1/2} + q_i^x trace\Sigma_k^{1/2}] \cdot \frac{[E^{-1}]^T_{ith-row}E^T_{ith-col}X_s(2,2)}{[\Sigma_k^{1/2}]_{ii}}\} \;, \tag{38b}$$

where E is the matrix of eigenvectors of the matrix $K_e(2,2)X_s(2,2)$.

The proof of Theorem 2 is similar to that of the main theorem in [7] with minor modifications.

Remark 1 *Equation (37) reduces to the standard LQG design by setting* $\beta_i^x = \infty$ *(or equivalently* $q_i^x = 0$, *i.e.* $E_x = 0$*) and* $\delta = 0$. *In this case, the 1-1 block of* X_s *in (37) reduces to the Kalman filter Riccati equation, and the 2-2 block of* K_s *in (37) reduces to the control Riccati equation.*

Remark 2 *Equation (37) reduces to the* LQG_{FW} *design in [7] by deleting equations (37g) and (37h). Hence, (37) is a generalization of results in [7], and also provides a way to select the wordlengths for the A/D and D/A converters.*

Now, we have the following LQG_{FWSC} controller design algorithm:

The LQG_{FWSC} *Algorithm*

Step 1 Design the standard LQG controller gain G with given weighting matrices Q_p and R.

Step 2 Solve for G, q_i^x, q_i^u and q_i^z from equations (37). (Initialize with G from standard LQG).

Step 3 Compute $\underline{T}_c = U_x^T \Sigma_x^{1/2} U_k^T U_t \Pi_t V_t^T$ by solving U_x, Σ_x, U_k, U_t, Π_t, V_t from (34), using the G and q_i^x obtained in Step 1.

In Step 2 of the above algorithm, a steepest descent method can be applied to find solutions for G, q_i^x, q_i^u and q_i^z satisfying (37). Note that for the infinite wordlength computer, A/D and D/A converters, the initial LQG controller stabilizes the given plant with skewing in control channels. The procedure to solve for G, q_i^x, q_i^u and q_i^z in Step 2 can be described as follows:

i) Solve (37a) and (37b) for X_s and K_e with given G, q_i^x, q_i^u and q_i^z.

ii) Solve (38) for $\nabla_x(2,2)$ and $\nabla_k(2,2)$, and form ∇_x and ∇_k.

iii) Solve (37c) and (37d) for K_s and K_T.

iv) Solve (37f), (37g) and (37h) for q_i^x, q_i^u and q_i^z.

v) Compute the gradient of G as follows

$$\begin{aligned}\Delta G \;=\;& (H^T Q H + B^T K_s B) G (M X_s M^T + J W_s J^T) \\ &+ B^T (K_s A X_s + K_e A K_T) M^T \\ &+ (B^T K_e B + H^T Q H) G M K_T M^T \;.\end{aligned} \tag{39a}$$

vi) Obtain a new control gain

$$G = G - \alpha \Delta G \tag{40}$$

where α is a step size parameter, which is chosen to be small enough such that the closed loop system remains stable.

The iterative process must be repeated until a controller which satisfies the necessary conditions (37) is obtained.

<u>*The Special Case of Equal Wordlengths*</u>

In many practical control problems, the wordlengths of the controller states, and the A/D converters and D/A converters are chosen to be equal, that is,

$$\rho_i^x = \rho_x \;;\; q_i^x = q_x \;;\; i = 1,\, 2,\, \ldots,\, n_c \tag{41a}$$
$$\rho_i^u = \rho_u \;;\; q_i^u = q_u \;;\; i = 1,\, 2,\, \ldots,\, n_u \tag{41b}$$
$$\rho_i^z = \rho_z \;;\; q_i^z = q_z \;;\; i = 1,\, 2,\, \ldots,\, n_z \;. \tag{41c}$$

The following corollary provides the necessary conditions for the above case.

Corollary 1 *The necessary conditions for G to be the solution of the optimal controller design problem (36), when conditions (41) hold, are similar to those stated in Theorem 2, except that equations (37f), (37g) and (37h) need to be replaced by*

$$q_x = \rho_x^{1/2} n_c / trace \sqrt{\Sigma_k} \;; \tag{42a}$$
$$q_u = \{\rho_u n_u / trace[D^T K_s D]\}^{1/2} \;; \tag{42b}$$
$$q_z = \{\rho_z n_z / trace[J^T G^T (H^T Q H + B^T K_s B) G J]\}^{1/2} \;, \tag{42c}$$

and (38) by

$$\nabla_x(2,2) = \frac{q_x}{n_c} trace \Sigma_k^{1/2} \{ \sum_{i=1}^{n_c} \frac{K_e(2,2)[E^{-1}]_{ith-row}^T E_{ith-col}^T}{[\Sigma_k^{1/2}]_{ii}} \} ; \quad (43a)$$

$$\nabla_k(2,2) = \frac{q_x}{n_c} trace \Sigma_k^{1/2} \{ \sum_{i=1}^{n_c} \frac{[E^{-1}]_{ith-row}^T E_{ith-col}^T X_s(2,2)}{[\Sigma_k^{1/2}]_{ii}} \} . \quad (43b)$$

It is clear that the necessary conditions for equal wordlength distribution in the control computer, the A/D and the D/A converter can be obtained from the corresponding necessary conditions of the unequally distributed wordlength case by averaging the corresponding unequal wordlengths.

5 Computational Example

The JPL LSCL facility [13] is shown in Figure 1. The main component of the apparatus consists of a central hub to which 12 ribs are attached. The diameter of the dish-like structure is slightly less than 19 feet. The ribs are coupled together by two rings of wires which are maintained under nearly constant tension. The ribs, being quite flexible and unable to support their own weight without excessive droop, are each supported at two locations along their free length by levitators. The hub is mounted to the backup structure through a gimbal arrangement so that it is free to rotate about two perpendicular axes in the horizontal plane. A flexible boom is attached to the hub and hangs below it.

Actuation of the structure is as follows. Each rib can be individually manipulated by a rib-root actuator (RA1, RA4, RA7 and RA10 in Figure 1) mounted on that rib near the hub. In addition, two actuators (HA1 and HA10) are provided which torque the hub about its two gimbal axes. The placement of these actuators guarantees good controllability of the flexible modes of motion. The locations of the actuators are shown in Figure 1.

The sensor locations are also shown in Figure 1. Each one of the 24 levitators (LS1-LS24) is equipped with a sensor which measures the relative angle of the levitator pulley. The levitator sensors provide the measurement of the vertical position of the corresponding ribs at the points where the levitators are attached. Four position sensors (RS1, RS4, RS7 and RS10), which are co-located with the rib root actuators, measure rib-root displacement. Sensing for the hub consists of two rotation sensors (HS1 and HS10) which are mounted directly at the gimbal bearing.

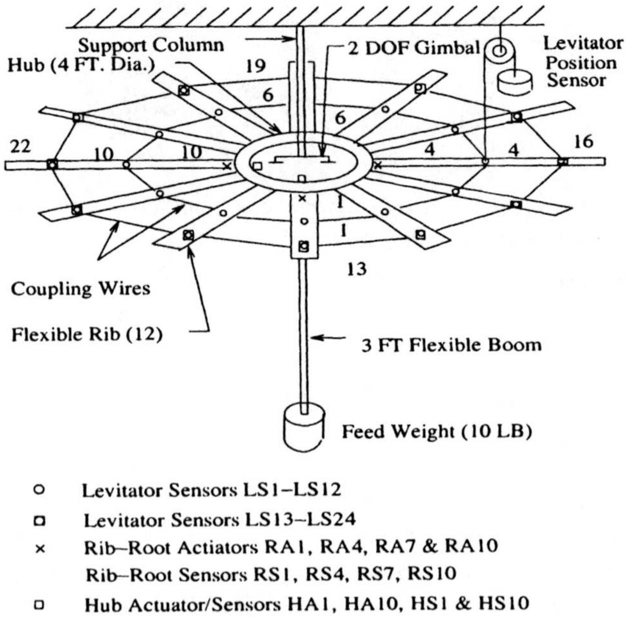

Figure 1: The JPL LSCL Structure

JPL created a finite element model with 30 degrees of freedom, (60 state variables). A 24th order reduced order model obtained from the 60th order finite element model is used for design. By augmenting the plant model with the actuator and sensor dynamics, we obtain a continuous system model of order 34 in the form of (1), where the performance outputs are $LS1$, $LS4$, $LS7$, $LS10$, $LS13$, $LS16$, $LS19$, $LS22$, $HS1$ and $HS10$ ($n_y = 10$). The measurements share the same outputs as y_p ($n_z = 10$); the controls consist of $HA1$, $HA10$, $RA1$, $RA4$, $RA7$ and $RA10$ ($n_u = 6$). Finally, the system disturbance w_p enters through the same channels as u ($n_w = 6$).

The continuous plant is discretized at 40 Hz with skewing $\delta = 0.01$ seconds in the control channels. The order of the discretized plant is 40 because of the augmentation of the plant state (order 34) with the control input vector u (dimension 6).

We consider the following output and input weighting matrices Q_p and R, and system noise and measurement noise covariance matrices W_p and V

$$Q_p = block\ diag\ [1.1430 I_8,\ 0.698 I_2]\ ;\ \ R = 0.125 I_6 \tag{44}$$

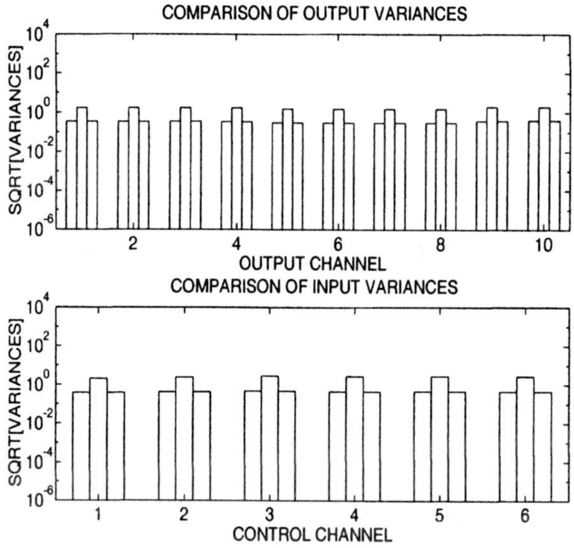

Figure 2: Comparison of Output/Input Variances for Case 1

$$W_p = 0.0016 I_6 \ ; \ V = 0.0001 I_1 0 \ . \tag{45}$$

The weighting matrices Q_p and R were selected using an output covariance constraint approach [12] to provide a good initial LQG design. Following the procedure described in section 4, an initial LQG controller (with no finite wordlength considerations) G_{LQG} is designed. Then, the optimal finite wordlength controller G_{LQGFW} is designed by applying the LQG_{FWSC} algorithm of Section 4, where we assume the following scalar weightings for the wordlengths of the control computer and the A/D and D/A converters.

$$\rho_i^x = 10^{-14} \ ; \ i = 1, \ 2, \ \ldots, \ n_c \tag{46a}$$

$$\rho_i^u = 10^{-14} \ ; \ i = 1, \ 2, \ \ldots, \ n_u \tag{46b}$$

$$\rho_i^z = 10^{-14} \ ; \ i = 1, \ 2, \ \ldots, \ n_z \ . \tag{46c}$$

The results of the optimal finite wordlength design are presented in Table 1 and Figure 2.

Table 1 Optimal Wordlength

$\beta_{25}^{x} - \beta_{40}^{x}$	$\beta_{22}^{x} - \beta_{24}^{x}$	β_{21}^{x}	$\beta_{14}^{x} - \beta_{15}^{x}, \beta_{17}^{x} - \beta_{20}^{x}$
3	4	5	6
$\beta_{1}^{x} - \beta_{13}^{x}, \beta_{16}^{x}$	$\beta_{1}^{u}, \beta_{5}^{z} - \beta_{9}^{z}$	$\beta_{2}^{u} - \beta_{6}^{z}, \beta_{1}^{z} - \beta_{4}^{z}$	$\beta_{9}^{z} - \beta_{10}^{z}$
7	8	9	10

Table 1 provides the optimal number of bits allocated by the optimization algorithm, to the control computer (β_i^x) and the A/D and D/A converters (β_i^z and β_i^u). The largest number of bits assigned to the controller state variables is 7, and the number of bits assigned for A/D and D/A conversion are between 8 and 10. Figure 2 presents the output and input (control) variances for each output and control channel, where the first column corresponds to the output/input variances of the initial LQG controller G_{LQG} with infinite wordlength implementation; the second column represents the variances of G_{LQG} with finite wordlength implementation (for the wordlengths shown in Table 1); and the third column corresponds to the optimal finite wordlength controller G_{LQGFW} with finite wordlength implementation. It is clear that the variances in the second column are much larger than those in the first column (the difference is about 2 orders of magnitude), which indicates that if one implements the LQG controller with finite wordlength, the performance will be much worse than what predicted by standard LQG theory. We observe that in the first and third columns, the variances are very close. Hence, the optimal finite wordlength controller G_{LQGFW}, provides closed loop system performance which is very closed to the original LQG design with infinite wordlength implementation. This suggests the following design procedure for finite wordlength optimal LQG controllers: i) Design a standard LQG controller G_{LQG} to satisfy the desired closed loop performance objectives (e.g. using the results of [12]) ii) Design the optimal finite wordlength controller with the same LQG weighting matrices Q_p and R as in i), using the G_{LQG} controller designed in i) as the initial controller. Then, the resulting optimal finite wordlength controller G_{LQGFW} will have closed loop performance (output/input variances) close to the expected one.

Next, the optimal finite wordlength controller with equal wordlength at the control computer, A/D and D/A converters, respectively, is designed using the same initial LQG controller as above (using the results of Corollary 1). The results are provided in Table 2 and Figure 3.

Table 2 Optimal Wordlength

β_x	β_u	β_z
7	9	9

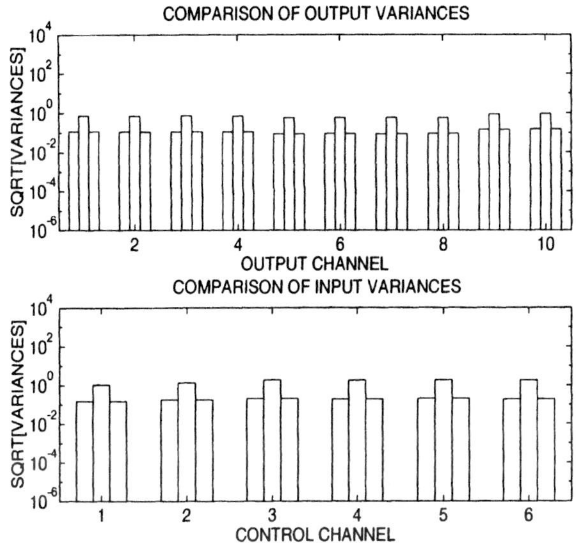

Figure 3: Comparison of Output/Input Variances for Case 2

We notice that the optimal cost for the case that allows each wordlength of state, A/D and D/A converters to be unequal was found to be

$$J_1 = 2.82697^{-3}\,, \tag{47}$$

and for the case of equally distributed wordlength (i.e., when the wordlength of all D/A channels is β_u, of all A/D channels if β_z, and of all controller state variables β_x, where the β's are as in Table 2),

$$J_1 = 2.82807^{-3}\,. \tag{48}$$

Note that J_1 and J_2 are approximately equal, but $J_1 < J_2$, hence we can achieve slightly better performance by allowing unequal wordlength distribution.

In this example, the optimally allocated wordlengths β_i^x in the control computer were not significantly different (3 bits-7 bits) to justify deleting any controller state. The same arguments hold for the sensors and actuators channels (A/D and D/A). Furthermore, similar performance was obtained by setting all β_i^x to a common value $i = 1, 2, \cdots, n_x$, and all β_i^u, $i = 1, \cdots, n_u$ to a common value, and all β_i^z, $i = 1, \cdots n_z$ to a common value to be optimized, with the advantage of greatly reduced computation. This example shows that for JPL's LSCL structure similar performance can be obtained with a 7 bit controller computer, and a 10 bit A/D and a 9 bit D/A, as the performance predicted with a standard "infinite precision" LQG solution. This performance is *not* achieved by quantizing the standard LQG gains, but by solving a new optimization problem.

The above example was repeated using 10^{-10} to replace 10^{-14}, the wordlength penalty scalars in (46). The increase in the penalty of wordlengths changes the wordlengths in Table 1 by approximately one third. However, the significant difference in the two examples is that the optimal coordinate transformation T dominates the performance in the case of ρ_i^x, ρ_i^u, $\rho_i^z = 10^{-14}$ and the optimal control gain dominates the performance in the case of ρ_i^x, ρ_i^u, $\rho_i^z = 10^{-10}$. Our general guideline therefore is that, if β is large, the optimal realization is more important than the optimization of the gain G. This is a useful guide because applying the optimal T to a standard LQG value of G is much easier than finding the optimal G using the LQG_{FWCS} algorithm.

6 Conclusions

This chapter solves the problem of designing an LQG controller to be optimal in the presence of finite wordlength effects (modeled as white noise sources whose variances are a function of computer, A/D and D/A wordlengths) and skewed sampling. The controller, A/D and D/A converter wordlengths are used as a control resource to be optimized in the design. This new controller, denoted LQG_{FWSC}, has two computational steps. First the gains are optimized, and then a special coordinate transformation must be applied to the controller. This transformation depends on the controller gains, so the transformation cannot be performed *a priori*. (Hence, there is no separation theorem.) The new LQG_{FWSC} controller design algorithm reduces to the standard LQG controller when an infinite wordlength is used for the controller synthesis and the sampling is synchronous, so this is a natural extension of the LQG theory. The selection of the LQG weights using Output Covariance Constraint (OCC) techniques in [12] will be investigated in future work. This work provides a mechanism for trading output performance (variances) with controller complexity.

References

1. G. F. Franklin, J. D. Powell, and M. L. Workman. *Digital Control of Dynamic Systems.* Addison and Wesley, 1990.

2. M. Gevers and G. Li. *Parametrizations in Control, Estimation and Filtering Problems.* Springer-Verlag, 1993.

3. K. Grigoriadis, K. Liu, and R. Skelton. Optimizing linear controllers for finite precision synthesis using additive quantization models. Proceedings of the International Symposium MTNS-91, Mita Press, 1992.

4. K. Grigoriadis, R. Skelton, and D. Williamson. Optimal finite wordlength digital control with skewed sampling and coefficient quantization. In *Proceedings of American Control Conference,* Chicago, Illinois, June 1992.

5. S. Hwang. Minimum uncorrelated unit noise in state-space digital filtering. *IEEE Trans. Acoust. Speech, Signal Processing,* 25(4), pp. 273-281, 1977.

6. K. Kadiman and D. Williamson. Optimal finite wordlength linear quadratic regulation. *IEEE Trans. Automat. Contr.*, 34(12), pp. 1218-1228, 1989.

7. K. Liu, R. E. Skelton, and K. Grigoriadis. Optimal controllers for finite wordlength implementation. *IEEE Trans. Automat. Contr.*, 37(9), pp.1294-1304, 1992.

8. C. Mullis and R. Roberts. Synthesis of minimum round-off noise fixed point digital fiters. *IEEE Trans. Circuits and Syst.*, 23(9), pp. 551-562, 1976.

9. A. Sripad and D. Snyder. A necessary and sufficient condition for quantization error to be uniform and white. *IEEE Trans. Acoust. Speech, Signal Processing,* 2(5), pp. 442-448, 1977.

10. D. Williamson. Finite word length design of digital kalman filters for state estimation. *IEEE Trans. Automat. Contr.*, 30(10), pp. 930-939, 1985.

11. D. Williamson. *Digital Control and Implementation: Finite Wordlength Considerations.* Prentice Hall, 1991.

12. G. Zhu, M. Rotea, and R. Skelton. A convergent feasible algorithm for the output covariance constraint problem. In *1993 American Control Conference*, San Francisco, June 1993.

13. G. Zhu and R. E. Skelton. *Integration of Model Reduction and Controller Design for Large Flexible Space Structure – An Experiment on the JPL LSCL Facility*. Purdue University Report, March 1992.

Optimal Pole Placement for Discrete-Time Systems[1]

Hal S. Tharp

Department of Electrical and Computer Engineering
University of Arizona
Tucson, Arizona 85721
tharp@ece.arizona.edu

I. INTRODUCTION

This chapter presents a technique for relocating closed-loop poles in order to achieve a more acceptable system performance. In particular, the technique provides a methodology for achieving exact shifting of nominal eigenvalues along the radial line segment between each eigenvalue and the origin. The pole-placement approach is based on modifying a standard, discrete-time, linear quadratic (LQ) regulator design [1, 2]. There are several reasons behind basing the pole-placement strategy on the LQ approach. First, the LQ approach is a multivariable technique. By using an LQ approach, the trade-offs associated with the relative weighting between the state vector and the input vector can be determined by the magnitudes used to form the state weighting matrix, Q, and the input weighting matrix, R [3]. In addition, for systems with multiple inputs, the LQ approach automatically distributes the input signal between the different input channels and automatically assigns the eigenvectors. Second, effective stability margins are automatically guaranteed with LQ-based full-state feedback designs [4]. Third, the closed-loop system that results from an LQ design can provide a target model response that can be used to define a reference model. This reference model could then be used in an adaptive control design, for example, or it could be used to establish a desired behavior for a closed-loop system. Similarly, the present LQ-based design could be used as a nominal stabilizing controller that might be required in more sophisticated control design techniques [5,6]. In all of these scenarios, the LQ design

[1] **Portions reprinted, with permission, from *IEEE Transactions on Automatic Control*, Vol. 37, No. 5, pp. 645-648, May 1992. ©1992 IEEE.**

has provided acceptable closed-loop eigenvalue locations automatically. Finally, the LQ approach is fairly straight-forward and easy to understand. It may not require as much time, effort, and money to achieve an acceptable control design based on an LQ approach versus a more advanced design approach like H_∞, adaptive, or intelligent control.

The pole-placement problem has been actively pursued over the past few decades. As a representative sample of this activity, the following three general techniques are mentioned. During the brief discussion of these three techniques, the main emphasis will be on how these techniques differ from the technique discussed in this chapter.

The first technique is the exact pole-placement technique [7]. As made clear by Moore in [8], a system with more than one input, allows flexibility in the pole-placement design process. In particular, this design freedom can be used to shape the eigenvectors associated with the particular eigenvalue locations that have been selected. Kautsky uses this freedom to orient the eigenvectors to make the design as robust to uncertainty as possible. One drawback of this approach is the issue of where the eigenvalues should be located. Of course, there are tables which contain eigenvalue locations which correspond to particular prototype system responses, e.g., the poles could correspond to Bessel polynomials [9]. However, if there are zeros in the system, their presence could complicate the process of selecting acceptable closed-loop poles.

The second technique is the related topic of eigenstructure assignment [10]. In contrast to the first technique, with the second technique, the eigenvectors are selected to provide shaping of the response. For example, the eigenstructure technique could be used to decouple certain subsystem dynamics in a system. Again, this strategy assumes that the designer knows where to locate the closed-loop eigenvalues. Also, it assumes that some knowledge about the closed-loop eigenvectors is available.

Finally, the topic of regional pole-placement is mentioned [11]. This technique allows the closed-loop poles to be located in a circular region inside the unit circle. With this regional placement approach, however, all of the closed-loop eigenvalues need to be located in the defined circular region. In certain situations, it may be unnecessary or undesirable to locate all of the eigenvalues inside this circular region.

All of the above techniques have their advantages. In some instances, it might be advantageous to utilize the desirable properties of the above techniques along with the present pole-placement strategy. Later in this chapter, it will be shown how the regional pole-placement technique can be combined with the present pole-placement technique. This chapter will also discuss how it might be beneficial to combine the presented technique with the robust, exact pole-placement technique.

The present pole-placement strategy differs from the above techniques in

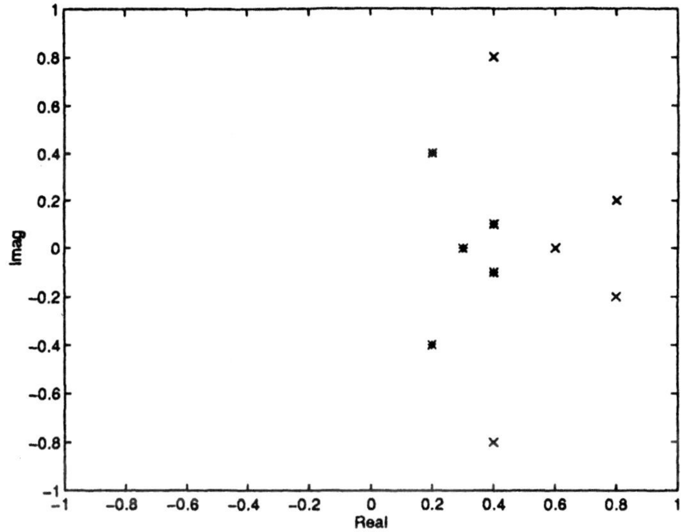

Figure 1: Relocation of closed-loop eigenvalues. x - Nominal Locations, * - Desired Locations.

that it relocates the closed-loop eigenvalues relative to an initial eigenvalue arrangement. This relocation can be accomplished either by relocating all of the eigenvalues by a particular multiplicative amount or by allowing the eigenvalues to be relocated independently. The relocated eigenvalues lie on a line segment joining the original eigenvalue location with the origin. Figure 1 illustrates how all of the nominal eigenvalues might be relocated with the present technique. In Figure 1, the desired eigenvalues equal a scaled version of the nominal eigenvalues, with the scaling factor equaling $\frac{1}{2}$.

By providing the opportunity to expand or contract the eigenvalue locations, the system time response can be effectively modified. Because these relocations are accomplished using an LQ approach, the movements are achieved while still preserving the trade-offs associated with the state and control weightings and the trade-offs in the distribution of the control energy among the different control channels.

This chapter contains the following material. Section II presents the theory behind how the closed-loop eigenvalues are moved or relocated. Both relocation techniques are presented. One technique is concerned with moving all of the eigenvalues by the same amount and the other technique is concerned with moving the eigenvalues independently. Section III discusses how the pole-shifting strategy can be combined with the regional pole-placement strategy. Examples illustrating this pole-placement technique are provided in Section IV. As a means of disseminating this technique, MATLAB m-files are included in Section VII.

II. POLE-PLACEMENT PROCEDURES

The pole-placement technique [12] that is being presented is really a pole-shifting technique as illustrated in Figure 1. To accomplish this eigenvalue shifting, the technique relies on two important facts associated with the LQ problem. One fact concerns the existence of more than one solution to the Discrete-time Algebraic Riccati Equation (DARE). The second fact concerns the invariance of a subset of the eigenvectors associated with two different Hamiltonian matrices, when particular structural differences between the two Hamiltonian matrices exist.

Before presenting the pole shifting theorems, the notation to be used is stated. Consider the following discrete-time system

$$x(k+1) = Ax(k) + Bu(k) . \tag{1}$$

Suppose a nominal closed-loop system is obtained by solving the LQ performance criterion

$$J = \frac{1}{2}\sum_{k=0}^{\infty}[x^T(k)Qx(k) + u^T(k)Ru(k)] . \tag{2}$$

As presented in [13], the closed-loop system is given by

$$F = A - BK, \tag{3}$$

where the state-feedback matrix, K, is defined by

$$K = (R + B^T M_s B)^{-1} B^T M_s A, \tag{4}$$

and M_s is defined to be the stable solution satisfying the DARE

$$M = Q + A^T MA - A^T MB(R + B^T MB)^{-1} B^T MA . \tag{5}$$

However, M_s is not the only solution of the DARE [14]. One way to obtain other solutions to Eq. (5) is by using the Hamiltonian matrix associated with the given performance criterion. The Hamiltonian matrix associated with the system in Eq. (1) and the performance criterion in Eq. (2), can be written as

$$H = \begin{bmatrix} A + BR^{-1}B^T A^{-T} Q & -BR^{-1}B^T A^{-T} \\ -A^{-T} Q & A^{-T} \end{bmatrix} . \tag{6}$$

In Eq. (6), the system matrix, A, has been assumed nonsingular. If the matrix A is singular, then at least one eigenvalue is located at the origin. Before beginning the following pole-shifting procedure, it is assumed that the system has been factored to remove these eigenvalues at the origin and consequently not consider them in the pole-shifting design process.

Recall that the Hamiltonian matrix associated with an LQ regulator problem contains a collection of 2n eigenvalues. In this collection of eigenvalues, n of the eigenvalues lie inside the unit circle and the other n eigenvalues are the reciprocal of the n eigenvalues inside the unit circle. The n eigenvalues inside the unit circle correspond to the stable eigenvalues and the n eigenvalues outside the unit circle correspond to the unstable eigenvalues.

The stable solution of the DARE, M_s, is constructed from the eigenvectors, $[X_s^T, Y_s^T]^T$, associated with the stable eigenvalues of the Hamiltonian matrix, i.e., $M_s = Y_s X_s^{-1}$. One way to obtain other DARE solutions is to construct a solution matrix, M, out of other combinations of the 2n eigenvectors associated with H. For example, the unstable DARE solution, M_u, is constructed using the eigenvectors associated with the unstable eigenvalues of the Hamiltonian matrix, i.e., $M_u = Y_u X_u^{-1}$. These two solutions, M_s and M_u, are constructed out of disjoint sets of eigenvectors. In fact, these two solutions use all the eigenvectors associated with the Hamiltonian matrix. With these definitions, the eigenvalue/eigenvector relationship associated with H can be written as

$$H \begin{bmatrix} X_s & X_u \\ Y_s & Y_u \end{bmatrix} = \begin{bmatrix} X_s & X_u \\ Y_s & Y_u \end{bmatrix} \begin{bmatrix} \Lambda_s & 0 \\ 0 & \Lambda_u \end{bmatrix} . \tag{7}$$

Other DARE solutions, M_i, can be constructed from a mixture of stable and unstable eigenvectors of H, as will be demonstrated and utilized in the ensuing development.

Suppose the pole locations resulting from the nominal performance criterion, given by Eq. (2), are not completely satisfactory. This unsatisfactory behavior could be because the system response is not fast enough. Using the nominal criterion as a starting point, the DARE in Eq. (6) will be modified to reposition the closed-loop eigenvalues at more desirable points on the radial line segments connecting the origin with each of the nominal closed-loop eigenvalues.

The first technique for shifting the nominal closed-loop eigenvalues, concerns contracting all of them by an amount equal to $\frac{1}{\rho^2}$. This technique is stated in the following theorem.

Theorem 1. Consider the closed-loop system (3) obtained by minimizing (2) for the system given in Eq. (1). The full state feedback matrix,

K_{ps}, obtained by minimizing the modified performance criterion

$$J_p = \frac{1}{2}\sum_{k=0}^{\infty}[\bar{x}^T(k)Q_p\bar{x}(k) + \bar{u}^T(k)R_p\bar{u}(k)], \tag{8}$$

for the modified system

$$\bar{x}(k+1) = A_p\bar{x}(k) + B_p\bar{u}(k), \tag{9}$$

where $Q_p = Q + \Delta Q$, $R_p = R + \Delta R$, $A_p = pA$, and $B_p = pB$, will result in the closed loop system

$$\overline{F} = A - BK_{ps} , \tag{10}$$

where

$$u(k) = -K_{ps}x(k) , \tag{11}$$

having each of its eigenvalues equal to the eigenvalues of F from Eq. (3) multiplied by $(\frac{1}{p^2})$, with $(p > 1)$. The adjustments to the appropriate matrices are given by

$$\Delta Q = (p^2 - 1) * (Q - M_u) \tag{12}$$

and

$$\Delta R = (p^2 - 1) * R . \tag{13}$$

The M_u matrix in Eq. (12) is the unstable solution to the DARE in Eq. (5).

In Theorem 1, the modified full-state feedback gain matrix is given by

$$K_{ps} = (R_p + B_p^T M_{ps} B_p)^{-1} B_p^T M_{ps} A_p . \tag{14}$$

The adjustments to the nominal system and the performance criterion are equivalent to minimizing the original system in Eq. (1) with the modified performance criterion

$$J_m = \frac{1}{2}\sum_{k=0}^{\infty}(p^{2k})[x^T(k)Q_px(k) + u^T(k)R_pu(k)]. \tag{15}$$

This criterion, J_m, has the following interpretation [15]. If Q_p and R_p were the original state and input weighting for the system in Eq. (1), then the p^{2k} term has the effect of moving the closed-loop poles inside a circle of radius $(\frac{1}{p})$. This causes the transient response associated with J_m to decay with a rate at least as fast as $(\frac{1}{p})^k$.

A consequence of Theorem 1 is that the nominal performance criterion, as given by Eq. (2), has been modified. As a result, Theorem 1 yields a feedback gain matrix, K_{ps}, which is suboptimal with respect to the nominal performance criterion. The degree of suboptimality is provided by the difference between the two Riccati equation solutions, M_{ps} and M_s, used to calculate the two feedback gain matrices K_{ps} and K. However, many times the LQ criterion is used more as a pole-placement design tool than as an objective that must be optimized explicitly. In such a design approach, the degree of suboptimality introduced by the application of Theorem 1 may be unimportant.

Before providing the proof of Theorem 1, a lemma dealing with M_u is introduced.

Lemma 1. The unstable solution, M_u, of Eq. (5) also satisfies the modified DARE associated with Eqs. (8) and (9), i.e., $M_{pu} = M_u$.

Proof of Lemma 1. The DARE associated with Eqs. (8) and (9) is given by

$$M = Q_p + A_p^T M A_p - A_p^T M B_p (R_p + B_p^T M B_p)^{-1} B_p^T M A_p. \tag{16}$$

Using the modified matrices defined in Theorem 1, Eq. (16) can be simplified to equal Eq. (5) when $M = M_u$. Therefore, since M_u satisfies Eq. (16), the unstable solution of Eq. (16) must equal M_u, i.e., $M_{pu} = M_u$.

Proof of Theorem 1. The eigenvalues of $F = A - BK$, denoted by Λ_s, and the eigenvalues of $F_u = A - BK_u$, denoted by Λ_u, with $K_u = (R + B^T M_u B)^{-1} B^T M_u A$, are reciprocals of one another. From Lemma 1, straightforward manipulation yields

$$K_{pu} = K_u \tag{17}$$

$$F_{pu} = A_p - B_p K_{pu} = pF_u \tag{18}$$

$$\lambda(F_{pu}) = p\lambda(F_u) = p\Lambda_u . \tag{19}$$

The reciprocal nature of the Hamiltonian eigenvalues provides the locations of the stable eigenvalues inside the Hamiltonian matrix, once the locations of the unstable eigenvalues of the Hamiltonian matrix are known. With $F_{ps} = A_p - B_p K_{ps}$, it is clear that the eigenvalues in F_{ps} and F_{pu} are reciprocals of one another. This allows the eigenvalues of F_{ps} and F to be related.

$$\lambda(F_{ps}) = \frac{1}{\lambda(F_{pu})} = \frac{1}{p\Lambda_u} = (\frac{1}{p})\Lambda_s \tag{20}$$

The abuse of notation in Eq. (20) is used to imply the reciprocal nature of each eigenvalue in the given matrices. To complete the proof, a relationship between the eigenvalues of F_{ps} and the eigenvalues of $\overline{F}$ must be established. In this discussion, $\bar{F}$ is the modified closed-loop system matrix formed from the original system matrix, input matrix, and the stable, full-state feedback gain given by Eq. (14).

$$\overline{F} = A - BK_{ps} \tag{21}$$

The equation for F_{ps} provides the needed relationship.

$$F_{ps} = (pA - pBK_{ps}) = p(A - BK_{ps}) = p\overline{F} \tag{22}$$

$$\lambda(\overline{F}) = (\frac{1}{p})\lambda(F_{ps}) = (\frac{1}{p^2})\Lambda_s \tag{23}$$

Equation (23) completes the proof.

Theorem 1 provides a strategy for moving all of the closed-loop eigenvalues by the same multiplicative factor, $\frac{1}{p^2}$. Suppose some of the nominal eigenvalues in Eq. (3) are in acceptable locations and only a subset of the nominal eigenvalues need to be moved an amount $(\frac{1}{p^2})$. Let Λ_1 contain the eigenvalues of F that are to be retained at their nominal locations and Λ_2 contain the eigenvalues of F to be contracted toward the origin by an amount $(\frac{1}{p^2})$, where

$$\Lambda_1 = \{\lambda_1, \lambda_2, ..., \lambda_k\} \tag{24}$$

and

$$\Lambda_2 = \{\lambda_{k+1}, \lambda_{k+2}, ..., \lambda_n\}. \tag{25}$$

The following theorem can be utilized to accomplish this selective movement.

Theorem 2. Let the nominal closed-loop system be given by Eq. (3) and let the eigenvalues from this system be partitioned into the two subsets given in Eqs. (24) and (25). Suppose all the eigenvalues in Λ_1 remain inside the unit circle when expanded by a factor p, i.e.,

$$|p\lambda_i| < 1.0, \quad \text{for } 1 \leq \text{i} \leq \text{k}. \tag{26}$$

The eigenvalues in Λ_2 can be contracted by $\frac{1}{p^2}$, if the nominal system and performance criterion are modified to produce a new optimization problem composed of new system, input, state weighting, and input weighting matrices with

$$Q_n = Q + (p^2 - 1) * (Q - M_i), \tag{27}$$

$$R_n = R + (p^2 - 1) * R, \tag{28}$$

$$A_n = pA, \tag{29}$$

and

$$B_n = pB. \tag{30}$$

$M_i = Y_i X_i^{-1}$ is constructed from eigenvectors of the nominal Hamiltonian matrix given in Eq. (6). The eigenvectors, from which Y_i and X_i are obtained, are associated with the eigenvalues in the set Λ_1 and the eigenvalues which are the reciprocals of the eigenvalues in Λ_2. The full state feedback matrix is given by

$$K_n = (R_n + B_n^T M_{ns} B_n)^{-1} B_n^T M_{ns} A_n. \tag{31}$$

Proof of Theorem 2. Since M_i is a solution to the nominal DARE, like M_u, this proof is the same as the proof of Theorem 1 with M_u replaced by M_i.

The restriction in the above theorem, concerning the magnitude of the eigenvalues in Λ_1 when they are multiplied by p, can be removed. Before discussing how to remove this restriction, a little more information behind the existence of the restriction is given.

When the modifications to the nominal problem are made (see Theorem 2), n of the nominal eigenvalues in Eq. (7) are expanded by the factor p in the modified Hamiltonian matrix. The n eigenvalues that are expanded correspond to the n eigenvectors used to form M_i in Eq. (27). The modified Hamiltonian matrix, which can be used to form the solutions to Eq. (16), contains these n expanded eigenvalues. In addition, the remaining n eigenvalues of the modified Hamiltonian matrix equal the reciprocal of these n expanded eigenvalues. If any of the nominally stable eigenvalues in Λ_1 are moved outside the unit circle when they are multiplied by p, then their reciprocal eigenvalue, which is now inside the unit circle, will be the eigenvalue whose corresponding eigenvector is used to form the stable Riccati solution, M_{ns}. This will result in that eigenvalue in Λ_1 not being retained in

$$F_n = A - BK_n . \tag{32}$$

To allow for exact retention of all the eigenvalues in the set Λ_1, when $|p\lambda_i| > 1$ for some λ_i in Λ_1, the eigenvectors associated with those eigenvalues that move outside the unit circle, after they are multiplied by p, must be used when constructing the solution to the modified DARE. Suppose this solution is called M_{ni}. Using this solution of the DARE to construct the feedback gain matrix given by

$$K_{ni} = (R_n + B_n^T M_{ni} B_n)^{-1} B_n^T M_{ni} A_n. \tag{33}$$

results in the desired eigenvalues in Λ_1 being retained. All the remaining eigenvalues in Λ_2 will be shifted by the factor $\frac{1}{p^2}$.

On the other hand, the stable solution to the modified DARE, M_{ns}, is constructed from the eigenvectors associated with the stable eigenvalues in the modified Hamiltonian matrix. This means the stable solution, M_{ns}, will be constructed from eigenvectors associated with $\frac{1}{p\lambda_j}$, when $p\lambda_j$ lies outside the unit circle. Thus, the resulting closed loop eigenvalue found in $A - BK_{ns}$, with

$$K_{ns} = (R_n + B_n^T M_{ns} B_n)^{-1} B_n^T M_{ns} A_n \ , \tag{34}$$

will be located at $(\frac{1}{p})(\frac{1}{p\lambda_j})$. As can be seen, this eigenvalue is not equal to λ_j.

The Hamiltonian matrix in each of these theorems above relies on the fact that the system matrix, A, has an inverse. If this is not the case, i.e., some of the open-loop eigenvalues are at the origin, then this collection of eigenvalues at the origin can be factored out of the system matrix before applying the above theorems.

When complex eigenvalues are encountered in the application of the above theorems, their complex eigenvectors, if necessary, can be converted into real vectors when forming the DARE solutions $M_i = Y_i X_i^{-1}$. This conversion is accomplished by converting one of the two complex conjugate eigenvectors into two real vectors, where the two real vectors consist of the real part and the imaginary part of the chosen complex eigenvector.

III. REGIONAL PLACEMENT WITH POLE-SHIFTING

To allow the above pole-shifting procedure to be even more useful, this section discusses how the above pole-shifting strategy can be combined with the regional pole-placement technique [11]. Very briefly, the regional pole placement technique allows the closed-loop eigenvalues, associated with a modified LQ problem, to be located inside a circular region of radius α and centered on the real-axis at a location β.

Accomplishing this regional pole-placement procedure is similar to the pole-shifting procedure, in the sense that, both techniques require a modified LQ regulator problem be solved. For the regional placement technique, the modifications amount to changing the system matrix and the input matrix while leaving the state weighting matrix and the input weighting matrix unchanged. The modified system matrix and input matrix, in terms of the nominal system and input matrices, A and B, are as shown below.

$$A_r = (\frac{1}{\alpha})[A - \beta I] \tag{35}$$

$$B_r = (\frac{1}{\alpha})B \tag{36}$$

The stable solution to the following DARE is then used to create the necessary full-state feedback gain matrix.

$$M_r = Q + A_r^T M_r A_r - A_r^T M_r B_r (R + B_r^T M_r B_r)^{-1} B_r^T M_r A_r \tag{37}$$

The full-state feedback gain matrix, K_r, that places the closed-loop poles inside the circular region of radius α with its center at β on the real axis can be calculated using the stable solution in Eq. (37), M_{rs}.

$$K_r = (R + B_r^T M_{rs} B_r)^{-1} B_r^T M_{rs} A_r \tag{38}$$

The closed-loop system matrix is found by applying K_r to the original system matrix and input matrix.

$$F_r = A - BK_r \tag{39}$$

Once K_r is determined, the pole-shifting technique can then be applied to relocate all or some of the closed-loop eigenvalues that have been positioned in the circular region given by the pair (α,β). To apply the pole-shifting technique, simply assign the appropriate matrices as the nominal matrices in Theorem 1 or 2. In particular, define the system matrix as F_r, use the nominal input matrix B, zero out the state weighting matrix, $Q = 0_{n \times n}$, and select the input weighting to be any positive definite matrix, e.g., $R = I$. Suppose that the pole-shifting has been accomplished and has given rise to a state-feedback gain matrix K_{tmp}, with the desired eigenvalues corresponding to the eigenvalues of $F_r - BK_{tmp}$. The final, full-state feedback gain that can be applied to the original (A,B) system pair is obtained by combining the gains from the regional placement and the pole-shifting technique.

$$K_f = K_r + K_{tmp} \tag{40}$$

K_f is the desired, full-state feedback gain matrix, with $A - BK_f$ containing the desired eigenvalues. As can be seen from the above development, the pole-shifting technique can be utilized to modify any nominal full-state feedback gain results that have been generated from some particular design environment.

IV. ILLUSTRATIVE EXAMPLES

This section has been included to help illustrate two of the design strategies that were presented in Section II. Both systems in these examples have previously appeared in the literature.

IV.A. Chemical Reactor Design

The first example helps illustrate the movement of all nominal eigenvalues by the same amount. This example is a two-input, fourth-order chemical reactor model [7,9]. The objective of this design is to determine a full-state feedback gain matrix that provides an acceptable initial condition response and is robust to uncertainty in the system matrices.

A discrete-time description of this system is given below.

$$x(k+1) = Ax(k) + Bu(k) \tag{41}$$

where

$$A = \begin{bmatrix} 1.1782 & 0.0015 & 0.5116 & -.4033 \\ -.0515 & 0.6619 & -.0110 & 0.0613 \\ 0.0762 & 0.3351 & 0.5606 & 0.3824 \\ -.0006 & 0.3353 & 0.0893 & 0.8494 \end{bmatrix} \tag{42}$$

and

$$B = \begin{bmatrix} 0.0045 & -.0876 \\ 0.4672 & 0.0012 \\ 0.2132 & -.2353 \\ 0.2131 & -.0161 \end{bmatrix} . \tag{43}$$

The open-loop eigenvalues of A are

$$1.2203 \quad 0.6031 \quad 1.0064 \quad 0.4204 \ .$$

Using a state-weighting matrix of $Q = diag([1, 1000, 1000, 1000])$ and an input-weighting matrix of $R = I_{2\times 2}$, the nominal, closed-loop eigenvalues are given by

$$0.8727 \quad 0.5738 \quad 0.0088 \quad 0.0031 \ .$$

Via an initial condition response, with $x(0) = [1, 0, 0, 0]^T$, the transient performance was concluded to be too fast and overly demanding. As an attempt to improve the response, the nominal closed-loop eigenvalues were actually moved away from the origin by a factor that placed the slowest eigenvalue at $\lambda = 0.95$. Table I contains the m-file script that was used to perform this design. The m-file functions called from the script in Table I are included in Section VII. The transient response for this example is given in Figure 2. As seen in Figure 2, the deviation in the four state components for this response are at least as good as the state response obtained using the robust, pole-placement technique in [9].

Depending on the robustness characteristics of the resulting design, it may be desirable to use these closed-loop eigenvalues in an approach like in [7,9]. Again, if this LQ design is being used to suggest nominal stabilizing controllers for more sophisticated controller designs, then this resulting design may be more than adequate as a nominal design.

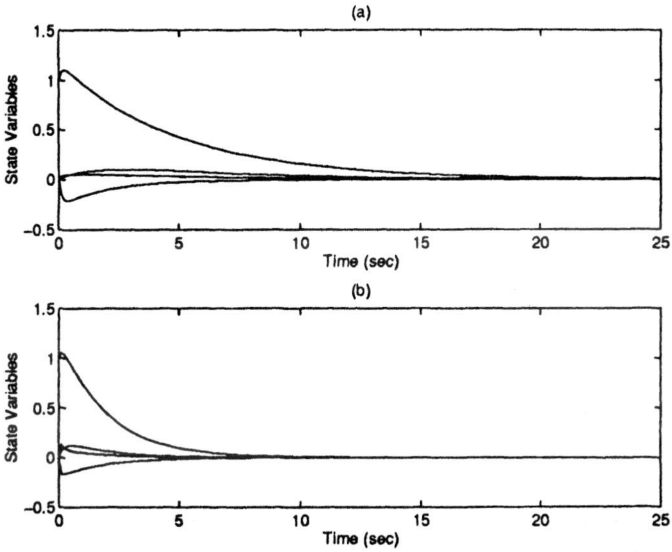

Figure 2: Chemical reactor initial condition response: (a) Gain matrix associated with robust pole-location technique; (b) Gain matrix associated with pole-shifting technique.

Table I: Script m-file for chemical reactor design.

```
% File – react.m –
% Script file to perform pole-shifting on reactor.
% Model from Vaccaro, "Digital Control: ... ", pp. 394-397.
%
% Enter the continuous-time system and input matrices.
ac=[1.3800,-0.2077,6.7150,-5.6760;
-0.5814,-4.290,0.0,0.6750;
1.0670, 4.2730,-6.6540,5.8930;
0.0480, 4.2730, 1.3430,-2.1040];
bc=[0.0,0.0;
5.6790, 0.0;
1.1360,-3.1460;
1.1360, 0.0];
ts=0.1;
[a,b]=c2d(ac,bc,ts), % Discrete sys. w/ 10.0 Hz sampling.
ea=eig(a), % Display open-loop eigenvalues for discrete sys.
%
lopt=[-0.1746,0.0669,-0.1611,0.1672;
-1.0794,0.0568,-0.7374,0.2864],% Vaccaro gain.
fopt=a-b*lopt;
ef=eig(fopt), % Closed-loop eigenvalues using Vaccaro gain.
% Calculate feedback gain using pole-shifting.
q=diag([1,1000,1000,1000]),r=eye(2), % Nominal weightings.
dham; % Solve the nominal LQ problem.
ff=a-b*ks;eff=eig(ff), % Display the nominal closed-loop eigenvalues.
p=0.9584527; % Shift all eigenvalues to slow down response.
dhamp; % Calculate the feedback gain for the desired eigenvalues.
ffp=a-b*kps;effp=eig(ffp), % Display the desired eigenvalues.
x0=[1;0;0;0];
t=0:0.1:25;
u=zeros(length(t),2);
c=[1,0,0,0];d=[0,0];
[yv,xv]=dlsim(fopt,b,c,d,u,x0); % I.C. Response for Vaccaro gain.
[yd,xd]=dlsim(ffp,b,c,d,u,x0); % I.C. Response for pole-shifting gain.
subplot(211);
plot(t,xv); % State variables with Vaccaro gain.
```

```
title('(a)')
ylabel('State Variables')
xlabel('Time (sec)')
subplot(212);
plot(t,xd); % State variables with pole-shifting gain.
title('(b)')
ylabel('State Variables')
xlabel('Time (sec)')
```

This first example has also illustrated the fact that the nominal eigenvalues can actually by moved away from the origin. The restriction that must be observed when expanding eigenvalues is to make sure that the magnitudes of these eigenvalues being moved remain less than one when expanded by $\frac{1}{p}$.

IV.B. Two Mass Design

The second example is from a benchmark problem for robust control design [16]. The system consists of two masses coupled by a spring, with the input force applied to the first mass and the output measurement obtained from the position of the second mass. The equations of motion are given by

$$\dot{x} = \begin{bmatrix} 0 & 0 & 1 & 0 \\ 0 & 0 & 0 & 1 \\ -k/m_1 & k/m_1 & 0 & 0 \\ k/m_2 & -k/m_2 & 0 & 0 \end{bmatrix} x + \begin{bmatrix} 0 \\ 0 \\ 1/m_1 \\ 0 \end{bmatrix} u \tag{44}$$

$$y = [0, 1, 0, 0]x \ . \tag{45}$$

In this system, x_1 and x_2 are the position of masses one and two, respectively. x_3 and x_4 are the respective velocities. The output, y, corresponds to the position of the second mass and the force input, u, is applied to the first mass.

For illustration purposes, the control objective is to design a linear, time-invariant controller (full-state, estimator feedback), such that the settling time of the system, due to an impulse, is 20 seconds for all values of the spring constant, k, between 0.5 and 2.0. Table II contains the m-file script associated with the design. The m-file functions in Table II that are not a part of MATLAB or the Control System Design Toolbox are included in Section VII.

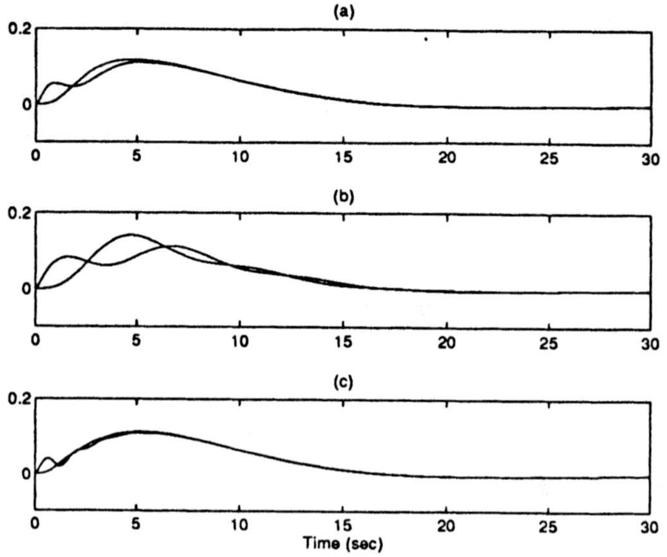

Figure 3: Two-mass positional responses: (a) Nominal system with $k = 1.0$; (b) Perturbed system with $k = 0.5$; and (c) Perturbed system with $k = 2.0$.

For this particular system, moving all of the nominal, state-feedback controller eigenvalues the same amount is ineffective at producing a controller that stabilizes the three different systems. Herein, the three different systems are associated with the systems that result when $k = 0.5$, $k = 1.0$, and $k = 2.0$, with a sampling period of $T_s = 0.1$. An acceptable design was achieved by moving only the slowest nominal eigenvalue closer to the origin by a factor of $(1/p^2)$, with $p = 1.05$. This value of p corresponds to a contraction of the slowest eigenvalue by a factor of approximately 0.907.

As shown in Table II, the full-order observer eigenvalues were found by contracting a nominal set of eigenvalues by a factor of 0.20. Thus, the observer eigenvalues were about five times as fast as the controller eigenvalues. The pole-shifting technique in Theorem 1 was used to design the observer gain. The transient response, associated with the two mass positions, when an impulse is applied at mass one, is shown in Figure 3.

All three different systems, corresponding to $k = 0.5$, $k = 1.0$, and $k = 2.0$, exhibit acceptable behavior. If desired, more sophisticated designs can be continued from this baseline full-order controller design.

Table II: Script m-file for two-mass design.

```
% File – two_mass.m –
%
% This file uses the system from J. Guid. Contr. Dyn.,
% Vol. 15, No. 5, pp. 1057-1059, 1992.
%
% Read in the dynamic system descriptions for the nominal
% matrices.
mass_spring
% Convert these systems into a discrete-time representation.
ts=0.10;
ac=a;
bc=b;
[a,b]=c2d(ac,bc,ts)
% Store these matrices for later recall.
atemp=a;
btemp=b;
% Read in the perturbed matrices.
mass_spring_p
%Perform the design on the design matrices (addes,bddes).
% These matrices are defined in 'mass_spring_p.m'
q=0.01*eye(4);
r=1;
ahat=addes;bhat=bddes;
ko=dlqr(ahat,bhat,q,r)
qtemp=q;
ff=a-b*ko;
% Shift these eigenvalues
q=0*eye(4);r=1;
ftmp=ff;b=b;
kt=ko;
dhind
ko=ko+ktmp
ff=atemp-b*ko;
eff=eig(ff)
efl=eig(adpl-bdpl*ko)
efu=eig(adpu-bdpu*ko)
pause
```

```
% Design the observer!
%
q=qtemp;
cm=[0,1,0,0];
l1t=dlqr(ahat',cm',q,r);
fonom=addes-l1t'*cm;
efonom=eig(fonom)
pause
% Move these nominal observer eigenvalues closer to the origin.
% Let (1/p^2)=1/5=0.2. ==> p=2.236
%
% First create the nominal lq solution (q=0*eye(4), r=1).
q=0*eye(4);r=1;
a=fonom';b=cm';
dham
p=2.236068, % A factor of 0.2 reduction.
dhamp
l=l1t'+kps';
fo=addes-l*cm;efo=eig(fo) % Check the eigenvalues.
% Create augmented controlled system.
t=0:0.1:30;
[m,n]=size(t);
a=atemp;b=btemp;
cbig=[eye(2),0*ones(2,6)];dbig=[0;0];
abig=[a,-b*ko;l*cm,addes-bddes*ko-l*cm];bbig=[b;0*ones(4,1)];
[ybig,xbig]=dimpulse(abig,bbig,cbig,dbig,1,n);
subplot(311)
plot(t,ybig)
title('(a)')
%
% Check the response of controller with lower limit system.
abigl=[adpl,-bdpl*ko;l*cm,addes-bddes*ko-l*cm];
bbigl=[bdpl;0*ones(4,1)];
[ybigl,xbigl]=dimpulse(abigl,bbigl,cbig,dbig,1,n);
subplot(312)
plot(t,ybigl)
title('(b)')
%
% Check the response of controller with upper limit system.
abigu=[adpu,-bdpu*ko;l*cm,addes-bddes*ko-l*cm];
bbigu=[bdpu;0*ones(4,1)];
[ybigu,xbigu]=dimpulse(abigu,bbigu,cbig,dbig,1,n);
subplot(313)
plot(t,ybigu)
title('(c)')
```

V. CONCLUSIONS

This chapter has presented a technique to increase the damping in discrete-time closed-loop systems by modifying a nominal LQ performance criterion. These modifications are not complex or difficult to implement after the nominal performance criterion is specified. In addition, a pole shifting technique that can be used to independently position the closed-loop eigenvalues has been presented. This independent pole-shifting technique can be applied to any nominal set of pole locations that might arise in any full-state feedback design strategy.

VI. REFERENCES

1. B.D.O. Anderson and J.B. Moore, *Optimal Control: Linear Quadratic Methods*, Prentice Hall, Englewood Cliffs, N.J., 1990.
2. P. Dorato, C. Abdallah, and V. Cerone, *Linear-Quadratic Control: An Introduction*, Prentice Hall, Englewood Cliffs, N.J., 1995.
3. H. Kwakernaak and R. Sivan, *Linear Optimal Control Systems*, Wiley-Interscience, New York, 1972.
4. M.G. Safonov and M. Athans, "A Multiloop Generalization of the Circle Criterion for Stability Margin Analysis," *IEEE Trans. Automat. Contr.*, AC-26, No. 2, pp. 415-422, 1981.
5. J.C. Doyle, B.A. Francis, and A.R. Tannenbaum, *Feedback Control Theory*, Macmillan Publishing Company, New York, 1992.
6. M. Vidyasagar, *Control System Synthesis: A Factorization Approach*, The MIT Press, Cambridge, Massachusetts, 1987.
7. J. Kautsky, N.K. Nichols, and P. Van Dooren, "Robust Pole Assignment in Linear State Feedback," *Int. J. Contr.*, Vol. 41, pp. 1129-1155, 1985.
8. B.C. Moore, "On the Flexibility Offered by State Feedback in Multivariable Systems Beyond Closed-Loop Eigenvalue Assignment," *IEEE Trans. Automat. Contr.*, Vol. AC-21, No. 5, October 1976, pp. 689-692.
9. R.J. Vaccaro, *Digital Control: A State-Space Approach*, McGraw-Hill, Inc., New York, 1995.

10. B.A. White, "Eigenstructure Assignment for Aerospace Applications," in *MATLAB Toolboxes and Applications for Control*, (A.J. Chipperfield and P.J. Fleming, ed.), pp. 179-204, Peter Peregrinus Ltd., United Kingdom, 1993.

11. T. Lee and S. Lee, "Discrete Optimal Control with Eigenvalue Assigned Inside a Circular Region," *IEEE Trans. Automat. Contr.*, Vol. AC-31, No. 10, October 1986, pp. 958-962.

12. H.S. Tharp, "Optimal Pole-Placement in Discrete Systems," *IEEE Trans. Automat. Contr.*, Vol. 37,No. 5, pp. 645-648, 1992.

13. K. Ogata, *Discrete-Time Control Systems,* Prentice-Hall, Inc., Englewood Cliffs, New Jersey, 1987.

14. K. Martensson, "On the Matrix Riccati Equation," *Inform. Sci.*, Vol. 3, No. 1, 1971, pp. 17-50.

15. G.F. Franklin, J.D. Powell, and M.L. Workman, *Digital Control of Dynamic Systems*, Addison-Wesley, Reading, Massachusetts, 1990.

16. B. Wie and D.S. Bernstein, "Benchmark Problems for Robust Control Design," *J. Guid., Contr., Dyn.*, Vol. 15, No. 5, pp. 1057-1059, 1992.

VII. EIGENVALUE MOVEMENT ROUTINES

This section contains MATLAB m-files that can be used to relocate closed-loop eigenvalues. Table III's m-file performs a nominal LQ controller design by using a Hamiltonian matrix approach. The m-file in Table IV relies on the results from Table III and must be executed after implementing the code in Table III. Before executing the m-file in Table IV, the scaling factor p must be defined. This m-file in Table IV allows all of the nominal eigenvalues to be relocated by a multiplicative factor of $\frac{1}{p^2}$. Table V contains an m-file that can be used to relocate eigenvalues individually. Before executing 'dhind', however, a nominal system must exist within MATLAB. The names associated with the nominal system matrices are indicated in the comments inside the m-file 'dhind'. Tables VI and VII contain the data used in the two-mass design in Section IV.

Table III: Function m-file for nominal LQ design.

```
% File – dham.m –
% Discrete-time Hamiltonian Matrix Creator.
%
% Matrices 'a', 'b', 'q', and 'r' must exist.
[n,m]=size(a);
brb=b/r*b';
ait=inv(a');
ba=brb*ait;
h=[a+ba*q,-ba;-ait*q,ait];
[xh,eh]=eig(h);
[tmp,indx]=sort(abs(diag(eh)));
xs=xh(1:n,indx(1:n));
ys=xh(n+1:2*n,indx(1:n));
xu=xh(1:n,indx(n+1:2*n));
yu=xh(n+1:2*n,indx(n+1:2*n));
ms=real(ys/xs);
mu=real(yu/xu);
ks=(r+b'*ms*b)\b'*ms*a;
```

Table IV: Function m-file to shift all eigenvalues.

```
%File – dhamp.m –
% Discrete-time Hamiltonian Maker for Perturbed System.
%
% (The scalar p must be predefined.)
% (This routine follows the 'dham' routine.)
%
ap=p*a;
bp=p*b;
p2m1=(p^2 - 1);
dr=p2m1*r;
rp=r+dr;
dq=p2m1*(q-mu);
qpp=q+dq;
brbp=bp/rp*bp';
apit=inv(ap');
bap=brbp*apit;
hp=[ap+bap*qpp,-bap;-apit*qpp,apit];
[xhp,ehp]=eig(hp);
% Find the stable roots in ehp.
[tmpp,indxp]=sort(abs(diag(ehp)));
xsp=xhp(1:n,indxp(1:n));
ysp=xhp(n+1:2*n,indxp(1:n));
mps=real(ysp/xsp);
kps=(rp+bp'*mps*bp)\ bp'*mps*ap;
```

Table V: Individual eigenvalue routine.

```
% File – dhind.m –
% Individual eigenvalue movement using an LQ
% Hamiltonian Matrix technique.
%
% A nominal design should have already been found.
% This nominal design should have as its system matrix,
% ftmp=a-b*kt.
% 'dhind.m' will update 'ftmp' and 'kt'.
%
% The matrices 'ftmp', 'b', and 'kt' must already be defined.
%
% Create the nominal Hamiltonian.
%
[n,m]=size(ftmp);
[rb,cb]=size(b);
q=0*eye(n);
r=eye(cb);
brb=b/r*b';
fit=inv(ftmp');
ba=brb*fit;
h=[ftmp+ba*q,-ba;-fit*q,fit];
[xh,eh]=eig(h);
[tmp,indx]=sort(abs(diag(eh)));
deh=diag(eh);
lam=deh(indx(1:n));
num=1:n;
dlam=[num',lam,tmp(1:n)];
disp(' Number Eigenvalue Magnitude')
disp(dlam)
disp(' ')
disp('Which eigenvalues are to be moved?')
disp(' ')
disp('Enter a row vector containing the number(s)')
mov=input('associated with the eigenvalue(s) to move >');
movp=((2*n)+1)-mov;
```

```
% Retain all of the eigenvalues not listed in 'mov'.
ret=num;
ret(:,mov)=[ ];
% Create the index vector that selects the correct columns from 'xh'.
iindx=[indx(movp);indx(ret)];
xi=xh(1:n,iindx);
yi=xh(n+1:2*n,iindx);
mu=real(yi/xi);
disp(' ')
%
disp('Distance Factor = (1/p^2)')
p=input('Enter the distance to move the eigenvalues, p = ');
%
% Find the eigenvalues that will move outside the unit circle.
% Store these eigenvalues in the vector 'tchkp'
tmp=find(p*abs(deh(indx(ret)))>1.0);
% 'tmp' contains element location of the e-val.
tindx=indx(ret(tmp)); % 'tindx' indexes these e-val. relative to 'eh'.
tchk=deh(tindx); % 'tchk' contains the actual values of these e-val.
tchkp=p*tchk;
[sizt,ct]=size(tchkp);
%
ap=p*ftmp;
bp=p*b;
p2m1=(p^2-1);
dr=p2m1*r;
rp=r+dr;
dq=p2m1*(q-mu);
qpp=q+dq;
brbp=bp/rp*bp';
apit=inv(ap');
bap=brbp*apit;
hp=[ap+bap*qpp,-bap;-apit*qpp,apit];
[xhp,ehp]=eig(hp);
```

```
% Find the desired roots in ehp.
[tmpp,indxp]=sort(abs(diag(ehp)));
dehp=diag(ehp);
rindxp=indxp(1:n);
sdehp=sort(dehp);
ret2=[ ];
for i=1:sizt;...
tmp2=find((abs(dehp-tchkp(i)))<0.001);...
tmp3=find((abs(sdehp-(1/tchkp(i))))<0.001);...
ret2=[ret2;tmp2];...
rindxp(tmp3,1)=0;...
end
tmp4=find(rindxp==0);
rindxp(tmp4,:)=[ ];
iiindx=[rindxp;ret2];
xsp=xhp(1:n,iiindx);
ysp=xhp(n+1:2*n,iiindx);
%
mps=real(ysp/xsp);
ktmp=(rp+bp'*mps*bp)\bp'*mps*ap;
% Update closed-loop system matrix and feedback gain matrix.
% This allows this routine to be repeated to modify other
% eigenvalue locations.
%
ftmp=ftmp-b*ktmp;
kt=kt+ktmp;
% a-b*kt will contain the desired closed-loop eigenvalues.
```

Table VI: Nominal mass-spring system.

```
% ************ mass_spring.m *************
%
% Nominal system model for the mass-spring system
% from J. Guidance, Control, Dynamics, Vol. 15, No. 5,
% 1992, pp. 1057-1059.
%
a=[0,0,1,0;0,0,0,1;-1,1,0,0;1,-1,0,0];
b=[0;0;1;0];
```

Table VII: Perturbed mass-spring systems.

```
% ********* mass_spring_pm ***********
%
% Perturbed mass spring system from J.G,C,D. 1992.
% The sampling period, ts, must be defined.
% The cont.-time input matrix, 'bc' must exist.
%
apl=[0,0,1,0;0,0,0,1;-.5,.5,0,0;.5,-.5,0,0]
apu=[0,0,1,0;0,0,0,1;-2,2,0,0;2,-2,0,0]
[adpl,bdpl]=c2d(apl,bc,ts)
[adpu,bdpu]=c2d(apu,bc,ts)
% Try the following matrices during the design.
ades=[0,0,1,0;0,0,0,1;-1.00,1.00,0,0;1.00,-1.00,0,0]
[addes,bddes]=c2d(ades,bc,ts)
```

On Bounds for the Solution of the Riccati Equation for Discrete-Time Control Systems

Nicholas Komaroff

Department of Electrical and Computer Engineering
The University of Queensland
Australia 4072

I. INTRODUCTION

This chapter is on the evolution, state and research directions of bounds for the solution of the discrete algebraic Riccati equation (DARE). The background and generalities of bounds are outlined. A collection and classification of mathematical tools or inequalities that have been used to derive bounds is presented for the first time. DARE bounds that have been published in the literature are summarized. From this list trends are discerned. Examples and discussions illustrate research directions.

The DARE [1]-[2] is

$$P = A'PA - A'PB(I + B'PB)^{-1}B'PA + Q, \; Q = Q' \geq 0 \,. \qquad (1.1)$$

Here matrices A, P, Q $\in R^{nxn}$, B $\in R^{nxr}$, I = unit matrix, and (′) ≥ denote transpose and positive semi-definiteness respectively. It is assumed that (A,B) is stabilizable and (A,Q) is detectable. Under these conditions, the solution P > 0 i.e., P is positive definite.

The version

$$P = A'(P^{1} + R)^{-1}A + Q, \; R = BB' \tag{1.2}$$

of (1.1) is usually employed to obtain bounds on the solution P. Equation (1.1) is converted to (1.2) by an application of the matrix identity

$$(X^{-1} + YZ)^{-1} = X - XY(I + ZXY)^{-1}ZX \;. \tag{1.3}$$

The DARE (1.1) plays a fundamental role in a variety of engineering fields such as system theory, signal processing and control theory. More specifically, it is central in discrete-time optimal control, filter design, stability analysis and transient behaviour evaluation. It is indispensable in the design and analysis of linear dynamical systems.

The organization of the chapter is as follows.

Section II is on motivation and reasons for obtaining bounds, their nature and mathematical content, notation, and quality criteria. Section III lists and classifies the inequalities that have been employed to bound the solution of the DARE. Published bounds for the DARE are in section IV. Examples that derive bounds are in section V. The overall status of bounds and research directions are discussed here, as well as in other sections. A conclusion is in section VI.

II. ON APPROXIMATE SOLUTIONS

This section discusses approximate solutions or bounds for the DARE under four headings. Subsection A deals with motivations and reasons, B with notation, C with formulation and expressions and D with their quality and usefulness.

A. Motivation for Approximations

The computation of the solution P of (1.1) is not always simple. For instance, it becomes a problem of some difficulty when the dimension n of the matrices is high. To attest to this is the quantity of literature that proposes computational schemes that ideally are simple, stable, and not expensive in terms of computer time and memory storage.

It would therefore appear that estimates of the solution P that do not require heavy computational effort, could be desirable. Indeed they are for two reasons. Whilst not exact solutions, they are articulated in terms of the characteristics of the independent matrices A, Q, R of (1.1). This relationship throws direct light on the dynamics and performance of the represented system. The second application of approximations is to facilitate numerical computations. This is because the efficiency of evaluating algorithms depends and often strongly, on how close the algorithm's starting value P_0 is to the final solution P. As expected, the smaller the magnitude of the difference $P-P_0$ is, the less computer time is required to reach an acceptable solution.

B. Notation

The real numbers we employ are often arranged in nonascending order of magnitude.

Thus for the set x_i, i=1,2,...n of real numbers

$$x_1 \geq x_2 \geq \dots \geq x_n. \tag{2.1}$$

The subscript i is reserved to indicate this ordering. It applies to Re λ_i (X) the real parts of eigenvalues of a matrix X, to the magnitudes $|\lambda_i (X)|$, to the singular values $\sigma_i (X) = [\lambda_i (XX')]^{1/2}$ of X, and to $\delta_i(X)$ the real diagonal elements of X. Integer $n \geq 1$ is also the dimension of $n \times n$ matrices.

Note that Re λ_1 (X) [Re λ_n (X)] refer to the maximum [minimum] of the Re λ_i (X). Also, $\lambda_i (-X) = -\lambda_{n-i+1} (X)$ and $\lambda_i (X^{-1}) = \lambda^{-1}{}_{n-i+1} (X)$ for real eigenvalues.

Other subscripts j, k, where j,k = 1,2,...,n do not indicate magnitude ordering. They can refer to an arbitrary member x_j of a sequence of numbers, or x_j is the component in the j-th position of a vector. These numbers can represent matrix eigenvalues or elements.

The trace of a matrix X is tr (X). Its determinant is $|X|$. $X \geq (>)\ 0$ means X is positive semidefinite (positive definite).

Many results for (1.1) are for $\Sigma_1^k \lambda_i (P)$ i.e., for summations of the k largest λ_i (P) including λ_1 (P) the maximal eigenvalue. Also used is the "reverse" numbering $\Sigma_1^k \lambda_{n-i+1}$ (P) for the k smallest λ_i (P) including λ_n (P) the minimal eigenvalue. Similarly, results exist for products $\Pi_1^k \lambda_i (P)$ and for $\Pi_1^k \lambda_{n-i+1} (P)$.

All matrices in (1.1) have real elements only. Only real-element matrices are considered in this chapter.

C. Nature of the Approximations

An approximation to the matrix P of (1.1), or an estimation of the matrix P, involves the size of P. The majority of the measures of the size or the extent of the solution P that have been presented are given by eigenvalues. Only the most recent results use matrices themselves as measurement of differences between estimates and the solution.

The eigenvalue technique is to obtain bounds, upper and lower, on the λ_i (P). These bounds as estimates of the solution should be as close as possible to the solution values.

The earliest results were for the extreme eigenvalues λ_1 (P) (maximum of the λ_i (P)) and λ_n (P) (minimum of the λ_i (P)). They provide extreme measures of P. The other λ_i (P) are assumed to be between these limits.

Next were developed bounds on functions of the λ_i (P). The first functions were tr(P) and $|P|$. These provide average and geometric-mean estimates of the λ_i (P). The average is greater than the geometric-mean value, as shown by the often-used arithmetic-mean geometric-mean inequality. They are the most useful single scalar results about eigenvalues.

Advancing in complexity and increasing in information are bounds for $\Sigma_1^k \lambda_i$ (P), summations of eigenvalues, and for $\Pi_1^k \lambda_i$ (P), products of eigenvalues, where k = 1,2,...,n. These, in general, have been derived, chronologically, after the bounds on tr(P) and $|P|$.

The most useful bounds but where very few results exist, are provided by matrix bounds. Here bounds on the eigenvalues λ_i (P) are replaced by bounds on P itself, given by matrix-valued functions of matrices. As an example, $P \geq Q$ is a well-known result for (1.1). Also, iteration solution schemes for (1.1) [3] can be regarded in this light. Assume P_j is calculated from P_{j-1}, starting with P_0 ($P_o > 0$), an initial estimate for P. The sequence is

$$P \leq \ldots \leq P_j \leq P_{j-1} \leq \ldots P_0, \; j = 1,2,\ldots$$

and as $j \rightarrow \infty$, convergence to P obtains. P_{j-1} is then an upper bound for P_j. The closer P_0 is to P, the better the scheme should work. The tightest (or sharpest) estimate P_0 as an upper bound on P, is what is required.

Because λ_i (P) $\leq$ (<) λ_i (P_0) follows from P $\leq$ (<) P_0, matrix bounds automatically contain the information in eigenvalue bounds. What is more, matrix bounds include eigenvectors as part of their makeup, a second and significant advantage over eigenvalue bounds.

D. Quality of Bounds

There are four criteria.

The first and the most important is tightness. How close is the bound to the actual solution? To illustrate we know that the solution $P > 0$ in (1.1). Thus an evaluation of $\lambda_i (P) > 0$ is not of value. However if one calculated lower bound is $\lambda_i (P) \geq b_1 > 0$ and a second is $\lambda_i(P) \geq b_2 > 0$, where $b_1 \geq b_2$, then the first bound is preferable: it is the tighter, or sharper. A similar statement applies to upper bounds. A combination of lower and upper bounds gives a range within which the solution lies.

Unfortunately it is usually not possible to compare two bounds. This is because of the mathematical complexity of most bound expressions. Also, the tightness of a given bound depends on the independent matrices A,R, Q in (1.1) i.e. small changes in A for instance, may result in large changes in the bound. An inequality has its "condition numbers".

The second criterion is quantity of computation. A bound that requires specialised or expensive-in-time algorithms for its evaluation may be too costly to evaluate. An example of a simple calculation is the trace of a matrix. A knowledge of numerical analysis aids in comparative judgement, and provides direction in the design of bounds.

The third factor influencing quality are the restrictions or side conditions that must be satisfied for the bound to be valid. Thus, some bounds for (1.1) require that λ_n (R) > 0, where λ_n (R) is the smallest-in-magnitude eigenvalue of R. Therefore the bound is not applicable if R is a singular matrix. The side condition R > 0 must be stipulated in order for the bound to be active.

The final factor affecting accuracy of the prediction of (or bound on) the solution is given by the number of independent variables involved. For example, a bound that is a function of but one of the n eigenvalues of a matrix contains less information than one that depends on all n eigenvalues. By this token a bound depending on the determinant of a matrix is preferable to one using but one extreme eigenvalue.

III. SUMMARY OF INEQUALITIES

This section presents inequalities that have been used to construct bounds on the solution of the DARE and other matrix equations. No such summary of inequalities exists in the literature.

The list cannot be complete and should serve only as a guide. There is a continuing search for inequalities that have not been used before, to obtain new approximations to the solution of the DARE. Therefore this list is no substitute for a study of the mathematical theory of inequalities, as found in [4] - [7].

There are three subsections A,B,C which contain algebraic, eigenvalue and matrix inequalities, respectively.

A. Algebraic Inequalities

The inequalities here refer to real numbers such as x_j, y_j, $j = 1,2,\dots n$. When ordered in magnitude, the subscript i is used - see (2.1).

One of the earliest results is the two variable version of the arithmetic - mean geometric - mean inequality. This was known in Greece more than two thousand years ago. It states $(x_1x_2)^{1/2} \le (x_1+x_2)/2$, for $x_1, x_2 \ge 0$. Its n-dimensional version is often used when the x_j are identified with matrix eigenvalues. For example, an upper bound for $(x_1+x_2)/2$ is automatically an upper bound for $(x_1x_2)^{1/2}$.

Inequalities exist for the summations $\sum_1^k x_i$ as well as for $\sum_1^k x_{n-i+1}$, $k = 1,2, \dots n$, and for similar orderings in products of the x_i.

Given the bound $\sum_1^k x_i \le \sum_1^k y_i$ (x_i, y_i being real numbers) the question is for what functions ϕ is $\sum_1^k \phi(x_i) \le \sum_1^k \phi(y_i)$? Answers to such and similar questions are provided in the references [4]-[7] already cited. The theory of convex and related functions, of vital importance in inequalities, is not developed here.

Probably the most important of all inequalities is the arithmetic- mean geometric-mean inequality.

Theorem 1[5]: Let x_j, j = 1,2,...,n be a set of nonnegative numbers. Then

$$\left(\prod_1^n x_j\right)^{1/n} \le \left(\sum_1^n x_j\right)/n. \tag{3.1}$$

There is strict inequality unless the x_j are all equal.
Associated with the above is the harmonic-mean geometric-mean inequality.

Theorem 2[5]: Let x_j, j=1,2,...,n be a set of positive numbers. Then

$$n/\left(\sum_1^n x_j^{-1}\right) \le \left(\prod_1^n x_j\right)^{1/n} \tag{3.2}$$

There is strict inequality unless the x_j are all equal.

Theorem 3 [5]: Let x_j, y_j , j=1,2,...,n be two sets of real numbers. Then

$$\left(\sum_1^n x_j y_j\right)^2 \le \left(\sum_1^n x_j^2\right)\left(\sum_1^n y_j^2\right) \tag{3.3}$$

There is equality if and only if the x_j,, y_j are proportional.
This is Cauchy's Inequality.

Theorem 4[4],[5]: Let x_i, y_i be two sets of real numbers. Then

$$(1/n)\sum_{1}^{n} x_{n-i+1}y_i \le (1/n)\left(\sum_{1}^{n} x_i\right)(1/n)\sum_{1}^{n} y_i \le (1/n)\sum_{1}^{n} x_i y_i . \tag{3.4}$$

This is Chebyshev's Inequality.

Remark 3.1: If more than two sequences of numbers are considered in Chebyshev's Inequality, then all numbers must be restricted to be nonnegative.

Theorem 5 [6,p.142]: Let x_i, y_i be two sets of nonnegative numbers. Then

$$\prod_{1}^{n}(x_i+y_i) \le \prod_{1}^{n}(x_j+y_j) \le \prod_{1}^{n}(x_i+y_{n-i+1}). \tag{3.5}$$

Remark 3.2: The index j in (x_j+y_j) means that a random selection of x_j, y_j is made; subscript i indicates ordered numbers are used in the terms (x_i+y_i) and (x_i+y_{n-i+1}).

Theorem 6 [6, p.95]: Let x_i, y_i be two sets of real numbers, such that for k=1,2,...n

$$\sum_{1}^{k} x_i \le \sum_{1}^{k} y_i \tag{3.6}$$

and let u_i be nonnegative real numbers. Then

$$\sum_{1}^{k} u_i x_i \le \sum_{1}^{k} u_i y_i . \tag{3.7}$$

Theorem 7: Let x_i, y_i be two sets of real numbers, such that for k = 1,2,...n

$$\sum_{1}^{k} x_{n-i+1} \ge \sum_{1}^{k} y_{n-i+1} \tag{3.8}$$

and let u_i be nonnegative real numbers. Then

$$\sum_1^k u_i x_{n-i+1} \geq \sum_1^k u_i y_{n-i+1} . \tag{3.9}$$

Proof: Inserting negative signs, (3.8) becomes :

$$\sum_1^k -x_{n-i+1} \leq \sum_1^k -y_{n-i+1} \tag{3.10}$$

where each term is now ordered as in (3.6). Use of (3.7) and removal of the negative signs produces (3.9).

Theorem 7 has not appeared in the literature.

Theorem 8 [6,pp.117-118]: Let x_i, y_i be two sets of real nonnegative numbers such that for k = 1,2,...n

$$\sum_1^k x_{n-i+1} \geq \sum_1^k y_{n-i+1} . \tag{3.11}$$

Then

$$\prod_1^k x_{n-i+1} \geq \prod_1^k y_{n-i+1} . \tag{3.12}$$

and

$$\sum_1^k x_{n-i+1}^{-1} \leq \sum_1^k y_{n-i+1}^{-1} . \tag{3.13}$$

Theorem 9[8]: Let x_i, y_i be two sets of real nonnegative numbers such that

$$x_1 \le y_1,\ x_1x_2 \le y_1y_2,\ \ldots,\ x_1 \ldots x_n \le y_1 \ldots y_n \tag{3.14}$$

then

$$\sum_1^k x_i^s \le \sum_1^k y_i^s, \quad k = 1,2,\ldots n \tag{3.15}$$

for any real exponent s > 0.

Theorem 10 [4,p.261]: Let x_j, y_j, j=1,2,...,n be two sets of real numbers. Then

$$\sum_1^n x_{n-i+1}y_i \le \sum_1^n x_j y_j \le \sum_1^n x_i y_i \ . \tag{3.16}$$

Theorem 11[9]: Let x_i, y_i be two sets of nonnegative real numbers. Then for k=1,2,...n

$$\sum_1^k x_i y_{n-i+1} \le \sum_1^k x_j y_j \le \sum_1^k x_i y_i . \tag{3.17}$$

Theorem 12[10]: Let x_i, y_i be two sets of real numbers with $x_i \ge 0$. Then for k = 1,2,...n

$$\sum_1^k x_i y_j \le \sum_1^k x_i y_i \ . \tag{3.18}$$

Note: The left hand side terms cannot be changed to x_j y_i.

Theorem 13[10]: Let x_i y_i be two sets of real numbers with $x_i \ge 0$. Then for k=1,2,...n

$$\sum_{1}^{k} x_i y_{n-i+1} \le \sum_{1}^{k} x_j y_j \,. \tag{3.19}$$

Remark 3.2: Theorem 10 applies to sign-indefinite numbers x_i y_i. If the single restriction $x_i \geq 0$ is made, the simultaneous inequalities of Theorems 12 and 13 exist. With the double restriction x_i, $y_i \geq 0$ are the simultaneous inequalities of Theorem 11.

B. Eigenvalue Inequalities

Inequalities between the eigenvalues and elements of related matrices are presented. Firstly relationships between the diagonal elements $\delta_i(X)$ and eigenvalues $\lambda_i(X)$ of a matrix $X \in R^{nxn}$ are shown. There are both summation and product expressions. Secondly, inequalities between $\lambda_i(X)$ and the eigenvalues of matrix-valued functions such as $X+X'$ and XX' are given. Finally, summation and product inequalities between $\lambda_i(X)$, $\lambda_i(Y)$ and λ_i $(X+Y)$, λ_i (XY) are listed - $X,Y \in R^{nxn}$. Many results automatically extend to more than two matrices.

The inequalities listed have been applied to discover how the $\lambda_i(P)$ are related to the eigenvalues of A,Q,R of (1.1).

Remark 3.3: Theorems 14-15 link diagonal elements $\delta_i(X)$ with $\lambda_i(X)$ for a symmetric matrix X.

Theorem 14[6,p.218]: Let matrix $X \in R^{nxn}$ be symmetric . Its diagonal elements are $\delta_i(X)$. Then for $k = 1,2,\ldots n$

$$\sum_1^k \delta_i(X) \leq \sum_1^k \lambda_i(X) \tag{3.20}$$

with equality when k = n. Because of this equality

$$\sum_1^k \delta_{n-i+1}(X) \geq \sum_1^k \lambda_{n-i+1}(X) \ . \tag{3.21}$$

Corollary 14.1 :

$$\sum_1^k \lambda_{n-i+1}(X) \leq \sum_1^k \delta_{n-i+1}(X) \leq \frac{k}{n}tr(X) \leq \sum_1^k \delta_i(X) \leq \sum_1^k \lambda_i(X) \ . \tag{3.22}$$

Theorem 15 [6,p.223]: Let matrix $X \in R^{nxn}$, and $X \geq 0$. Then for k=1,2,...,n

$$\prod_1^k \delta_{n-i+1}(X) \geq \prod_1^k \lambda_{n-i+1}(X) \tag{3.23}$$

When k=n, (3.23) is called Hadamard's inequality.

Note: To prove (3.23) apply result (3.12) on (3.21), where all elements are nonnegative, because $X \geq 0$.

Theorem 16 [6,p.498]: Let matrix $X \in R^{nxn}$ be symmetric. Then

$$X = T'\Lambda T \tag{3.24}$$

where T is orthogonal i.e. $T'T=I$ and $\Lambda = \text{diag.}\ (\lambda_1, \lambda_2, \ldots,\lambda_n)$ where the λ_i are the real characteristic roots of X, ordered $\lambda_1 \geq \lambda_2 \geq \ldots \geq \lambda_n$.

Remark 3.4: Theorems 17-22 relate the $\lambda_i\ (X)$ with $\sigma_i(X)$ and $\lambda_i(X+X')$.

Theorem 17[8]: Let matrix $X \in R^{nxn}$. Then for k = 1,2,...,n

$$\sum_1^k |\lambda_i(X)| \leq \sum_1^k \sigma_i(X) \tag{3.25}$$

and

$$\sum_1^k |\lambda_i(X)|^2 \leq \sum_1^k \sigma_i^2(X) . \tag{3.26}$$

Theorem 18[8]: Let matrix $X \in R^{nxn}$. Then for k=1,2,...n

$$\prod_1^k |\lambda_i(X)| \leq \prod_1^k \sigma_i(X) \tag{3.27}$$

with equality when k = n. Because of this equality

$$\prod_1^k |\lambda_{n-i+1}(X)| \geq \prod_1^k \sigma_{n-i+1}(X) . \tag{3.28}$$

Theorem 19[11]: Let matrix $X \in R^{nxn}$. Then for k=1,2,...n

$$\sum_1^k 2Re\lambda_i(X) \leq \sum_1^k \lambda_i(X'+X) \tag{3.29}$$

with equality when k = n. Because of this equality

$$\sum_1^k 2Re\lambda_{n-i+1}(X) \geq \sum_1^k \lambda_{n-i+1}(X'+X) . \tag{3.30}$$

Theorem 20[11]: Let matrix $X \in R^{nxn}$. Then for k=1,2,...,n

$$\sum_1^k 2Im\lambda_i(X) \leq \sum_1^k Im\lambda_i(X-X') \tag{3.31}$$

with equality when k = n. Because of this equality

$$\sum_1^k 2Im\lambda_{n-i+1}(X) \geq \sum_1^k Im\lambda_{n-i+1}(X-X') \ . \tag{3.32}$$

Remark 3.5: Theorems 19 and 20 together relate the real and imaginary parts of the eigenvalues of X and (X+X′), (X-X′) respectively.

Theorem 21 [13]: Let matrix $X \in R^{n\times n}$. Then

$$\lambda_i(X'+X) \leq 2\sigma_i(X) \ . \tag{3.33}$$

Also with k = 1,2,...n

$$\sum_1^k \sigma_i(X+X') \leq 2\sum_1^k \sigma_i(X). \tag{3.34}$$

Theorem 22[14]: Let matrix $X \in R^{nxn}$. If $\lambda_i(X+X') \geq 0$, then for k = 1,2,...,n

$$\prod_1^k \lambda_{n-i+1}(X+X') \leq \prod_1^k 2Re\lambda_{n-i+1}(X). \tag{3.35}$$

Theorem 23[6,p.216]: Let matrices $X,Y \in R^{nxn}$. Then

$$\lambda_i(XY) = \lambda_i(YX) \ . \tag{3.36}$$

Theorem 24[12]: Let symmetric matrices $X,Y \in R^{nxn}$, and let $1 \leq i, j \leq n$. Then

$$\lambda_{i+j-1}(X+Y) \leq \lambda_j(X) + \lambda_i(Y), \quad i+j \leq n+1 \tag{3.37}$$

and

$$\lambda_{i+j-n}(X+Y) \geq \lambda_j(X) + \lambda_i(Y), \quad i+j \geq n+1 \ . \tag{3.38}$$

Theorem 25[12]: Let matrices X,Y $\in R^{nxn}$ where X,Y ≥ 0, and let $1 \leq i, j \leq n$. Then

$$\lambda_{i+j-1}(XY) \leq \lambda_j(X)\lambda_i(Y), \quad i+j \leq n+1 \tag{3.39}$$

and

$$\lambda_{i+j-n}(XY) \geq \lambda_j(X)\lambda_i(Y), \quad i+j \geq n+1 \ . \tag{3.40}$$

Remark 3.6: Theorems 24-25 relate single eigenvalues of XY or X+Y with one eigenvalue from each of X and Y. Theorems 26-43 express summations and products of k, k=1,2,...n eigenvalues of XY or X+Y with k eigenvalues from each of X and Y. Usually, they are the k largest or k smallest of the ordered (subscript i used) eigenvalues.

Theorem 26[15]: Let symmetric matrices X,Y $\in R^{nxn}$. Then for k=1,2,...n

$$\sum_1^k \lambda_i(X+Y) \leq \sum_1^k [\lambda_i(X) + \lambda_i(Y)] \tag{3.41}$$

with equality when k = n.

Theorem 27[16]: Let matrices X,Y $\in R^{nxn}$. Then for k=1,2,..,n

$$\sum_{1}^{k} \sigma_i(X+Y) \leq \sum_{1}^{k} [\sigma_i(X) + \sigma_i(Y)] \; . \tag{3.42}$$

Theorem 28(6,p.245]: Let symmetric matrices $X,Y \in R^{nxn}$. Then for k = 1, 2,...,n

$$\sum_{1}^{k} \lambda_i(X+Y) \geq \sum_{1}^{k} [\lambda_i(X) + \lambda_{n-i+1}(Y)] \tag{3.43}$$

with equality when k=n.

Theorem 29: Let symmetric matrices $X,Y \in R^{nxn}$. Then for k=1,2,...,n

$$\sum_{1}^{k} \lambda_{n-i+1}(X+Y) \geq \sum_{1}^{k} [\lambda_{n-i+1}(X) + \lambda_{n-i+1}(Y)] \tag{3.44}$$

with equality when k = n.

Proof: Write -X(-Y) instead of X(Y) in (3.41), and use the identity $\lambda_i(-X) = -\lambda_{n-i+1}(X)$.

Theorem 30 [6,p.245]: Let symmetric matrices $X,Y \in R^{nxn}$. If $[\lambda_n(X) + \lambda_n(Y)] \geq 0$, then for k = 1,2,...,n

$$\prod_{1}^{k} \lambda_{n-i+1}(X+Y) \geq \prod_{1}^{k} [\lambda_{n-i+1}(X) + \lambda_{n-i+1}(Y)] \; . \tag{3.45}$$

Proof: This follows from (3.44), in view of (3.12).

Theorem 31 [17] : Let matrices $X, Y \in R^{nxn}$ and $X, Y \geq 0$. Then

$$\prod_1^n[\lambda_i(X) + \lambda_i(Y)] \leq |X+Y| \leq \prod_1^n[\lambda_i(X) + \lambda_{n-i+1}(Y)]. \tag{3.46}$$

Remark 3.7 : Inequality (3.46) resembles (3.5).

Theorem 32, [6, p.475] : Let matrices X, Y $\in R^{nxn}$ and X,Y ≥ 0. Then for k = 1, 2, ... , n

$$[\prod_1^k\lambda_{n-i+1}(X+Y)]^{1/k} \geq [\prod_1^k\lambda_{n-i+1}(X)]^{1/k} + [\prod_1^k\lambda_{n-i+1}(Y)]^{1/k} . \tag{3.47}$$

Theorem 33 [11], [6, p.476] : Let matrices X, Y $\in R^{nxn}$ and X, Y ≥ 0. Then for k = 1, 2, ... , n

$$\prod_1^k\lambda_{n-i+1}((X+Y)/2) \geq [\prod_1^k\lambda_{n-i+1}(X)]^{1/2} [\prod_1^k\lambda_{n-i+1}(Y)]^{1/2}. \tag{3.48}$$

Theorem 34 [19] : Let matrices X, Y $\in R^{nxn}$. Then for k = 1, 2, ... , n

$$\prod_1^k\sigma_i(XY) \leq \prod_1^k\sigma_i(X)\sigma_i Y \tag{3.49}$$

with equality when k = n. Because of this equality

$$\prod_1^k\sigma_{n-i+1}(XY) \geq \prod_1^k\sigma_{n-i+1}(X)\sigma_{n-i+1}(Y). \tag{3.50}$$

Theorem 35 [6, p. 247] : When matrices X, Y ≥ 0, then for k = 1, 2, ... , n

$$\prod_1^k \lambda_i(XY) \le \prod_1^k \lambda_i(X)\lambda_i(Y) \tag{3.51}$$

with equality when k=n. Because of this equality

$$\prod_1^k \lambda_{n-i+1}(XY) \ge \prod_1^k \lambda_{n-i+1}(X)\lambda_{n-i+1}(Y). \tag{3.52}$$

Theorem 36 [20, (11)] : Let matrices X, Y $\in R^{nxn}$. Then for k = 1, 2, ..., n

$$\sum_1^k \sigma_i(XY) \le \sum_1^k \sigma_i(X)\sigma_i(Y). \tag{3.53}$$

Theorem 37 [6, p. 249] : Let matrices X, Y $\in R^{nxn}$ with X, Y ≥ 0. Then for k = 1, 2, ... , n

$$\sum_1^k \lambda_i(X)\lambda_{n-i+1}(Y) \le \sum_1^k \lambda_i(XY). \tag{3.54}$$

Theorem 38 [18, Theorem 2.2] : Let matrices X, Y $\in R^{nxn}$ and X, Y ≥ 0. Then for k = 1, 2, ... , n

$$\sum_1^k \lambda_i^s(XY) \le \sum_1^k \lambda_i^s(X)\lambda_i^s(Y) \tag{3.55}$$

where real exponent s > 0.

Theorem 39 [18, Theorem 2.1] : Let matrices X, Y $\in R^{nxn}$ with Y symmetric. Then

$$\sum_{1}^{n} \lambda_{n-i+1}(X+X')\lambda_i(Y) \leq 2tr(XY)$$
$$\leq \sum_{1}^{n} \lambda_i(X+X')\lambda_i(Y). \tag{3.56}$$

Theorem 40 [20, Theorem 2.1] : Let matrices X, Y $\in R^{nxn}$ with Y symmetric. Then for k = 1, 2, ... , n

$$\sum_{1}^{k} Re\lambda_i(XY) \leq \sum_{1}^{k} \sigma_i(X)|\lambda_i(Y)| \ . \tag{3.57}$$

Note : When k = n, a tighter bound is given by the upper bound in (3.56).

Theorem 41 [21] : Let matrices X, Y $\in R^{nxn}$ be symmetric. Then

$$\sum_{1}^{k} \lambda_i(X)\lambda_{n-i+1}(Y) \leq tr(XY)$$
$$\leq \sum_{1}^{k} \lambda_i(X)\lambda_i(Y) \ . \tag{3.58}$$

Theorem 42 [10] : Let matrices X, Y $\in R^{nxn}$ with $X \geq 0$. Then for k = 1, 2, ... , n

$$\sum_{1}^{k} \lambda_{n-i+1}(XY) \leq \sum_{1}^{k} \lambda_i(X)\lambda_i(Y) \ . \tag{3.59}$$

Theorem 43 [10] : Let matrices X, Y $\in R^{nxn}$ with $X \geq 0$. Then for k = 1, 2, ..., n

$$\sum_{1}^{k} \lambda_i(X)\lambda_{n-i+1}(Y) \leq \sum_{1}^{k} \lambda_i(XY). \tag{3.60}$$

Note : Matrices X, Y cannot be interchanged in the left hand side of (3.60).

C. Matrix Inequalities

Algebraic and eigenvalue inequalities use ordering of their elements. So do matrix inequalities. One of these is the Loewner ordering [22] for matrix valued functions of symmetric matrices. Let X, Y $\in R^{nxn}$ be two such matrices. Then X > (≥)Y means X-Y is positive (semi)-definite. A matrix-increasing function ϕ exists if given Y < (≤) X, then ϕ (Y) < (≤) ϕ (X). For matrix-montone and matrix-convex functions, consult the references in [6, pp.462-475].

The following are several results.

Theorem 44 [22]: Let symmetric matrices X,Y $\in R^{nxn}$, and X > (≥) Y. Then

$$\lambda_i(X) > (\geq) \lambda_i(Y) \ . \tag{3.61}$$

Note: The converse is not necessarily true i.e., given (3.61), it does not follow that X > (≥) Y. A consideration of two diagonal matrices in two dimensions will demonstrate this.

Theorem 45 [23, pp. 469-471]: Let symmetric matrices X, Y $\in R^{nxn}$, and X > (≥) Y, and let Z $\in R^{nxn}$. Then

$$Z'XZ > (\geq)\ Z'YZ \tag{3.62}$$

with strict inequality if Z is nonsingular and X > Y.

Theorem 46 [22] [6,pp. 464-465]: Let symmetric matrices X, Y $\in R^{nxn}$ and X > (≥) Y≥ 0. Then for $0 \leq r \leq 1$, $-1 \leq s < 0$.

$$X^r > (\geq)\ Y^r, \qquad Y \geq 0 \tag{3.63}$$

and

$$X^s < (\leq)\ Y^s, \qquad Y > 0\ . \tag{3.64}$$

IV. BOUNDS FOR THE DARE

Bounds for the DARE, last summarized in [24], 1984, are listed in this section.

First in subsection A is an exposition of notation. The bounds themselves are in subsections B-K, associated with titles. For example, under **D. λ_1(P)** are lower and upper bounds for $\lambda_1(P)$, the maximum of the $\lambda_i(P)$.

A. Notation

The notation in [24] is followed in major part.

As abbreviations, write $\lambda_i(Q) = \beta_i$, $\lambda_i(R) = \rho_i$, $|\lambda_i(A)| = a_i$, $|\lambda_i(A)|^2 = a_i^2$, $\sigma_i(A)$

$= \gamma_i$, $\sigma_i^2(A) = \gamma_i^2$. Then $\beta_1, \rho_1, a_1, a_1^2, \gamma_1, \gamma_1^2$ ($\beta_n, \rho_n, a_n, a_n^2, \gamma_n, \gamma_n^2$) are the maximal (minimal) values of $\beta_i, \rho_i, a_i, a_i^2, \gamma_i, \gamma_i^2$. No abbreviations are used for the $\lambda_i(P)$. An arbitrary eigenvalue has subscript k e.g., $\lambda_k(Q) = \beta_k$.

$\|X\|$ is the matrix norm induced by the Euclidean norm i.e., $\|X\| = \sigma_1(X)$.

The scalar function

$$f(x,y,z) = \frac{-x+(x^2+yz)^{1/2}}{y} = \frac{z}{x+(x^2+yz)^{1/2}}, \quad y \neq 0 \qquad (4.1)$$

is used to write the bounds. The terms x, y, z in (4.1) include β_i etc., singly or in summation and product combinations.

G(g) refers to a matrix (scalar) expression.

Matrix bounds for P such as $P \geq (>) X$, $P \leq (<) Y$ have the interpretation $(P - X) \geq (>) 0$, $(P - Y) \leq (<) 0$, respectively.

B. $\lambda_n(P)$

$\geq f(g, 2\rho_1, 2\beta_n)$, $g = 1 - \gamma_n^2 - \rho_1\beta_n$ (Yasuda and Hirai, 1979 [25]) (4.2)

$\leq f(g, 2\rho_n, 2\beta_1)$, $g = 1 - a_n^2 - \rho_n\beta_1$ (Yasuda and Hirai, 1979 [25]) (4.3)

$\leq f(g, 2\rho_n, 2tr(Q)/n)$, $g = 1 - tr(AA')/n - \rho_n tr(Q)/n$

(Komaroff, 1992 [26]) (4.4)

C. $\lambda_k(P)$

$\leq f(g, 2k\rho_n, 2\Sigma_1^k\beta_i)$, $g = k - \Sigma_1^k\gamma_i^2 - \rho_n\Sigma_1^k\beta_i$

(Komaroff, 1992 [26]) (4.5)

D. $\lambda_1(P)$

$\geq f(\|G\|, 2\|R(A')^{-1}\|, |\lambda_1(H)|)$, $G = A^{-1} - A - R(A')^{-1}Q$,

$2H = Q A^{-1} + (A')^{-1}Q$, $|H| \neq 0 \neq |A|$.

(Kwon and Pearson, 1977[27]) (4.6)

$\geq f(g, 2tr (R), 2 tr (Q))$, $g = n - tr(RQ) - \Sigma_1^n a_i^2$

(Patel and Toda, 1978 [28]) (4.7)

$\geq f(g, 2\rho_1, 2\beta_n)$, $g = 1 - a_1^2 - \rho_1\beta_n$ (Yasuda and Hirai, 1979 [25]) (4.8)

$\geq f(g, 2\Sigma_1^k\rho_i, 2\Sigma_1^k\beta_{n-i+1})$, $g = k - \Sigma_1^k a_i^2 - \rho_n\Sigma_1^k\beta_{n-i+1}$, $R \neq 0$

(Garloff, 1986 [29]) (4.9)

$\leq f(g, 2(\rho_n, 2\beta_1)$, $g = 1 - \gamma_1^2 - \rho_n\beta_1$ (Yasuda and Hirai, 1979 [25]) (4.10)

$\leq (\gamma_1^2 \rho_n^{-1} + 4\beta_1)(4 - \gamma_1^2)^{-1}$, $\gamma_1 < 2$ (Garloff, 1986 [29] (4.11)

E. $tr(P)$

$\geq f(-g, 2\rho_1, 2n^2\beta_n)$, $g = \Sigma_n^1 a_i^2 + tr(RQ) - n$ (Kwon *et al*, 1985 [30]) (4.12)

$\geq f(-g, 2\rho_1, 2tr(Q))$, $g = \gamma_n^2 - 1 + \rho_1 tr(Q)$

(Mori *et al*, 1987 [31]) (4.13)

$\geq f(-g, 2n\rho_1, 2(tr(Q^{½}))^2)$, $g = \Sigma_1^n a_i^2 + \rho_1 (tr(Q^{½}))^2 - n$

(Komaroff and Shahian, 1992 [32])(4.14)

$\geq tr(Q) + tr(AA')g(1 + \rho_1 g)^{-1}$, $g = f(1 - \gamma_n^2 - \rho_1\beta_n), 2\rho_1, 2\beta_n)$

(Kim *et al*, 1993 [33]) (4.15)

$\geq tr(Q) + \gamma_k^2\Sigma^n_{j=n-k+1} \beta_j(1 + \beta_j\gamma_1)^{-1}$, $\gamma_k \neq 0$, $\gamma_i = 0$, $i = k+1,...,n$

(Kim *et al*, 1993 [33]) (4.16)

$\leq f (g, 2\rho_n/n, 2tr(Q))$, $g = 1 - \gamma_1^2 - \beta_1\rho_n$ (Komaroff, 1992 [26]) (4.17)

F. $|P|$

$\leq [f(g, 2\rho_n, 2tr(Q)/n]^n$, $g = 1 - \gamma_1^2 - \beta_1\rho_n$ (Komaroff, 1992 [26]) (4.18)

G. $\Sigma_1^k \lambda_i(P)$

$\geq \Sigma_1^k \beta_{n-i+1} + \Sigma_1^k a_i^2 (\beta_n^{-1} + \rho_1)^{-1}$ (Garloff, 1986 [29]) (4.19)

$\geq f(-g, 2\rho_1, 2k^2\beta_n)$, $g = \Sigma_1^k (a_1^2 + \lambda_{n-i+1} (RQ) - 1)$

(Komaroff and Shahian, 1992 [32]) (4.20)

$\leq \Sigma_1^k \beta_i[1 - \gamma_1^2 (1 + \beta_k \rho_n)^{-1}]^{-1}$, $\gamma_1^2 > 1 + \beta_k \rho_n$

(Garloff, 1986 [29]) (4.21)

$\leq f(g, 2\rho_n /k, 2\Sigma_1^k \beta_i)$, $g = 1 - \gamma_1^2 - \beta_1\rho_n$ (Komaroff, 1992 [26]) (4.22)

H. $\Sigma_1^k \lambda_{n-i+1} (P)$

$\geq \Sigma_1^k \beta_{n-i+1} [1 + \gamma_n^2 (1 + \beta_{n-i+1} \rho_1)^{-1}]$ (Garloff, 1986 [29]) (4.23)

I. $\Pi_1^k \lambda_i (P)$

$\leq [f(g, 2\rho_n, 2\Sigma_1^k \beta_i/k)]^k$, $g = 1 - \gamma_1^2 - \beta_1 \rho_n$

(Komaroff, 1992 [26]) (4.24)

J. $\Pi_1^k \lambda_{n-i+1}(P)$

$\geq (\Pi_1^k \beta_{n-i+1}) [1+ (\Pi_1^k \gamma^2_{n-i+1})^{1/k} \Pi_1^k (1 + \beta_{n-i+1} \rho_1)^{-1/k}]^k$

(Garloff, 1986 [29]) (4.25)

$\geq 2^k \Pi_1^k \gamma_{n-i+1} \Pi_1^k \beta_{n-i+1} \Pi_1^k (1 + \beta_{n-i+1} \rho_1)^{-1/2}$, $|A| \neq 0$

(Garloff, 1986 [29]) (4.26)

K. **P**

$\geq f(\|G\|, 2\| R(A')^{-1} \|, |\lambda_n(H)|)I$, $G = A^{-1} - A - R (A')^{-1} Q$,

$2H = QA^{-1} + (A')^{-1} Q$, $|H| \neq 0 \neq |A|$

(Kwon and Pearson, 1977 [27]) (4.27)

$\geq Q$ (strict inequality if $|A| \neq 0$) (Garloff, 1986 [29]) (4.28)

$\geq A' (Q^{-1} + R)^{-1} A + Q$ (strict inequality if $|A| \neq 0$).

(Komaroff, 1994 [34]) (4.29)

$\leq A' R^{-1} A + Q$ (strict inequality if $|A| \neq 0$).

(Komaroff, 1994 [34]) (4.30)

V. EXAMPLES AND RESEARCH

The trend in research, as the results of section IV illustrate, is to employ matrix-valued functions to bound the solution of (1.1). This section demonstrates the power of matrix bounds. Two examples and research suggestions are given.

The first example, in subsection A, shows a relationship between a matrix bound and eigenvalue function bounds. In B, the second example applies matrix bounds to analyse an iterative convergence scheme to solve (1.1). It is suggested, in subsection C, that matrix inequalities be designed to take the place of scalar inequalities employed in the fertile literature on the solution of scalar nonlinear equations.

A. Example 1

This example shows how a matrix inequality can include the information in eigenvalue inequalities. Specifically, (4.19) (written as (5.2) below) is derived from (4.29) (written as (5.1) below). Both inequalities, at the outset, rely on (4.28).

Theorem 5.1: Given the inequality

$$P \geq A'(Q^{-1} + R)^{-1} A + Q \tag{5.1}$$

strict if $|A| \neq 0$, for the solution P of (1.2), it follows that

$$\sum_1^k \lambda_i(P) \geq \sum_1^k \lambda_{n-i+1}(Q) + \sum_1^k |\lambda_i(A)|^2 [\lambda_n^{-1}(Q) + \lambda_1(R)]^{-1} \tag{5.2}$$

where k = 1, 2, ..., n.

Proof: Write $X = Q^{-1} + R$. Then from (5.1)

$$A'X^{-1} A \leq P - Q \tag{5.3}$$

which, multiplied on the left and on the right by $X^{1/2}$, gives

$$X^{1/2} A' X^{-1} A X^{1/2} \leq X^{1/2} (P - Q) X^{1/2} . \tag{5.4}$$

To the left hand side apply (3.26), and to the right hand side apply (3.55); then

$$\sum_1^k |\lambda_i(A)|^2 \leq \sum_1^k \lambda_i(X) \lambda_i(P - Q), \quad k = 1,2,\ldots,n \tag{5.5}$$

since $\lambda_i (X^{1/2}(P - Q) X^{1/2}) = \lambda_i (X(P - Q))$ by (3.36), and $(P - Q) \geq 0$ by (4.28), which permits the use of (3.55).

We now bound the right hand side of (5.5) given k = 1, 2, ..., n as

$$\sum_{1}^{k} |\lambda_i(A)|^2 \le \sum_{1}^{k} \lambda_1(X)\,\lambda_i(P-Q) \tag{5.6}$$

$$\le \sum_{1}^{k} \lambda_1(X)[\lambda_i(P) + \lambda_i(-Q)] \tag{5.7}$$

$$= \sum_{1}^{k} \lambda_1(X)\,[\lambda_i(P) - \lambda_{n-i+1}(Q)] \tag{5.8}$$

where to obtain (5.7) we used (3.41). Next,

$$\lambda_1(X) \le \lambda_1(Q^{-1}) + \lambda_1(R) \tag{5.9}$$

$$= \lambda_n^{-1}(Q) + \lambda_1(R) \tag{5.10}$$

where (3.41) is again used, with k = 1.

It remains to solve (5.8) for $\sum_{1}^{k} \lambda_i(\mathrm{P})$, which immediately produces (5.2), having employed (5.10).

Remark 5.1: Two inequalities were used to derive (5.5), and one for each of (5.6), (5.7) and (5.9). This totals five, to derive (5.2) from (5.1). Besides, each used (4.28).

B. Example 2

This example shows how rates of convergence in matrix iteration schemes can be compared through use of matrix inequalities. The two iteration schemes to solve (1.1) in [34] are investigated.

As the first step, both schemes employ the equation

$$P_1 = A'(P_0^{-1} + R)^{-1} A + Q \tag{5.11}$$

where P_0 is the initial estimate (or bound) for P, and P_1 is the resulting first iterate. In the first variant $P_0 < P_1 < P$ and in the other $P_0 > P_1 > P$.

The object is to compare the rate of convergence of the step (5.11) for the two cases. To distinguish between these cases write P_{01} and P_{02} as the two values for P_0:

$$P_{01} = P - D, \; P_{02} = P + D, \; D > 0 \tag{5.12}$$

This states that the initial values P_{01}, P_{02} are equidistant from P. It follows that

$$Q < P_{01} < P < P_{02} \tag{5.13}$$

where the inclusion of Q which also places a limit on D in (5.12), is necessary to ensure convergence of the algorithm of which the first step is (5.11) [34].

The following lemma will be used [34].

Lemma 5.1: Let matrices A, X, R, Y $\in R^{nxn}$ with $X > 0$, Y, $R \geq 0$ and $X > (\geq)$ Y. Then

$$A'(X^{-1} + R)^{-1} A > (\geq) A'(Y^{-1} + R)^{-1} A \tag{5.14}$$

with strict inequality if A is nonsingular, and $X > Y$.

Theorem 5.1 : With initial estimates P_{01} and P_{02} for the solution P of (1.1) defined by (5.12), let

$$P_{11} = A'(P_{01}^{-1} + R)^{-1} A + Q \tag{5.15}$$

and

$$P_{12} = A'(P_{02}^{-1} + R)^{-1} A + Q \tag{5.16}$$

Then

$$(P - P_{11}) > (P_{12} - P) . \tag{5.17}$$

Proof : From [34, Theorem 3.2],

$$P > P_{11} > P_{01} \tag{5.18}$$

and from [34, Theorem 3.4]

$$P < P_{12} < P_{02} \tag{5.19}$$

Remembering (5.12) and using (5.14), (5.15)

$$A'[(P - D)^{-1} + R]^{-1}A < A'[(P + D)^{-1} + R]^{-1} A \tag{5.20}$$

which means that the difference between P_{01} and P_{11} is less than the difference between P_{02} and P_{12}, which is stated by (5.17).

Remark 5.2 : Theorem 5.1 shows that if the iteration (5.11) starts a distance $D = P - P_{01}$ below the solution, convergence to the solution P is slower than for the scheme that starts iteration a distance $D = P_{02} - P$ above the solution.

The numerical example in [34] supports this. It uses the scalar equation

$$P = (P^{-1} + 0.5)^{-1} + 1, A = 1, R = 0.5, Q = 1$$

version of (1.2); its solution P = 2. For $P_{10} = 1$ and $P_{02} = 3$ (D = 1 in (5.12)), $P_{11} = 1.667$, $P_{12} = 2.2$, showing that $(P - P_{11}) = 0.33 > (P_{12} - P) = 0.2$.

In practice, there is another factor involved in the choice of the two schemes. If the available upper bound (or P_{02}) for the solution is much more pessimistic than the lower bound (or P_{01}), the above advantage may be negated. For, the closeness of P_0 (obtained from a bound in the literature) to P determines the number of iterative steps to achieve convergence to the solution.

C. On Research

In previous sections it was stressed that bounds for the DARE evolved from eigenvalue to matrix bounds. The inequalities used to obtain these bounds are classified in accordance with this evolution.

The number of matrix inequalities is few. However the information they convey inherently contains all eigenvalue, and what is additional, all eigenvector estimates of P. This increasing trend to employ matrix inequalities to bound solutions of matrix equations is not only of engineering (hardware applications, software implementations) significance, but adds impetus to the development of mathematical matrix inequalities.

Matrices are a generalization of scalars. Direct use of matrices in inequalities makes it possible to exploit the rich literatures devoted to the numerical solution of nonlinear equations. For example, the average $(P_{12} - P_{11})/2$ (see (5.15), (5.16)) may provide a better approximation for P than P_{12} or P_{11} alone. The Regula Falsi and associated schemes can be "matricized". Likewise scalar methods that compare the degree of convergence of iteration methods can be modified for matrix equations.

Such research directions are natural once matrices are identified with scalars.

VI. CONCLUSIONS

Bounds on the solution P of the DARE have been presented for the period 1977 to the present.

The reasons for seeking bounds, their importance and applications have been given. Mathematical inequalities, intended to be a dictionary of tools for deriving bounds for the DARE, have been collected for the first time. The collection is not and cannot be complete; it is not a substitute for a study of inequalities in the cited references.

The listing of bounds for the DARE updates the previous summary in 1984. It shows the trend of deriving bounds, directly or indirectly, for an increasing number of the eigenvalues of P, and the latest results are for matrices that bound solution matrix P. The bibliography of mathematical inequalities used to obtain these bounds had been expressly categorized to mirror this evolution of results for the DARE. The two listings together show "tools determine the product".

Two examples illustrate the derivation of bounds, and show some implications of matrix versus eigenvalue bounds. Research directions and suggestions, a cardinal aim of the exposition, are to be found in various parts of the chapter.

VII. REFERENCES

[1] B.C. Kuo, "*Digital Control Systems*", 2nd Ed., Orlando, FL: Saunders College Publishing, Harcourt Brace Jovanovich, 1992.

[2] F.L. Lewis, "*Applied Optimal Control and Estimation*", New Jersey: Prentice Hall, 1992.

[3] D. Kleinman, "Stabilizing a discrete, constant, linear system with application to iterative methods for solving the Riccati equation," *IEEE Trans. Automat. Contr.*, vol. AC-19, pp. 252-254, June 1974.

[4] G.H. Hardy, J.E. Littlewood and G. Polya, *Inequalities*, 2nd. Ed., Cambridge: Cambridge University Press, 1952.

[5] D.S. Mitrinovic, *Analytic Inequalities*, New York: Springer-Verlag, 1970.

[6] A.W. Marshall and I. Olkin, *Inequalities: Theory of Majorization and its Applications*, New York: Academic, 1979.

[7] R.A. Horn and G.A. Johnson, *Topics in Matrix Analysis*, Cambridge: Cambridge University Press, 1991.

[8] H. Weyl, "Inequalities between the two kinds of eigenvalues of linear transformation, *Proc. Nat. Acad. Sci.*, vol. 35, pp. 408-411, 1949.

[9] P.W. Day, "Rearrangement inequalities", *Canad. J. Math.*, 24, pp. 930-943, 1972.

[10] N. Komaroff, "Rearrangement and matrix product inequalities", *Linear Algebra Appl.*, 140, pp. 155-161, 1990.

[11] K. Fan, "On a theorem of Weyl concerning eigenvalues of linear transformations II", *Proc. Nat. Acad. Sci.*, vol. 36, pp. 31-35, 1950.

[12] A.R. Amir-Moez, "Extreme properties of eigenvalues of a Hermitian transformation and singular values of the sum and product of linear transformations", *Duke Math. J.*, vol. 23, pp. 463-467, 1956.

[13] K. Fan and A. Hoffman, "Some metric inequalities in the space of matrices", *Proc. Amer. Math. Soc.*, vol. 6, pp. 111-116, 1955.

[14] K. Fan, "A minimum property of the eigenvalues of a Hermitian transformation", *Amer. Math. Monthly*, vol. 60, pp. 48-50, 1953.

[15] K. Fan, "On a theorem of Weyl concerning eigenvalues of linear transformations I", *Proc. Nat. Acad. Sci.*, vol. 35, pp. 652-655, 1949.

[16] K. Fan, "Maximum properties and inequalities for the eigenvalues of completely continuous operators", *Proc. Nat. Acad. Sci.*, vol. 37, pp. 760-766, 1951.

[17] M. Fiedler, "Bounds for the determinant of the sum of Hermitian matrices", *Proc. Amer. Math. Soc.*, vol. 30, pp. 27-31, 1971.

[18] N. Komaroff, "Bounds on eigenvalues of matrix products with an application to the algebraic Riccati equation", *IEEE Trans. Automat. Contr.*, vol. AC-35, pp. 348-350, Mar. 1990.

[19] A. Horn, "On the singular values of a product of completely continuous operators", *Proc. Nat. Acad. Sci.*, vol. 36, pp. 374-375, 1950.

[20] N. Komaroff, "Matrix inequalities applicable to estimating solution sizes of Riccati and Lyapunov equations", *IEEE Trans. Automat. Contr.*, vol. AC-34, pp.97-98, Jan. 1989.

[21] L. Mirsky, "On the trace of matrix products", *Math. Nachr.*, vol. 20, pp. 171-174, 1959.

[22] C. Loewner, "Üeber monotone Matrixfunktionen", *Math. Z.*, 38, pp. 177-216, 1934.

[23] R.A. Horn and C.R. Johnson, *Matrix Analysis*, Cambridge: Cambridge University Press, 1985.

[24] T. Mori and I.A. Derese, "A brief summary of the bounds on the solution of the algebraic matrix equations in control theory", *Int. J. Contr.*, vol. 39, pp. 247-256, 1984.

[25] K. Yasuda and K. Hirai, "Upper and lower bounds on the solution of the algebraic Riccati equation", *IEEE Trans. Automat. Contr.*, vol. AC-24, pp. 483-487, June 1979.

[26] N. Komaroff, "Upper bounds for the solution of the discrete Riccati equation", *IEEE Trans. Automat. Contr.*, vol. AC-37, pp. 1370-1373, Sept. 1992.

[27] W.H. Kwon and A.E. Pearson, "A note on the algebraic matrix Riccati

equation", *IEEE Trans. Automat. Contr.*, vol. AC-22, pp. 143-144, Feb. 1977.

[28] R.V. Patel and M. Toda, "On norm bounds for algebraic Riccati and Lyapunov equations", *IEEE Trans. Automat. Contr.*, vol. AC-23, pp. 87-88, Feb. 1978.

[29] J. Garloff, "Bounds for the eigenvalues of the solution of the discrete Riccati and Lyapunov matrix equations", *Int. J. Contr.*, vol. 43, pp. 423-431, 1986.

[30] B.H. Kwon, M.J. Youn and Z. Bien, "On bounds of the Riccati and Lyapunov matrix equation", *IEEE Trans. Automat. Contr.*, vol. AC-30, pp. 1134-1135, Nov. 1985.

[31] T. Mori, N. Fukuma and M. Kuwahara, "On the discrete Riccati equation", *IEEE Trans. Automat. Contr.*, vol. AC-32, pp. 828-829, Sep. 1987.

[32] N. Komaroff and B.Shahian, "Lower summation bounds for the discrete Riccati and Lyapunov equations", *IEEE Trans. Automat. Contr.*, vol. AC-37, pp. 1078-1080, July 1992.

[33] S.W. Kim, P.G. Park and W.H. Kwon, "Lower bounds for the trace of the solution of the discrete algebraic Riccati equation", *IEEE Trans. Automat. Contr.*, vol. AC-38, pp. 312-314, Feb. 1993.

[34] N. Komaroff, "Iterative matrix bounds and computational solutions to the discrete algebraic Riccati equation", *IEEE Trans Automat. Contr.*, vol. AC-39, pp. 1676-1678, Aug. 1994.

ANALYSIS OF DISCRETE-TIME LINEAR PERIODIC SYSTEMS

Sergio Bittanti and Patrizio Colaneri
Politecnico di Milano
Dipartimento di Elettronica e Informazione
Piazza Leonardo da Vinci 32
20133 Milano (Italy)

FAX ++39.2.23993587

Emails
bittanti@elet.polimi.it
colaneri@elet.polimi.it

Abstract
This paper is intended to provide an updated survey on the main tools for the analysis of discrete-time linear periodic systems. We first introduce classical notions of the periodic realm: *monodromy matrix*, *structural properties* (reachability, cotrollability etc...), *time-invariant reformulations* (lifted and cyclic) and *singularities* (zeros and poles). Then, we move to more recent developments dealing with the *system norms* (H_2, H_∞, Hankel), the *symplectic pencil* and the *realization problem*.

1. INTRODUCTION

The long story of periodic systems in signals and control can be traced back to the sixties, see (Marzollo,1972) for a coordinated collection of early reports or (Yakubovich and Starzhinskii, 1975) for a pioneering volume on the subject. After two decades of study, the 90's have witnessed an exponential growth of interests, mainly due to the pervasive diffusion of digital techniques in signals (Gardner, 1994) and control. Remarkable applications appeared in chemical reactor control, robot guidance, active control of vibrations, flight fuel consumption optimization, economy management, etc. Among other things, it has been recognized that the performances of time-invariant plants can be upgraded by means of periodic controllers. Even more so, the consideration of periodicity in control laws have led to the solution of problems otherwise unsolvable in the time-invariant realm.

This paper is intended to be a tutorial up-to-date introduction to the analysis of periodic discrete-time systems, see (Bittanti, 1986) for a previous relevant survey. The organization is as follows. The basic concepts of monodromy matrix, stability and structural properties are outlined in Sect. 2. A major tool of analysis consists in resorting to suitable time invariant reformulations of periodic systems; the two most important reformulations are the subject of Sect. 3. In this way, it is possible to give a frequency-domain interpretation to periodic systems. The concepts of adjoint system and symplectic pencil are dealt with in Sect. 4. As is well known, they are most useful in analysis and design problems in both H_2 and H_∞ contexts. Thanks to the time invariant reformulations, the notions of poles and zeros of periodic systems are defined in Sect. 5. In particular, the zero blocking property is properly characterized by means of the so-called exponential periodic signals. The main definitions of norm of a system (L_2, L_∞ and Hankel) are extended in Sect. 6, where the associated input-output interpretations are also discussed. Finally, the issue of realization is tackled in Sect. 7 on the basis of recent results. For minimality, it is necessary to relax the assumption of time-invariance of the dimension of the state space; rather, such a dimension must be periodic in general.

2. BASICS ON LINEAR DISCRETE-TIME PERIODIC SYSTEMS

In this paper, we consider systems over discrete time ($t \in Z$) described by

$$x(t+1) = A(t)x(t) + B(t)u(t) \tag{1.a}$$
$$y(t) = C(t)x(t) + D(t)u(t) \tag{1.b}$$

where $u(t) \in R^m$, $x(t) \in R^n$, $y(t) \in R^p$, are the input, state and output vectors, respectively. Matrices $A(\cdot)$, $B(\cdot)$, $C(\cdot)$ and $D(\cdot)$ are real matrices, of appropriate dimensions, which depend periodically on t:

$$A(t+T) = A(t); \quad B(t+T) = B(t); \quad C(t+T) = C(t) \quad D(t+T) = D(t).$$

T is the period of the system.

2.1 Monodromy matrix and stability

The state evolution from time τ to time $t > \tau$ of system (1.a) is given by the Lagrange formula

$$x(t) = \Psi_A(t,\tau)x(\tau) + \sum_{j=\tau+1}^{t} \Psi_A(t,j)B(j-1)u(j-1), \tag{2}$$

where $\Psi_A(t,\tau) = A(t-1)A(t-2) \ldots A(\tau)$ is the transition matrix of the system. It is easily seen that the periodicity of the system entails the "biperiodicity" of matrix $\Psi_A(t,\tau)$, namely:

$$\Psi_A(t+T, \tau+T) = \Psi_A(t,\tau). \tag{3}$$

The transition matrix over one period, viz. $\Phi_A(t) = \Psi_A(t+T,t)$, is named *monodromy matrix* at time t, and is T-periodic. Apparently, this matrix determines the system behaviour from one period (starting at t) to the subsequent one (starting at $t+T$). In particular, the T-sampled free motion is given by $x(t+kT) = \Phi_A(t)^k x(t)$. This entails that the system, or equivalently matrix $A(\cdot)$, is (asymptotically) stable if and only if the eigenvalues of $\Phi_A(t)$ belong to the open unit disk. Such eigenvalues, referred to as *characteristic multipliers* are independent of t, see e.g. (Bittanti, 1986).

The characteristic multipliers are all different from zero iff matrix $A(t)$ is nonsingular for each t. In such a case, the system is *reversible*, in that the state $x(\tau)$ can be recovered from $x(t)$, $t>\tau$ (assuming that input $u(\cdot)$ over the interval $[\tau, t-1]$ be known).

Remark 1

A more general family of discrete-time periodic systems is constituted by the so-called *descriptor periodic* systems, which are characterized by the modified state equation $E(t)x(t+1)= A(t)\, x(t) + B(t)\, u(t)$, where $E(t)$ is also T-periodic and singular for some t. The analysis of such systems goes beyond the scope of the present paper.

2.2 Structural properties

In this section we deal first with the notions of *reachability* and *observability*. As is well known, (Kalman, 1960), reachability deals with the possibility of driving $x(t)$ to any desired point in the state-space by a proper input sequence, while observability is connected with the possibility of uniquely estimating $x(t)$ from the observation of the future outputs. When any state at any time can be reached in an interval of length k, we speak of *k-step reachability*. If not all points in the state-space can be reached over any finite length interval, one can define the *reachability subspace* as the set of states which are reachable. Analogous concepts can be introduced for *observability*. In the periodic case, a thorough body of results regarding these notions for periodic systems is available, see e.g. (Bittanti, Bolzern, 1985 and 1986), (Bittanti, Colaneri, De Nicolao, 1986) and references quoted there. Among the various characterizations, the following ones are worth mentioning.

Reachability Criterion

System (1) is k-step reachable at time t iff *rank* $[R_k(t)] = n, \forall t$, where

$$R_k(t) = \left[B(t-1) \quad \Psi_A(t,t-1)B(t-2) \cdots \Psi_A(t,t-k+1)B(t-k)\right] \tag{4}$$

Moreover, system (1) is reachable at time t iff it is nT-step reachable.■

Observability Criterion

System (1) is k-step observable at time t iff *rank* $[O_k(t)] = n, \forall t$, where

$$O_k(t) = \left[C(t)' \quad \Psi_A(t+1,t)'C(t+1)' \cdots \Psi_A(t+k-1,t)'C(t+k-1)'\right]' \tag{5}$$

Moreover, system (1) is observable at time t iff it is nT-step observable.■

Notice that $R_k(t)R_k(t)'$ and $O_k(t)'O_k(t)$ are known as *Grammian reachability* and *Grammian observability* matrices, respectively.

Attention is drawn to the following fact. Even if $R_{nT}(t)$ [or $O_{nT}(t)$] has maximum rank for some t, it may fail to enjoy the same property at a different time point. This corresponds to the fact that the dimensions of reachability and unobservability subspaces of system (1) are, in general, time-varying. A notable exception is the reversible case, where the subspaces have constant dimension.

In the following, we will say that the pair $(A(\cdot),B(\cdot))$ is reachable [$(A(\cdot),C(\cdot))$ is observable] if system (1) is reachable [observable] at any t.

An alternative characterization of reachability and observability of periodic systems refers to the characteristic multipliers as follows:

Reachability Modal Characterization
A characteristic multiplier λ of $A(\cdot)$ is said to be $(A(\cdot),B(\cdot))$-unreachable at time τ, if there exists $\eta \neq 0$, such that

$$\Phi_A(\tau)'\eta = \lambda\eta, \quad B(j-1)'\Psi_A(\tau, j)'\eta = 0, \quad \forall j \in [\tau - T + 1, \tau] \tag{6}$$

A characteristic multiplier which is not unreachable is said to be reachable. System (1) is reachable if all characteristic multipliers are reachable.■

Observability Modal Characterization
A characteristic multiplier λ of $A(\cdot)$ is said to be $(A(\cdot),C(\cdot))$-unobservable at time τ, if there exists $\xi \neq 0$, such that

$$\Phi_A(\tau)\xi = \lambda\xi, \quad C(j)\Psi_A(j,\tau)\xi = 0, \quad \forall j \in [\tau, \tau + T - 1] \tag{7}$$

A characteristic multiplier which is not unobservable is said to be observable. System (1) is observable if all characteristic multipliers are observable.■

These "modal" notions correspond to the so-called PBH (Popov-Belevitch-Hautus) characterization in the time invariant case, see e.g. (Kailath, 1970). It should be noted that if a characteristic multiplier $\lambda \neq 0$ is unreachable [resp. unobservable] at time t, it is also unreachable [resp. unobservable] at any time point. On the contrary, a null characteristic multiplier may be reachable [resp. observable] at a time point and unreachable [resp. unobservable] at a different time instant, see (Bittanti, Bolzern 1985) for more details.

Two further important structural properties are *controllability* and *reconstructibility*. The former deals with the possibility of driving the system state to the origin in finite time by a proper choice of the input sequence, while the latter concerns the possibility of estimating the current state from future output observations. If not all points in the state-space can be controlled one can define the *controllability subspace* as the set of states which are controllable. Analogously, one can introduce the *reconstructability* and *unreconstructability* subspaces. The characterization of controllability and reconstructibility in terms of Grammians is somewhat involved, see e.g. (Bittanti, Bolzern, 1985 and 1986). Here, we will focus on modal characterizations only.

Controllability Modal Characterization
A characteristic multiplier $\lambda \neq 0$ of $A(\cdot)$ is said to be $(A(\cdot),B(\cdot))$-uncontrollable, if there exists $\eta \neq 0$, such that, for some τ, eq. (6) holds. A null characteristic multiplier or a characteristic multiplier $\lambda \neq 0$ which is not uncontrollable is said to be controllable. System (1) is controllable if all characteristic multipliers are controllable.■

Reconstructibility Modal Characterization
A characteristic multiplier $\lambda \neq 0$ of $A(\cdot)$ is said to be $(A(\cdot),C(\cdot))$-unreconstructible, if there exists $\zeta \neq 0$, such that, for some τ, eq. (7) holds. A null characteristic multiplier or a characteristic multiplier $\lambda \neq 0$ which is not unreconstructible is said to be reconstructible. System (1) is reconstructible if all characteristic multipliers are reconstructible.■

Note that, in the above definitions, the role of τ is immaterial. Indeed, with reference to the uncontrollability notion, one can prove that if a characteristic multiplier $\lambda \neq 0$ of $A(\cdot)$ is such that $\Phi_A(\tau)'\eta = \lambda\eta$, and $B(j\text{-}1)'\,\Psi_A(\tau, j)'\eta = 0$, $\forall j \in [\tau\text{-}T\text{+}1, \tau]$, then the same is true for any other time point τ. Analogous considerations hold true for the unreconstructibility notion.

As already mentioned, the dimensions of the reachabibility and observability subspaces may (periodically) vary with time. On the contrary, the controllability and reconstructibility subspaces have constant dimensions. The reason of this difference lies in the peculiar behaviour of the null characteristic multipliers of $A(\cdot)$. Indeed all null multipliers are obviously controllable; however they may not correspond to reachable modes. Precisely, in general the reachability subspace [observability subspace] at time t is contained in or equals the controllability subspace [reconstructibility subspace] at time t; for instance, the difference between the dimension of the controllability and reachability subspaces at a time point t equals the number of null characteristic multipliers which are unreachable at t (Bittanti and Bolzern, 1984). Notice that such a number is possibly time-varying, see (Bittanti, 1986) for more details.

Obviously, if $\det A(t) \neq 0, \forall t$ (reversibility), then the reachability subspace [observability subspace] at time t coincides with the controllability subspace [reconstructibility subspace] at the same time point t.

As in the time-invariant case, the state representation of a periodic system can be canonically decomposed. In view of the above seen properties of the structural subspaces, in order to come out with four constant dimensional subsystems, reference must be made to controllability and reconstructibility only. This amounts to saying that there exists a nonsingular periodic change of basis $T(t)$ such that matrix $T(t+1)A(t)T(t)^{-1}$ is block-partitioned into four submatrices accounting for the controllable and unreconstructible, controllable and reconstructible, uncontrollable and reconstructible, uncontrollable and unreconstructibility parts.

The notions of stabilizability and detectability of periodic systems can then be introduced.

Stabilizability Decomposition-based Characterization
System (1) is said to be stabilizable if its uncontrollable part is stable. ■

Detectability Decomposition-based Characterization
System (1) is said to be detectable if its unreconstructible part is stable.■

Other equivalent characterizations are the following modal ones.

Stabilizability Modal Characterization
A characteristic multiplier λ of $A(\cdot)$, with $|\lambda| \geq 1$, is said to be $(A(\cdot),B(\cdot))$-unstabilizable if there exists $\eta \neq 0$, such that, for some τ, eq. (6) holds. A characteristic multiplier λ is stabilizable if either $|\lambda| < 1$ or $|\lambda| \geq 1$ with λ not unstabilizable. System (1) is stabilizable if all characteristic multipliers are stabilizable.■

Detectability Modal Characterization
A characteristic multiplier λ of $A(\cdot)$, with $|\lambda| \geq 1$, is said to be $(A(\cdot),C(\cdot))$-undetectable if there exists $\xi \neq 0$, such that, for some τ, eq. (7) holds. A characteristic multiplier λ is detectable if either $|\lambda| < 1$ or $|\lambda| \geq 1$ with λ not undetectable. System (1) is detectable if all characteristic multipliers are detectable. ■

Finally, in the context of control and filtering problems, the above notions take the following form.

Stabilizability Control Characterization
System (1) is stabilizable if there exists a T-periodic matrix $K(\cdot)$ such that $A(\cdot)+B(\cdot)K(\cdot)$ is stable. ■

Detctability Estimation Characterization
System (1) is detectable if there exists a T-periodic matrix $L(\cdot)$ such that $A(\cdot)+L(\cdot)C(\cdot)$ is stable.■

3. TIME INVARIANT REFORMULATIONS

A main tool of analysis and design of periodic systems exploits the natural correspondence between such systems and time-invariant ones. There are two popular correspondences named *lifted reformulation* (Jury, 1959), (Mayer and Burrus, 1975) (Khargonekar, Poola and Tannenbaum, 1985) and *cyclic reformulation* (Verriest,1988) (Flamm, 1989).

3.1 Lifted reformulation

The rationale underlying such a reformulation is to sample the system state with sampling interval coincident with the system period T, and to pack the input and output signals over one period so as to form input and output signals of enlarged dimensions. Precisely, let τ be a sampling tag and define the sampled state as

$$x_\tau(k) = x(kT + \tau). \tag{8.a}$$

Moreover, introduce the "packed input" and "packed output" segments as follows:

$$u_\tau(k) = [u(\tau + kT)' \quad u(\tau + kT + 1)' \cdots u(\tau + kT + T - 1)']' \tag{8.b}$$

$$y_\tau(k) = [y(\tau + kT)' \quad y(\tau + kT + 1)' \cdots y(\tau + kT + T - 1)']'. \tag{8.c}$$

In the *lifted reformulation*, the state $x_\tau(k+1) = x(\tau + (k+1)T)$ is related to $x_\tau(k) = x(\tau + kT)$ by means of the "packed input" segment $u_\tau(k)$. As for the "packed output" segment $y_\tau(k)$, it can be obtained from $x_\tau(k)$ and $u_\tau(k)$. More precisely, define $F_\tau \in R^{n\times n}$, $G_\tau \in R^{n\times mT}$, $H_\tau \in R^{pT\times n}$, $E_\tau \in R^{pT\times mT}$ and $u_\tau(k) \in R^{mT}$ as:

$$F_\tau = \Phi_A(\tau),$$

$$G_\tau = [\Psi_A(\tau + T, \tau + 1)B(\tau) \quad \Psi_A(\tau + T, \tau + 2)B(\tau + 1) \cdots B(\tau + T - 1)]$$

$$H_\tau = [C(\tau)' \quad \Psi_A(\tau + 1, \tau)'C(\tau + 1)' \cdots \Psi_A(\tau + T - 1, \tau)'C(\tau + T - 1)']'$$

$$E_\tau = \{(E_\tau)_{ij}\}, \qquad i, j = 1,2,\cdots,T,$$

$$(E_\tau)_{ij} = \begin{cases} 0 & i < j \\ D(\tau + i - 1) & i = j \\ C(\tau + i - 1)\Psi_A(\tau + i - 1, \tau + j)B(\tau + j - 1) & i > j \end{cases}$$

Thanks to these definitions, the lifted reformulation can be introduced:

$$x_\tau(k+1) = F_\tau x_\tau(k) + G_\tau u_\tau(k) \tag{9.a}$$

$$y_\tau(k) = H_\tau x_\tau(k) + E_\tau u_\tau(k) \tag{9.b}$$

In view of (2), it is easy to see that, if $u_\tau(\cdot)$ is constructed according to (8.b) and $x_\tau(0)$ is taken equal to $x(\tau)$, then $x_\tau(k) = x(kT + \tau)$ and $y_\tau(\cdot)$coincides with the segment defined in (8.c).

From (8.a) it is apparent that the time-invariant system (9) can be seen as a state-sampled representation of system (1), fed by an augmented input vector and producing an augmented output vector. Such vectors $u_\tau(k)$ and $y_\tau(k)$ are obtained by stacking the values of the input $u(\cdot)$ and the output $y(\cdot)$ over each period as pointed out by (8.b) and (8.c).

Obviously, one can associate a transfer function $W_\tau(z)$ to the time-invariant system (9):

$$W_\tau(z) = H_\tau(zI - F_\tau)^{-1}G_\tau + E_\tau. \qquad (10)$$

This transfer function will be named the *lifted transfer function* of system (1) at time τ.

Two important properties of $W_\tau(z)$ are:

- $W_\tau(z)$ has a block-triangular structure at infinity ($z \to \infty$) since $W_\tau(z)\big|_{z\to\infty} = E_\tau$
- as τ varies, $W_\tau(z)$ has a recurrent structure as specified by the following equation:

$$W_{\tau+1}(z) = \Delta'_p(z^{-1})W_\tau(z)\Delta_m(z) \qquad (11)$$

where

$$\Delta_k(z) = \begin{bmatrix} 0 & z^{-1}I_k \\ I_{k(T-1)} & 0 \end{bmatrix},$$

see e.g. (Colaneri, Longhi, 1995). Interestingly enough, $\Delta_k(z)$ is inner, i.e. $\Delta'_\tau(z^{-1})\Delta_k(z) = I_{kT}$.

The lifted reformulation shares the same structural properties as the original periodic system. In particular, system (9) is reachable (observable) if and only if system (1) is reachable (observable) at time τ; system (9) is controllable (reconstructable, detectable, stabilizable) if and only if system (1) is controllable (reconstructable, detectable, stabilizable).
Moreover, system (9) is stable if and only if system (1) is stable.

3.2 Cyclic reformulation

In the lifted reformulation, only the output and input vectors were enlarged, whereas the dimension of the state space was preserved. In the cyclic reformulation, the state space is enlarged too. To explain the basic underlying idea, consider a signal $v(t)$, $t=\tau,\tau+1,\tau+2,...$, where the initial tag τ belongs to $[0,T\text{-}1]$, and define the enlarged signal

$$\bar{v}_\tau(t) = vec(\bar{v}_{\tau 1}(t), \bar{v}_{\tau 2}(t), \ldots, \bar{v}_{\tau T}(t)) \qquad (12.a)$$

$$\bar{v}_{\tau i}(t) = \begin{cases} v(t), & t = \tau + i - 1 + kT,\ k = 0, \pm 1, \pm 2, \cdots \\ 0 & otherwise \end{cases} \qquad (12.b)$$

In other words, the signal $v(t)$ cyclically shifts along the row blocks of $\bar{v}_\tau(t)$:

$$\bar{v}_\tau(t) = \begin{bmatrix} v(t) \\ 0 \\ \vdots \\ 0 \end{bmatrix}, t = \tau + kT, \quad \bar{v}_\tau(t) = \begin{bmatrix} 0 \\ v(t) \\ \vdots \\ 0 \end{bmatrix}, t = \tau + 1 + kT, \cdots, \quad \bar{v}_\tau(t) = \begin{bmatrix} 0 \\ 0 \\ \vdots \\ v(t) \end{bmatrix}, t = \tau + T - 1 + kT$$

The cyclic reformulation is now defined as

$$\bar{x}_\tau(t+1) = \bar{F}_\tau \bar{x}_\tau(t) + \bar{G}_\tau \bar{u}_\tau(t) \tag{13.a}$$

$$\bar{y}_\tau(t) = \bar{H}_\tau \bar{x}_\tau(t) + \bar{E}_\tau \bar{u}_\tau(t), \tag{13.b}$$

where:

$$\bar{F}_\tau = \begin{bmatrix} 0 & 0 & \cdots & 0 & A(\tau+T-1) \\ A(\tau) & 0 & \cdots & 0 & 0 \\ 0 & A(\tau+1) & \cdots & 0 & 0 \\ \cdots & \cdots & \ddots & \cdots & \cdots \\ 0 & 0 & \cdots & A(\tau+T-2) & 0 \end{bmatrix}$$

$$\bar{G}_\tau = \begin{bmatrix} 0 & 0 & \cdots & 0 & B(\tau+T-1) \\ B(\tau) & 0 & \cdots & 0 & 0 \\ 0 & B(\tau+1) & \cdots & 0 & 0 \\ \cdots & \cdots & \ddots & \cdots & \cdots \\ 0 & 0 & \cdots & B(\tau+T-2) & 0 \end{bmatrix}$$

$$\bar{H}_\tau = blockdiag\,\{C(\tau), C(\tau+1), \cdots, C(\tau+T-1)\}$$

$$\bar{E}_\tau = blockdiag\,\{D(\tau), D(\tau+1), \cdots, D(\tau+T-1)\}.$$

The dimensions of the state, input and output spaces of the cyclic reformulation are those of the original periodic systems multiplied by T.

Remark 2

For the simple case $T=2$, one has:

$$\bar{x}_0(t) = \begin{cases} \begin{bmatrix} x(t) \\ 0 \end{bmatrix} & t = even \\ \begin{bmatrix} 0 \\ x(t) \end{bmatrix} & t = odd \end{cases} \qquad \bar{x}_1(t) = \begin{cases} \begin{bmatrix} 0 \\ x(t) \end{bmatrix} & t = even \\ \begin{bmatrix} x(t) \\ 0 \end{bmatrix} & t = odd \end{cases}$$

and the signals $\bar{u}_0(t)$, $\bar{u}_1(t)$, $\bar{y}_0(t)$ and $\bar{y}_1(t)$ are defined analogously. Moreover, the system matrices for $\tau=0$ and $\tau=1$, are given by

$$\overline{F}_0 = \begin{bmatrix} 0 & A(1) \\ A(0) & 0 \end{bmatrix}, \quad \overline{F}_1 = \begin{bmatrix} 0 & A(0) \\ A(1) & 0 \end{bmatrix}, \quad \overline{G}_0 = \begin{bmatrix} 0 & B(1) \\ B(0) & 0 \end{bmatrix}, \quad \overline{G}_1 = \begin{bmatrix} 0 & B(0) \\ B(1) & 0 \end{bmatrix}$$

$$\overline{H}_0 = \begin{bmatrix} C(0) & 0 \\ 0 & C(1) \end{bmatrix}, \quad \overline{H}_1 = \begin{bmatrix} C(1) & 0 \\ 0 & C(0) \end{bmatrix}, \quad \overline{E}_0 = \begin{bmatrix} D(0) & 0 \\ 0 & D(1) \end{bmatrix}, \quad \overline{E}_1 = \begin{bmatrix} D(1) & 0 \\ 0 & D(0) \end{bmatrix}$$

■

Obviously, the transfer function of the cyclic reformulation is given by

$$\overline{W}_\tau(z) = \overline{H}_\tau(zI - \overline{F}_\tau)^{-1}\overline{G}_\tau + \overline{E}_\tau. \tag{14}$$

This transfer function will be called the *cyclic transfer function* of system (1) at time τ.

Two important properties of $\overline{W}_\tau(z)$ are:

- $W_\tau(z)$ has a block-diagonal structure at infinity ($z \to \infty$) since $\overline{W}_\tau(z)\big|_{z\to\infty} = \overline{E}_\tau$
- as τ varies, $\overline{W}_\tau(z)$ changes only throughout a permutation of its input and output vectors. Precisely,

$$\overline{W}_{\tau+1}(z) = \overline{\Delta}_p' \overline{W}_\tau(z) \overline{\Delta}_m \tag{15}$$

where

$$\overline{\Delta}_k = \begin{bmatrix} 0 & I_k \\ I_{k(T-1)} & 0 \end{bmatrix}. \tag{16}$$

As in the case of the lifted reformulation, the structural properties of the cyclic reformulation are determined by the properties of the original system. However, there are some slight but notable differences due to the extended dimension of the state space in the cyclic reformulation. For example, if system (1) is reachable at time τ, system (9) is reachable too, whereas system (13) is not necessarily reachable. The appropriate statement is that system (1) is reachable (observable) at *each time* if and only if system (13) is reachable (observable) (at any arbitrary time τ, which parametrizes eq. (13)). This reflects the fact that if system (13) is reachable for a parametrization τ, it is reachable for any parametrization. As for the remaining structural properties, one can recognize that system (1) is controllable (reconstructable, detectable, stabilizable) if and only if system (13) is controllable (reconstructable, detectable, stabilizable). Furthermore, system (13) is stable if and only if system (1) is stable; indeed, the eigenvalues of $\overline{F}_\tau$ are the *T-th* roots of the characteristic multipliers of system (1).

Finally, transfer functions (10) and (14) are obviously related each other. Simple computations show that

$$\overline{W_\tau}(z) = \widehat{\Delta}_p(z) W_\tau(z^T) \widehat{\Delta}_m(z^{-1}) \tag{17}$$

where

$$\widehat{\Delta}_k(z) = diag\{I_k, z^{-1}I_k, \cdots, z^{-T+1}I_k\}.$$

Notice that

$$\widehat{\Delta}_k(z)\widehat{\Delta}_k(z^{-1}) = I_k. \tag{18}$$

4. ADJOINT SYSTEM AND PERIODIC SYMPLECTIC PENCIL

As seen above, system (1) admits the lifted reformulation at time τ given by (9). Such a time invariant reformulation has the transfer function $W_\tau(z)$. As is well known, the adjoint of system (9) is the following descriptor system:

$$F_\tau{}'\lambda_\tau(k+1) = \lambda_\tau(k) - H_\tau{}'v_\tau(k) \tag{19.a}$$

$$\varsigma_\tau(k) = G_\tau{}'\lambda_\tau(k+1) + E_\tau{}'v_\tau(k) \tag{19.b}$$

the transfer function of which is $W_\tau(z^{-1})'$.

Consider now the periodic system in descriptor form:

$$A(t)'\lambda(t+1) = \lambda(t) - C(t)'v(t) \tag{20.a}$$

$$\varsigma(t) = B(t)'\lambda(t+1) + D(t)'v(t). \tag{20.b}$$

It is easy to see that, if one sets

$$v_\tau(k) = [v(\tau+kT)' \quad v(\tau+kT+1)' \cdots v(\tau+kT+T-1)']'$$
$$\varsigma_\tau(k) = [\varsigma(\tau+kT)' \quad \varsigma(\tau+kT+1)' \cdots \varsigma(\tau+kT+T-1)']'$$

then (19) is the lifted reformulation at time τ of system (20). This is why we are well advised to name (20) the *adjoint system* of (1).

We are now in a position to define the *periodic symplectic pencil* relative to system (1) and the associated adjoint system (20). Consider the symplectic pencil associated with the pair of time-invariant systems (9) and (19). Such a pencil is obtained by putting the two systems in a feedback configuration by letting

$$v_\tau(k) = y_\tau(k)$$
$$u_\tau(k) = \sigma\varsigma_\tau(k)$$

where σ is either -1 or +1. Correspondingly, the symplectic pencil is given by the descritor type equations:

$$\begin{bmatrix} I & G_\tau(\sigma^{-1}I+E_\tau'E_\tau)^{-1}G_\tau' \\ 0 & [F_\tau-G_\tau(\sigma^{-1}I+E_\tau'E_\tau)^{-1}E_\tau'H_\tau]' \end{bmatrix}\begin{bmatrix} x_\tau(k+1) \\ \lambda_\tau(k+1) \end{bmatrix}=\begin{bmatrix} F_\tau-G_\tau(\sigma^{-1}I+E_\tau'E_\tau)^{-1}E_\tau'H_\tau & 0 \\ -\sigma^{-1}H_\tau'(\sigma^{-1}I+E_\tau E_\tau')^{-1}H_\tau & I \end{bmatrix}\begin{bmatrix} x_\tau(k) \\ \lambda_\tau(k) \end{bmatrix}$$

On the other hand, it is easy to see that this equation can be obtained as a lifted reformulation of the following periodic system:

$$\begin{bmatrix} I & B(t)(\sigma^{-1}I+D(t)'D(t))^{-1}B(t)' \\ 0 & [A(t)-B(t)(\sigma^{-1}I+D(t)'D(t))^{-1}D(t)'H(t)]' \end{bmatrix}\begin{bmatrix} x(t+1) \\ \lambda(t+1) \end{bmatrix}=\begin{bmatrix} A(t)-B(t)(\sigma^{-1}I+D(t)'D(t))^{-1}D(t)'C(t) & 0 \\ -\sigma^{-1}C(t)'(\sigma^{-1}I+D(t)'D(t))^{-1}C(t) & I \end{bmatrix}\begin{bmatrix} x(t) \\ \lambda(t) \end{bmatrix} \tag{21}$$

which can be seen as the feedback configuration of systems (1) and (20) as indicated in Fig.1.

Notice that the exixtence of the inverses appearing in the above expressions is guaranteed only if $\sigma = +1$.

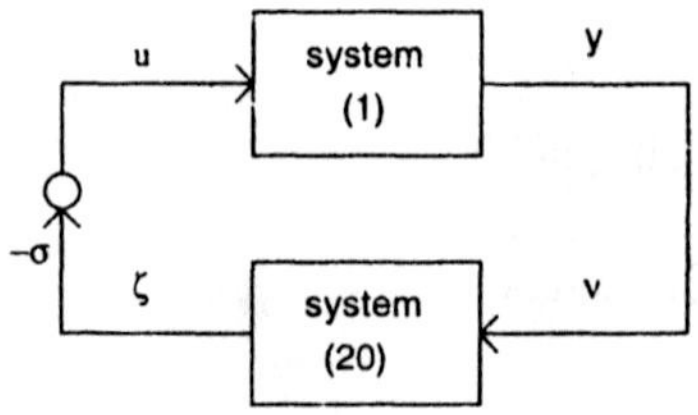

Fig. 1

By letting $\lambda(t) = P(t)x(t)$, the symplectic pencil (21) gives rise to a periodic Riccati equation in $P(t)$. If $\sigma=+1$, the usual (H_2) Riccati equation of optimal periodic control arises, whereas if $\sigma = -1$ the Riccati equation for the H_∞ analysis problem is recovered, see Sect. 3. For the periodic H_2 Riccati equation, the interested reader is referred to (Bittanti, Colaneri, De Nicolao, 1991).

Associated with the periodic pencil (21), we define the *characteristic polynomial equation* at τ:

$$\det\left\{\lambda\begin{bmatrix} I & G_\tau(\sigma^{-1}I+D_\tau'D_\tau)^{-1}G_\tau' \\ 0 & [F_\tau-G_\tau(\sigma^{-1}I+D_\tau'D_\tau)^{-1}D_\tau'H_\tau]' \end{bmatrix}-\begin{bmatrix} F_\tau-G_\tau(\sigma^{-1}I+D_\tau'D_\tau)^{-1}D_\tau'H_\tau & 0 \\ -\sigma^{-1}H_\tau'(\sigma^{-1}I+D_\tau D_\tau')^{-1}H_\tau & I \end{bmatrix}\right\}=0$$

The singularities of such an equation are the so-called *characteristic multipliers* at τ of the periodic pencil (21). In this connection, it is worthwhile pointing out that, if $\lambda \neq 0$ is a singularity for the characteristic polynomial equation at τ, it is also a singularity for the characteristic polynomial equation at $\tau+1$, and therefore at any other time point. Indeed, it is easy to show that $\lambda \neq 0$ is a solution of the above equation iff

$$\det\left[\frac{I}{\sigma}+W_\tau(\lambda^{-1})'W_\tau(\lambda)\right]=0.$$

In view if (11),

$$\frac{I}{\sigma}+W_{\tau+1}(\lambda^{-1})'W_{\tau+1}(\lambda)=\Delta_m(\lambda^{-1})'\left[\frac{I}{\sigma}+W_\tau(\lambda^{-1})'\Delta_m(\lambda)\Delta_m(\lambda^{-1})'W_\tau(\lambda)\right]\Delta_m(\lambda)$$
$$=\Delta_m(\lambda^{-1})'\left[\frac{I}{\sigma}+W_\tau(\lambda^{-1})'W_\tau(\lambda)\right]\Delta_m(\lambda)$$

so that

$$\det\left[\frac{I}{\sigma}+W_{\tau+1}(\lambda^{-1})'W_{\tau+1}(\lambda)\right]=0.$$

Hence, λ is also a characteristic multiplier at $\tau+1$ of the periodic pencil (21).

In conclusion, for the nonzero characteristic multipliers of the symplectic pencil (21), it is not necessary to specify an associated time point.
Moreover, as is well known due the symplectic property, if $\lambda \neq 0$ is a characteristic multipler of the symplectic pencil (21), then λ^{-1} is a characteristic multiplier as well.

5. ZEROS AND POLES

As is well known, the zeros of time-invariant systems can be characterized in terms of the blocking property: associated with any zero there is an input function of exponential type such that the output is identically zero for a suitable choice of the initial state. The definition of zeros and poles of a periodic system can be introduced starting from any of the two time-invariant reformulations introduced in Sect. 3, see (Bolzern, Colaneri, Scattolini, 1986) and (Grasselli, Longhi, 1988).

Definition 1

(i) The complex number z is an invariant [transmission] zero at time τ of system (1), if it is an invariant [resp. transmission] zero of the associated lifted system (9).
(ii) The complex number z is a pole at time τ of system (1), if it is a pole of the associated lifted system (9). ■

To better understand this definition, we elaborate further as follows.

Periodic zero blocking property

Let's begin with the invariant zeros of system (1) at time τ and first focus on tall systems ($p \geq m$). Based on Def. 1, consider an invariant zero λ of system (9). Then, consider the system matrix of the lifted system, namely:

$$\Sigma_\tau(\lambda) = \begin{bmatrix} \lambda I - F_\tau & -G_\tau \\ H_\tau & E_\tau \end{bmatrix}.$$

Then, there exist $\eta = [\eta_0' \quad \eta_1' \quad \cdots \quad \eta_{T-1}']'$ and $x_\tau(0)$, not simultaneously zero, such that

$$\Sigma_\tau(\lambda) \begin{bmatrix} x_\tau(0) \\ \eta \end{bmatrix} = 0,$$

i.e. $F_\tau x_\tau(0) + G_\tau \eta = \lambda x_\tau(0)$ and $H_\tau x_\tau(0) + E_\tau \eta = 0$. Furthermore, the well known blocking property for the invariant zeros of time-invariant systems entails that system (9) with input signal given by $u_\tau(k) = \eta \lambda^k$, $k \geq 0$, and initial state $x_\tau(0)$, results in the null output: $y_\tau(k) = 0$, $k \geq 0$. By recalling the definition of the lifted signals $u_\tau(k)$ and $y_\tau(k)$ in terms of $u(t)$ and $y(t)$, respectively, (see (8.b) and (8.c)), this implies that there exists an *Exponentially Periodic Signal* (EPS) $u(t+kT) = u(t)\lambda^k$, $t \in [\tau, \tau+T-1]$, with $u(\tau+i) = \eta_i$, $i \in [0, T-1]$, and initial state $x(\tau) = x_\tau(0)$ such that $y(t) = 0$, $t \geq \tau$.

Time-invariance of the zeros

It is easy to see that, if $\lambda \neq 0$ is an invariant zero at time τ, then it is also an invariant zero at time $\tau+1$. Actually it suffices to select as input $u(t+kT) = u(t)\lambda^k$, $t \in [\tau+1, \tau+T]$, with $u(\tau+i+1) = \eta_i$, $i \in [0, T-2]$ and $u(\tau+T) = \lambda \eta_0$, and initial state $x(\tau+1) = A(\tau)x_\tau(0) + B(\tau)\eta_0$ to ensure that $y(t) = 0$, $t \geq \tau+1$. Note that, for consistency, one has to check that

$\bar{\eta} = [\eta_1' \quad \cdots \quad \eta_{T-1}' \quad \lambda\, \eta_0']'$

and

$x(\tau+1) = A(\tau)x_\tau(0) + B(\tau)\eta_0$

are not simultaneously zero. If $\eta \neq 0$, this is obvious. If instead $\eta = 0$, then the system matrix interpretation leads to $F_\tau x_\tau(0) = \lambda x_\tau(0)$. Moreover, $\bar{\eta} = 0$ and $x(\tau+1) = A(\tau)x_\tau(0)$. To show that $x(\tau+1) \neq 0$, notice that $A(\tau)x_\tau(0) = 0$ would imply that $A(\tau+T-1)A(\tau+T-2)\cdots A(\tau)x_\tau(0) = 0$, i.e. $F_\tau x_\tau(0) = 0$. Therefore, it would turn out that $\lambda = 0$, in contradiction with our initial assumption.

As for the transmission zeros, again for tall systems, Def. 1 implies that λ is a transmission zero for a periodic system at time τ if there exists $\eta \neq 0$ such that $W_\tau(z)\big|_{z \to \lambda} \eta = 0$. From (12), since $\Delta_k(\lambda)$ is invertible if $\lambda \neq 0$, it is apparent that the nonzero transmission zeros of the periodic system do not change with τ as well.

The interpretation for the case of fat systems, i.e. $p<m$, both for invariant and transmission zeros for system (1) is easily derivable by transposition.

Time-invariance of the poles

Turning to the poles, it is apparent that they consistute a subset of the characteristic multipliers of system (1). Not differently from the zeros, the nonzero poles, together with their multiplicities, are in fact independent of τ. The proof of this statement, omitted here for the sake of brevity, can be worked out in terms of an impulse response characterization of the system.■

In analogy with the time invariant case, we define the notion of a minimum phase system as follows.

Definition 2

When all zeros and poles of a periodic system belong to the open unit disk, the system is said to be *minimum phase.* ■

Remark 3

A change of basis in the state-space amounts to transforming the state vector $x(t)$ into a new vector $\tilde{x}(t) = S(t)x(t)$, where $S(t)$ is T-periodic and nonsingular for each t. If one performs a change of basis on system (1), the new system is characterized by the quadruplet $(\tilde{A}(\cdot), \tilde{B}(\cdot), \tilde{C}(\cdot), \tilde{D}(\cdot))$:

$$\begin{aligned}
\tilde{A}(t) &= S(t+1)A(t)S(t)^{-1}, \\
\tilde{B}(t) &= S(t+1)B(t), \\
\tilde{C}(t) &= C(t)S(t)^{-1}, \\
\tilde{D}(t) &= D(t).
\end{aligned}$$

As usual, we will say that the original and the new quadruplet are *algebraically equivalent*. Stability, reachability, observability, etc., poles and zeros are not affected by a change of state coordinates.

6. L_2, L_∞ AND HANKEL NORMS OF PERIODIC SYSTEMS

6.1 L_2 norm

In this section, the notion and time-domain characterization of the L_2-norm for a periodic system is first introduced. To this purpose, it is first noted that the transfer functions of the lifted and cyclic reformulations $W_\tau(z)$ and $\overline{W}_\tau(z)$ have coincident L_2 norm. Indeed, from (17) and (18)

$$
\begin{aligned}
\left\|\overline{W}_\tau(z)\right\|_2 &:= \\
&= \left(\frac{1}{2\pi} tr \int_{-\pi}^{\pi} \overline{W}_\tau(e^{-j\vartheta})\overline{W}_\tau(e^{j\vartheta}) d\vartheta\right)^{1/2} = \\
&= \left(\frac{1}{2\pi} tr \int_{-\pi}^{\pi} \hat{\Delta}_m(e^{j\vartheta}) W_\tau(e^{-jT\vartheta}) W_\tau(e^{jT\vartheta}) \hat{\Delta}_p(e^{-j\vartheta}) d\vartheta\right)^{1/2} \\
&= \left(\frac{1}{2\pi} tr \int_{-\pi}^{\pi} W_\tau(e^{-jT\vartheta}) W_\tau(e^{jT\vartheta}) d\vartheta\right)^{1/2} = \\
&= \left(\frac{1}{2\pi T} tr \int_{-\pi T}^{\pi T} W_\tau(e^{-j\vartheta}) W_\tau(e^{j\vartheta}) d\vartheta\right)^{1/2} = \\
&= \left(\frac{1}{2\pi} tr \int_{-\pi}^{\pi} W_\tau(e^{-j\vartheta}) W_\tau(e^{j\vartheta}) d\vartheta\right)^{1/2} = \\
&=: \left\|W_\tau(z)\right\|_2
\end{aligned}
$$

Moreover, from (11) and (17) it is apparent that

$$tr(W'_{\tau+1}(e^{-j\vartheta})W_{\tau+1}(e^{j\vartheta})) = tr(W'_\tau(e^{-j\vartheta})W_\tau(e^{j\vartheta}))$$

$$tr(\overline{W}'_{\tau+1}(e^{-j\vartheta})\overline{W}'_{\tau+1}(e^{j\vartheta})) = tr(\overline{W}'_\tau(e^{-j\vartheta})\overline{W}_\tau(e^{j\vartheta}))$$

respectively, so that both $\left\|W_\tau(z)\right\|_2$ and $\left\|\overline{W}_\tau(z)\right\|_2$ are in fact independent of τ. These considerations are at the basis of the following:

Definition 3
Given system (1), the quantity

$$\left\|T_{yu}\right\|_2 := \left\|W_\tau(z)\right\|_2 = \left\|\overline{W}_\tau(z)\right\|_2, \quad \forall \tau$$

is named the L_2 norm of the periodic system. ■

Obviously, the above norm is bounded if the periodic system does not have unit modulus poles. Moreover, in case of stable systems, the so defined L_2 norm can be given interesting time-domain interpretations in terms of Lyapunov equations and impulse response.

Lyapunov equation interpretation
Consider the two periodic Lyapunov equations:

$$A(t)'P(t+1)A(t) + C(t)'C(t) = P(t) \tag{PLE1}$$

$$A(t)Q(t)A'(t) + B(t)B(t)' = Q(t+1) \tag{PLE2}$$

As is well known, see e.g. (Bolzern, Colaneri, 1987) the stability of $A(\cdot)$ entails the existence of a unique periodic solution of both these equations. From such solutions, the L_2 norm can be computed as follows:

$$\begin{aligned}\left\|T_{yu}\right\|_2 &= \left[tr\left(\sum_{i=0}^{T-1} B'(i)P(i+1)B(i)+D'(i)D(i)\right)\right]^{1/2} = \\ &= \left[tr\left(\sum_{i=0}^{T-1} C(i)Q(i)C'(i)+D(i)D'(i)\right)\right]^{1/2}\end{aligned} \tag{22}$$

Let's show the correctness of the first equality, the second being provable in a completely analogous way. Actually, from the time-invariant theory,

$$\left\|T_{yu}\right\|_2 = \left\|\overline{W}_\tau(z)\right\|_2 = \left[tr(\overline{G}_\tau'\overline{P}_\tau\overline{G}_\tau + \overline{E}_\tau'\overline{E}_\tau)\right]^{1/2},$$

where $\overline{P}_\tau$ is the unique solution of the Lyapunov equation:

$$\overline{F}_\tau{}'\overline{P}_\tau\overline{F}_\tau + \overline{H}_\tau'\overline{H}_\tau = \overline{P}_\tau \qquad \text{(ALEC1)}.$$

Due to the particular structure of the matrices in the cyclic reformulation, it is easy to recognize that

$$\overline{P}_\tau = diag\left\{P(\tau), P(\tau+1), \dots, P(\tau+T-1)\right\}$$

which immediately leads to our claim. Analogously, it can be easily checked that

$$\overline{Q}_\tau = diag\left\{Q(\tau), Q(\tau+1), \dots, Q(\tau+T-1)\right\}$$

is the unique solution of

$$\overline{F}_\tau\overline{Q}_\tau\overline{F}_\tau{}' + \overline{G}_\tau\overline{G}_\tau{}' = \overline{Q}_\tau \qquad \text{(ALEC2)}$$

Finally, it is immediately seen that the value at $t=\tau$ of the periodic solution of (PLE1) and (PLE2) are the constant solutions of the Lyapunov equations associated with the lifted reformulation of the periodic system at $t=\tau$:

$$F_\tau{}'\,P(\tau)F_\tau + H_\tau{}'\,H_\tau = P(\tau) \qquad \text{(ALEL1)}$$

$$F_\tau Q(\tau)F_\tau{}' + G_\tau G_\tau{}' = Q(\tau) \qquad \text{(ALEL2)}$$

Impulse response interpretation

Starting from the impulse response interpretation for the L_2 norm of time-invariant systems, one can derive the following interpretation of the L_2 norm of periodic systems:

$$\|T_{yu}\|_2 = \left[\sum_{t=\tau}^{\infty} \sum_{j=1}^{m} \sum_{i=0}^{T-1} \|h^{i,j}(t)\|^2 \right]^{1/2}$$

where $h^{i,j}(t)$ is the response of system (1) to an impulse applied to the j-th input component at time $\tau + i$ with initial condition $x(\tau) = 0$.

Remark 4
The L_2 norm is used many times in control problems. A typical problem is the following *disturbance attenuation problem*. Consider the periodic system:

$$x(t+1) = A(t)x(t) + B(t)u(t) + \bar{B}(t)v(t) \qquad (23.a)$$
$$y(t) = C(t)x(t) + D(t)u(t) + \bar{D}(t)v(t) \qquad (23.b)$$

where $u(t)$ is a disturbance term and $v(t)$ is the control input. The L_2 norm full information feedback control problem can be stated as the problem of finding a stabilizing periodic state control law

$$v(t) = K(t)x(t)$$

so as to minimize the L_2 norm (from $u(t)$ to $y(t)$) of the corrisponding closed-loop system.

For a given stabilizing periodic $K(\cdot)$, one can exploit characterization (22) of the L_2 norm where $\bar{P}(\cdot)$ is the unique T-periodic solution of the Lyapunov equation:

$$[A(t) + B(t)K(t)]'\bar{P}(t+1)[A(t) + B(t)K(t)] + C(t)'C(t) = \bar{P}(t).$$

We leave to the reader to verify that the above equation can be equivalently written as:

$$\begin{aligned}\bar{P}(t) &= A(t)'\bar{P}(t+1)A(t) + C(t)'C(t) \\ &\quad - [A(t)'\bar{P}(t+1)\bar{B}(t) + C(t)'\bar{D}(t)](\bar{D}(t)'\bar{D}(t) + \bar{B}(t)'\bar{P}(t+1)\bar{B}(t))^{-1}[A(t)'\bar{P}(t+1)\bar{B}(t) + C(t)'\bar{D}(t)]' \\ &\quad + [K(t) - K^0(t)]'(\bar{D}(t)'\bar{D}(t) + \bar{B}(t)'\bar{P}(t+1)\bar{B}(t))[K(t) - K^0(t)]\end{aligned} \qquad (24)$$

where

$$K^0(t) = -(\bar{B}(t)'\bar{P}(t+1)\bar{B}(t) + C(t)'\bar{D}(t))^{-1}[\bar{B}(t)'\bar{P}(t+1)A(t) + \bar{D}(t)'C(t)]. \qquad (25)$$

Now, if eq. (24) admits a solution $P(t)$ with $K(t) = K^0(t)$, then $P(t) \le \bar{P}(t), \forall t$, for any periodic solution $\bar{P}(t)$associated with any other periodic stabilizing matrix $K(\cdot)$. The proof of this statement is based on monotonicity arguments and is left to the patient reader. Therefore, $K(t) = K^0(t)$ leads to the minimum attainable value of performance index (22).

Eq. (24) with $K(t) = K^0(t)$ becomes the standard periodic Riccati equation:

$$P(t) = A(t)'P(t+1)A(t) + C(t)'C(t)$$
$$-[A(t)'P(t+1)\overline{B}(t) + C(t)'\overline{D}(t)](\overline{D}(t)'\overline{D}(t) + \overline{B}(t)'P(t+1)\overline{B}(t))^{-1}[A(t)'P(t+1)\overline{B}(t) + C(t)'\overline{D}(t)]'$$

(L_2-PRE)

Correspondingly, gain $K^0(t)$ is given by (25) with $\overline{P}(t)$ replaced by the solution $P(t)$ of the (L_2-PRE).

Equation (L_2-PRE) has been extensively studied in the last years. A necessary and sufficient condition for the existence of the (unique) periodic stabilizing solution is that the pair $(A(\cdot), B(\cdot))$ is stabilizable and the symplectic system

$$\begin{bmatrix} I & \overline{B}(t)(I+\overline{D}(t)'\overline{D}(t))^{-1}\overline{B}(t)' \\ 0 & [A(t)-\overline{B}(t)(I+\overline{D}(t)'\overline{D}(t))^{-1}\overline{D}(t)'H(t)]' \end{bmatrix} \begin{bmatrix} x_\tau(k+1) \\ \lambda_\tau(k+1) \end{bmatrix} = \begin{bmatrix} A(t)-\overline{B}(t)(I+\overline{D}(t)'\overline{D}(t))^{-1}\overline{D}(t)'C(t) & 0 \\ -C(t)'(I+\overline{D}(t)'\overline{D}(t))^{-1}C(t) & I \end{bmatrix} \begin{bmatrix} x_\tau(k) \\ \lambda_\tau(k) \end{bmatrix}$$

does not have characteristic multipliers on the unit circle. This statement is a generalization of a result given for the filtering case in (Bittanti, Colaneri, De Nicolao, 1988). Notice that this pencil is derived from pencil (21) with $\sigma = +1, D(t) \to \overline{D}(t), B(t) \to \overline{B}(t)$.

Under such a condition, as already said, $v(t) = K^0(t)x(t)$ is the optimal periodic control law in the the L_2 norm sense. Interestingly enough, this is exactly the optimal control law of the optimal periodic control problem with full information. Precisely, with reference to system (23) with $B(t) = 0, D(t) = 0, \forall t$, this problem consists in minimizing

$$J = \sum_{t=\tau}^{\infty} \|y(t)\|^2$$

with respect to $v(\cdot)$ for a given initial state $x(\tau)$, see e.g. (Bittanti,Colaneri, De Nicolao, 1991).

6.2 L_∞ norm

As in the previous case, it is first noted that the transfer functions of the lifted and cyclic reformulations $W_\tau(z)$ and $\overline{W}_\tau(z)$ have coincident L_∞ norm. Indeed, denoting by $\lambda_{\max}(A)$ the maximum eigenvalue of a matrix A, from (17) and (18)

$$
\begin{aligned}
\left\|\overline{W}_\tau(z)\right\|_\infty &= \left[\max_{\vartheta} \ \lambda_{\max}(\overline{W}_\tau'(e^{-j\vartheta})\overline{W}_\tau(e^{j\vartheta}))\right]^{1/2} = \\
&= \left[\max_{\vartheta} \ \lambda_{\max}(\hat{\Delta}_m'(e^{j\vartheta})W_\tau'(e^{-j\vartheta T})\hat{\Delta}_p(e^{-j\vartheta})\hat{\Delta}_p'(e^{j\vartheta})W_\tau(e^{j\vartheta T})\hat{\Delta}_m(e^{-j\vartheta}))\right]^{1/2} = \\
&= \left[\max_{\vartheta} \ \lambda_{\max}(\hat{\Delta}_m'(e^{j\vartheta})W_\tau'(e^{-j\vartheta T})W_\tau(e^{j\vartheta T})\hat{\Delta}_m(e^{-j\vartheta}))\right]^{1/2} = \\
&= \left[\max_{\vartheta} \ \lambda_{\max}(W_\tau'(e^{-j\vartheta})W_\tau(e^{j\vartheta}))\right]^{1/2} = \left\|W_\tau(z)\right\|_\infty
\end{aligned}
$$

Moreover, both norms do not depend on τ. Indeed, focusing on the lifted reformulation, from (11), it follows that

$$
\begin{aligned}
\left\|W_{\tau+1}(z)\right\|_\infty &= \left[\max_{\vartheta} \ \lambda_{\max}(W_{\tau+1}'(e^{-j\vartheta})W_{\tau+1}(e^{j\vartheta}))\right]^{1/2} = \\
&= \left[\max_{\vartheta} \ \lambda_{\max}(\Delta_m'(e^{-j\vartheta})W_\tau'(e^{-j\vartheta})\Delta_p(e^{j\vartheta})\Delta_p'(e^{-j\vartheta})W_\tau(e^{j\vartheta})\Delta_m(e^{j\vartheta}))\right]^{1/2} = \\
&= \left[\max_{\vartheta} \ \lambda_{\max}(\Delta_m'(e^{-j\vartheta})W_\tau'(e^{-j\vartheta})W_\tau(e^{j\vartheta})\Delta_m(e^{j\vartheta}))\right]^{1/2} = \\
&= \left[\max_{\vartheta} \ \lambda_{\max}(W_\tau'(e^{-j\vartheta})W_\tau(e^{j\vartheta}))\right]^{1/2} = \left\|W_\tau(z)\right\|_\infty
\end{aligned}
$$

Therefore the following definition makes sense:

Definition 4
Given system (1), the quantity

$$
\left\|T_{yu}\right\|_\infty := \left\|W_\tau(z)\right\|_\infty = \left\|\overline{W}_\tau(z)\right\|_\infty, \quad \forall\tau
$$

is called the L_∞ norm of the periodic system. ■

Obviously, this norm is bounded if the periodic system does not have unit modulus characteristic multipliers.

Input-output interpretation

From the well known input-output characterization of the L_∞ norm for stable time-invariant systems, the following input-output characterization for the stable periodic systems in terms of its L_2 - induced norm can be derived:

$$
\left\|T_{yu}\right\|_\infty := \sup_{u\in\Xi} \frac{\left\|y(\cdot)\right\|_2}{\left\|u(\cdot)\right\|_2} \quad , \quad \Xi = \{u : u \in L_2[\tau,\infty), u \neq 0\}
$$

where by the norm of a signal $q(\cdot) \in L_2[\tau,\infty)$ we mean:

$$\|q(\cdot)\|_2^2 = \sum_{t=\tau}^{\infty} q(t)'q(t).$$

Riccati equation interpretation

An important question is whether the infinite norm of a periodic system is bounded by some positive value γ. The reply is given in (Colaneri, 1991) and can be stated as follows: $A(\cdot)$ is stable and

$$\|T_{yu}\|_{\infty} < \gamma$$

if and only if there exists the T-periodic positive semidefinite solution of the Riccati equation :

$$\begin{aligned} P(t) &= A(t)'P(t+1)A(t)+C(t)'C(t) \\ &+[A(t)'P(t+1)B(t)+C(t)'D(t)](\gamma^2 I - D(t)'D(t)-B(t)'P(t+1)B(t))^{-1}[A(t)'P(t+1)B(t)+C(t)'D(t)]' \end{aligned}$$

(L$_\infty$-PRE)

such that

i) $\gamma^2 I - D(t)'D(t) - B(t)'P(t+1)B(t) > 0, \forall t$

ii) $A(t)+B(t)(\gamma^2 I - D(t)'D(t)-B(t)'P(t+1)B(t))^{-1}[A(t)'P(t+1)B(t)+C(t)'D(t)]'$ is stable.

In the present framework a solution of the (L$_\infty$-PRE) satisfying this last condition is said to be stabilizing. It can be proven that, if there exists such a solution, it is the unique stabilizing solution.

Remark 5

The solution $P(\cdot)$ of eq. (L$_\infty$-PRE) at $t=\tau$ (with properties i) and ii)), can be also related to the optmization problem for system (1) with non zero initial condition $x(\tau)$:

$$\sup_{u \in L_2[\tau,\infty)} \|y\|_2^2 - \gamma^2 \|u\|_2^2$$

It is easily seen that such a problem has the solution

$$\sup_{u \in L_2[\tau,\infty)} \|y\|_2^2 - \gamma^2 \|u\|_2^2 = x(\tau)'P(\tau)x(\tau)$$

Indeed, in view of system (1) and equation (L$_\infty$-PRE), it follows that

$$\begin{aligned} x(t)'P(t)x(t) - x(t+1)'P(t+1)x(t+1) &= x(t)'[P(t)-A(t)'P(t+1)A(t)]x(t) + x(t)'P(t+1)B(t)u(t) + \\ &+ u(t)'B(t)'P(t+1)x(t) = y(t)'y(t) - \gamma^2 u(t)'u(t) + q(t)'q(t) \end{aligned}$$

where

$$q(t)=V(t)^{-1/2}\left[B(t)'P(t+1)A(t)+D(t)'C(t)\right]-V(t)^{1/2}u(t)$$
$$V(t)=\left[\gamma^2 I-B(t)'P(t+1)B(t)-D(t)'D(t)\right]>0,\ \forall t$$

By taking the sum of both members from $t=\tau$ to $t=\infty$, we have

$$x(\tau)'P(\tau)x(\tau)=\|y\|_2^2-\gamma^2\|u\|_2^2+\|q\|_2^2$$

so that the conclusion easily follows by noticing that $q\equiv 0$ corresponds to the optimal input

$$u(t)=V(t)^{-1}\left[B(t)'P(t+1)A(t)+C(t)'D(t)\right]x(t)$$

belonging to $L_2[\tau,\infty)$, in view of the stabilizing property of $P(\cdot)$.

6.3 Hankel norm

As is well known, the Hankel operator of a stable system links the past input to the future output through the initial state of the system. Here we define the Hankel norm for a periodic system. For, assume that system (1) is stable, and consider the input

$$u(t)=0,\quad t>\tau-1,\quad u(\cdot)\in L_2(-\infty\ ,\tau-1] \tag{26}$$

Here, by $L_2(-\infty\ ,\tau-1]$ we mean the space of square summable signals over $(-\infty\ ,\tau-1]$. By assuming that the system state is 0 at $t=-\infty$, the state at $t=\tau$ is

$$x(\tau)=\sum_{k=-\infty}^{\tau-1}\Psi_A(\tau,k+1)B(k)u(k). \tag{27}$$

The output for $t\geq\tau$ is therefore

$$y(t)=C(t)\Psi_A(t,\tau)\sum_{j=-\infty}^{\tau-1}\Psi_A(\tau,j+1)B(j)u(j). \tag{28}$$

Thanks to the system stability, $y(\cdot)\in L_2[\tau,+\infty\ )$.

The Hankel operator at time τ of the periodic system is defined as the operator mapping the input over $(-\infty\ ,\tau-1]$ defined by (26) into the output over $[\tau,+\infty\ )$ given by (28). Such an operator can be related to the infinite Hankel matrix of the lifted reformulation of the periodic system at time τ. Recall the definition of the lifted input and output signals $y_\tau(k)$ and $u_\tau(k)$, and the associated lifted system (see Sect. 3). From (28) a simple computation shows that

$$\begin{bmatrix} y_\tau(0) \\ y_\tau(1) \\ y_\tau(2) \\ \vdots \end{bmatrix} = \begin{bmatrix} H_\tau G_\tau & H_\tau \Phi_A(\tau) G_\tau & H_\tau \Phi_A(\tau)^2 G_\tau & \cdots \\ H_\tau \Phi_A(\tau) G_\tau & H_\tau \Phi_A(\tau)^2 G_\tau & H_\tau \Phi_A(\tau)^3 G_\tau & \cdots \\ H_\tau \Phi_A(\tau)^2 G_\tau & H_\tau \Phi_A(\tau)^3 G_\tau & H_\tau \Phi_A(\tau)^4 G_\tau & \cdots \\ \vdots & \vdots & \vdots & \ddots \end{bmatrix} \begin{bmatrix} u_\tau(-1) \\ u_\tau(-2) \\ u_\tau(-3) \\ \vdots \end{bmatrix}$$

Therefore, the Hankel operator at τ of the periodic system is represented by the infinite Hankel matrix of its time invariant lifted reformulation at τ.

From previous considerations, it makes sense to define the Hankel norm at τ of the periodic system as the Hankel norm of its lifted reformulation at τ. Notice that such an operator is independent of the input-output matrix E_τ. From the time-invariant case, it is known that the Hankel norm can be computed as the square root of the largest eigenvalue of the product of the unique solutions of two Lyapunov equations. As such, the Hankel norm at τ of system (1) (assumed to be stable) is given by

$$\|T_{yu}(\tau)\|_H = [\lambda_{\max}(P(\tau)Q(\tau))]^{1/2}$$

where $P(\tau)$ and $Q(\tau)$ are the solutions of (ALEL1) and (ALEL2) respectively. Notice that, on the basis of the structure of the solutions of (ALEC1) and (ALEC2), the Hankel norm of the cyclic reformulation at τ is independent of τ and is given by $\max_\tau [\lambda_{\max}(P(\tau)Q(\tau))]^{1/2}$. This means that a proper definition of Hankel norm of a periodic system is induced from its cyclic reformulation, i.e.

$$\|T_{yu}\|_H = \max_\tau \|T_{yu}(\tau)\|_H = \max_\tau [\lambda_{\max}(P(\tau)Q(\tau))]^{1/2}$$

Remark 6

Let the eigenvalues of the matrix $P(\tau)Q(\tau)$ be ordered according to their values as follows:

$$\sigma_1(t)^2 \geq \sigma_2(t)^2 \geq \cdots \geq \sigma_n(t)^2$$

The i-th *Hankel singular value* of the periodic system (1) can be defined as

$$\bar{\sigma}_i = \max_\tau \sigma_i(\tau).$$

In analogy with the time-invariant case, one can pose the problem of finding an optimal Hankel norm approximation of reduced order of the given periodic system. The problem can be technically posed as follows: find a T-periodic system of a given order k<n so as to minimize the Hankel norm of the periodic system obtained by setting in parallel the original system and the reduced one and subtracting the outputs. This is a difficult problem. For a subclass of periodic systems, in (Colaneri, Maffe', 1995) it has been shown that the optimal Hankel norm difference is exactly $\bar{\sigma}_{k+1}$. The relevant algorithm is based on the lifted reformulation and appropriate periodic realization procedure (see Sect. 8 below).

7. REALIZATION ISSUES

The realization of time varying systems is reportedly a thorny issue, see e.g. (Bittanti, Bolzern and Guardabassi, 1985) or (Gohberg, Kaashoek and Lerer, 1992). For a proper discussion of *periodic realization*, it is advisable to enlarge the class of periodic systems so as to include also systems with *periodically time-varying dimension*. In other words, we will now admit that the dimension n of system (1) may actually be subject to periodic time variations, $n=n(t)$, $n(t+T)=n(t)$. Since $A(t) \in R^{n(t+1)\times n(t)}$, this means that matrix $A(t)$ may then be rectangular, with time varying dimension. Moreover, assuming that the dimensions of the input and output spaces are constant, $B(t) \in R^{n(t+1)\times m}$, $C(t) \in R^{p\times n(t)}$ and $D(t) \in R^{p\times m}$. Notice that the structural properties introduced in Sect. 2 can be straightforwardly extended to the time-varying dimensional case. Therefore we can speak of reachability of $(A(\cdot),B(\cdot))$ or observability of $(A(\cdot),C(\cdot))$ at a given time point even if matrix $A(t)$ is not square.

That being said, let's consider a multivariable rational matrix $W(z)$ of dimension $pT\times mT$. We will now address the question whether there exist a T-periodic system S - *possibly with time varying dimension* - whose lifted reformulation at a certain time point τ has $W(z)$ as transfer function. This is the *periodic realization problem* and S is a *periodic realization* of $W(z)$. Referring the reader to (Colaneri and Longhi, 1995) for the proofs and more details, we will present here a survey of the main results.

Existence of a periodic realization

Recalling the specific structure of a lifted reformulation of a periodic system (see Sect. 3), it is apparent that for a transfer $W(z)$ to be periodically realizable it is necessary that $W(\infty)$ has the lower block triangular structure. Precisely, define the class $\Xi(m,p,T)$ of $pT\times mT$ rational matrices $W(z)$ such that:

- $W(z)=\left[W_{ij}(z):\ W_{ij}(z)\in C^{p\times m}\quad i,j=1,\cdots,T\right],\quad W_{ij}(z)=0,\quad i<j,\quad i,j=1,\ldots T$
- the generic element of $W_{ij}(z)$ is a proper rational matrix.

Then the necessary condition for a periodic realization to exist is that $W(z)\in\Xi(m,p,T)$. Interestingly enough, this is also a sufficient condition, so leading to the following result

> Consider a multivariable rational matrix $W(z)$ of dimension $pT\times mT$.
> There exists a periodic realization of $W(z)$ iff $W(z)\in\Xi(m,p,T)$. ■

Minimal, quasi-minimal and uniform realizations

Not differently from the time invariant case, much importance is played by the issue of *minimality* of a periodic realization. Specifically, a periodic realization will be said to be

- *minimal* if its (time-varying) order is smaller (time by time) than the (time-varying) order of any other periodic realization.
- *quasi-minimal* if its (time-varying) order is smaller (at least at a certain time instant) than the (time-varying) order of any other periodic realization.
- *uniform* if its order is constant.

Under the obvious assumption that $W(z) \in \Xi(m,p,T)$, the picture of the main results on all these issues can be roughly outlined as follows. A periodic realization is minimal [quasi minimal] iff it is reachable and observable at any point [reachable and observable at a time point at least].
As for the existence issue, a minimal and uniform periodic realization may not exist in general. The basic reason is that, in order to guarantee reachability and observability at any time instant in a periodic system, it is in general required to let the dimension of the state space be time varying. However, it is possible to prove that from a transfer function $W(z) \in \Xi(m,p,T)$ one can always work out a realization which is uniform and quasi minimal. Precisely, let $n(t)$ be the order of a minimal realization. Note that all minimal realizations have the same order function (up to on obious shift in time). Consider the time instant t where $n(\cdot)$ is maximum. It is not difficult to show that a quasi-minimal and uniform realization of order $n(t)$ can be built from the (nonuniform) minimal one by suitably adding unreachable and/or unobservable dynamics. Moreover, if this slack dynamics has zero characteristic multipliers, the uniform quasi-minimal realization so achieved has the distintive feature of being completely controllable and reconstructible.

We finally address the important question of determining the order $n(t)$ of a minimal realization, from where we will easily move to the problem of the existence of a uniform and minimal realization. Consider $W(z) \in \Xi(m,p,T)$ and set $W_0(z) = W(z)$. For the subsequent analysis, it is advisable to introduce further T-1 rational functions, derived from $W_0(z)$, according to recursion (12):

$$W_h(z) = \begin{bmatrix} 0 & I_{p(T-h)} \\ zI_{ph} & 0 \end{bmatrix} W_0(z) \begin{bmatrix} 0 & z^{-1}I_{mh} \\ I_{m(T-h)} & 0 \end{bmatrix}, \qquad h=0,1,2,\ldots,T-1.$$

Now, let ρ_h be the degree of the least common multiple of all denominators in $W_h(z)$. From this expression, it is immediately seen that $|\rho_h - \rho_0| \le 1$, $\forall h$. This entails that the rank of the generic Hankel matrix $M_h(j)$ of dimension j, associated with $W_h(z)$ cannot increase for $j \ge \rho_0+1$. The order of a minimal realization at time h is exactly the rank of $M_h(\rho_0 + 1)$. From this, the conclusion below follows.

> Consider a multivariable rational matrix $W(z)$ of dimension $pT \times mT$. There exists a minimal and uniform periodic realization of $W(z)$ iff $W(z) \in \Xi(m,p,T)$ and rank($M_h(\rho_0 + 1)$) is constant with respect to h, $\forall h \in [0, T-1]$ ■

In the discussion preceding this theorem, it is apparent that $W_h(z)$ are nothing but the lifted refurmalations of the periodic state space system at time $t=h$. For the algorithmic aspects concerning the effective computation of a minimal realization, and for the proofs of the various statements given herein, see (Colaneri and Longhi, 1995).

Acknowledgement

Paper supported by the European Project HCM SIMONET, the Italian MURST Project "Model Identification, System Control, Signal Processing" and by the CNR "Centro di Teoria dei Sistemi" of Milan (Italy).

REFERENCES

Bittanti S., Deterministic and stochastic linear periodic systems, in *Time Series and Linear Systems*, S. Bittanti ed., Springer-Verlag, Berlin 1986, p. 141-182.

Bittanti S., P. Bolzern, Can the Kalman canonical decomposition be performed for a discrete-time linear periodic system?, Proc. *1° Congresso Latino-Americano de Automatica*, Campina Grande, 1984, p. 449-453, 1984.

Bittanti S., P. Bolzern, Discrete-time linear periodic systems: Gramian and modal criteria for reachability and controllability, *Int. J. Control*, 41, p. 909-928, 1985.

Bittanti S., P. Bolzern, On the structure theory of discrete-time linear systems, *Int. J. of Systems Science*, 17, p. 33-47, 1986.

Bittanti S., P. Bolzern, G. Guardabassi: Some critical issues concerning the state-representation of time-varying ARMA models. 7th IFAC Symposium on *Identification and System Parameter Estimation*, York (England), p. 1479-1484, 1985.

Bittanti S., P.Colaneri, Cheap control of discrete-time periodic systems, Proc. 2nd *European Control Conference*, p. 338-341, Groningen (NL), 1993.

Bittanti S., P. Colaneri, G. De Nicolao, Discrete-time linear periodic systems: a note on the reachability and controllability interval length, *Systems and Control Letters*, 8, p. 75-78, 1986.

Bittanti S., P. Colaneri, G, De Nicolao, The periodic Riccati equation, in *The Riccati Equation* (S. Bittanti, A.J. Laub, J.C. Willems eds.), Springer-Verlag, 1991.

Bittanti S., P. Colaneri, G. De Nicolao, The difference periodic Riccati equation for the periodic prediction problem, *IEEE Trans. Automatic Control*, vol. 33, 8, p. 706-712, 1988.

Bolzern P., P.Colaneri, The periodic Lypaunov equation, *SIAM Journal on Matrix Analysis and Application*, N.4, p. 499-512, 1988.

Bolzern P., P. Colaneri, R. Scattolini, Zeros of discrete-time linear periodic systems, *IEEE Trans. Automatic Control*, AC-31, p. 1057,1058, 1986.

Colaneri P., Hamiltonian systems and periodic symplectic matrices in H_2 and H_∞ control problems, 30th IEEE *Conf. Decision and Control*, Brighton (GB), 1991.

Colaneri P., S. Longhi, The realization problem for linear discrete-time periodic systems, *Automatica*, 1995.

Colaneri P., M. Maffe', Hankel norm approximation of periodic discrete-time systems, *Dipartimento di Elettronica e Informazione*, Politecnico di Milano, Int. Rep. 95-038, 1995.

Flamm D.S., A new shift-invariant representation for periodic systems, *Systems and Control Letters*, 17, p. 9-14, 1991.

Gardner W.A. (ed.), *Cyclostationarity in communications and signal processing*, IEEE Press, 1994.

Gohberg I, M.A.Kaashoek, L.Lerer, Minimality and realization of discrete time-varying systems, *Operator Theory: Advances and Application*, Vol. 56, p. 261-296, 1992.

Grasselli O.M., S. Longhi, Zeros and poles of linear periodic multivariable discrete-time systems, *IEEE Trans. on Circuit, Systems and Signal Processing*, 7, 361-380, 1988.

Jury E.J., F.J. Mullin, The analysis of sampled data control system with a periodically time varying sampling rate, *IRE Trans. Automatic Control*, vol. 5, p. 15-21, 1959.

Kailath T. , *Linear Systems*, Prentice-Hall, 1980.

Kalman R., P.L. Falb, M.A.Arbib, *Topics in Mathematical System Theory*, Englewood Cliffs, New York, 1969

Khargonekar P.P, K.Poolla, A.Tannenbaum, Robust control of linear time-invariant plants using periodic compensators, *IEEE Trans. on Automatic Control*, Vol. AC-30, p. 1088-1096, 1985.

Mayer R.A. and C.S. Burrus, Design and implementation of multirate digital filters, *IEEE Trans. Acoustics, Speech and Signal Processing*,1 p. 53-58, 1976.

Marzollo (ed), *Periodic optimization*, Springer-Verlag, 1972.

Park B., E.I. Verriest, Canonical forms on discrete-time periodically time varying systems and a control application, Proc. 28th *Conf. on Decision and Control*, p. 1220-1225, Tampa (USA),1989.

Yakubovich V.A., V.M.Starzhinskii, *Linear differential equations with periodic coefficients*, J. Wiley and sons, New York, 1975.

Alpha-Stable Impulsive Interference: Canonical Statistical Models and Design and Analysis of Maximum Likelihood and Moment-Based Signal Detection Algorithms

George A. Tsihrintzis [1]

Communication Systems Lab

Department of Electrical Engineering

University of Virginia

Charlottesville, VA 22903-2442

and

Chrysostomos L. Nikias

Signal and Image Processing Institute

Department of Electrical Engineering - Systems

University of Southern California

Los Angeles, CA 90089-2564

Abstract

Symmetric, alpha-stable distributions and random processes have, recently, been receiving increasing attention from the signal processing and communication communities as statistical models for signals and noises that contain impulsive components. This Chapter is intended as a comprehensive review of the fundamental concepts and results of signal processing with alpha-stable processes with emphasis placed on acquainting its readers with this emerging discipline and revealing potential applications. More specifically, we start with summarizing the key definitions and properties of symmetric, alpha-stable distributions and the corresponding random processes and proceed to

[1] Also with the Signal and Image Processing Institute, Department of Electrical Engineering - Systems, University of Southern California, Los Angeles, CA 90089-2564.

derive alpha-stable models for impulsive noise. The derivation serves as an illustration of the Generalized Central Limit Theorem, which states that the first-order distributions in all observed time series follow, to a higher or lesser degree, a stable law. We proceed to address two detection problems as examples of the methodologies required for designing robust processing algorithms for non-Gaussian, alpha-stable-distributed signals. These problems also build intuition on the differences between Gaussian and non-Gaussian, alpha-stable signal processing, as well as indicate the performance gains that are to be expected if the signal processing algorithms are designed on the basis of a non-Gaussian, alpha-stable assumption rather than on a Gaussianity assumption. A large number of references to the literature are included for the interested reader to study further.

I. Introduction

Statistical signal processing is mainly concerned with the extraction of information from observed data via application of models and methods of mathematical statistics. In particular, the procedure that generates the data is, in a first step, quantitatively characterized, either completely or partially, by appropriate probabilistic models and, in a second step, algorithms for processing the data are derived on the basis of the theory and techniques of mathematical statistics. Clearly, the choice of good statistical models is crucial to the development of efficient algorithms which, in the real world, will perform the task they are designed for at an acceptable performance level.

Traditionally, the signal processing literature has been dominated by Gaussianity assumptions for the data generation processes and the corresponding algorithms have been derived on the basis of the properties of

Gaussian statistical models. The reason for this tradition is threefold: (i) The well known Central Limit Theorem suggests that the Gaussian model is valid provided that the data generation process includes contributions from a large number of sources, (ii) The Gaussian model has been extensively studied by probabilists and mathematicians and the design of algorithms on the basis of a Gaussianity assumption is a well understood procedure, and (iii) The resulting algorithms are usually of a simple linear form which can be implemented in real time without requirements for particularly complicated or fast computer software or hardware. However, these advantages of Gaussian signal processing come at the expense of reduced performance of the resulting algorithms. In almost all cases of non-Gaussian environments, a serious degradation in the performance of Gaussian signal processing algorithms is observed. In the past, such degradation might be tolerable due to lack of sufficiently fast computer software and hardware to run more complicated, non-Gaussian signal processing algorithms in real time. With today's availability of inexpensive computer software and hardware, however, a loss in algorithmic performance, in exchange for simplicity and execution gains, is no longer tolerated. This fact has boosted the consideration of non-Gaussian models for statistical signal processing applications and the subsequent development of more complicated, yet significantly more efficient, nonlinear algorithms [38].

One physical process that is not adequately characterized in terms of Gaussian models is the process that generates "impulsive" signal or noise bursts. These bursts occur in the form of short duration interferences, attaining large amplitudes with probability significantly higher than the probability predicted by Gaussian distributions. The sources of impulsive intereference, natural and man-made, are abundant in nature: In underwater signals, impulsive noise is quite common and may arise from ice cracking in the arctic region, the presence of submarines and other underwater objects, and reflections from the seabed [58, 12, 60, 59, 61]. On the other hand, lightning in the atmosphere and accidental hits or transients from car ignitions result in impulsive interference in radar, wireless links, and telephone lines, respectively. Impulsive interference can be particularly annoying in the operation of communication receivers and in the performance of signal detectors. When subjected to impulsive interference, traditional communication devices, that have been built on Gaussianity assumptions, suffer degradation in their performance down to unacceptably low levels. However, significant gains in performance can be obtained if the design of the communication devices is based on more appropriate statistical-physical models for the impulsive interference [44, 37, 54, 52].

Classical statistical-physical models for impulsive interference have been proposed by Middleton [30, 31, 32, 33, 35, 34, 36] and are based on the filtered-impulse mechanism. The Middleton models can be categorized in three different classes of interference, namely A, B, and C. Interference in class A is "coherent" in narrowband receivers, causing a negligible amount of transients. Interference in class B, however, is "impulsive," consisting of a large number of overlapping transients. Finally, interference in class C is the sum of the other two. The Middleton model has been shown to describe real impulsive interferences with high fidelity; however, it is mathematically involved for signal processing applications. This is particularly true of the class B model, which contains seven parameters, one of which is purely empirical and in no way relates to the underlying physical model. Moreover, mathematical approximations need to be used in the derivation of the Middleton model, which [2] are equivalent to changes in the assumed physics of the noise and lead to ambiguities in the relation between the mathematical formulae and the physical scenario. In this Chapter, we review a recent alternative to the Middleton model, which is based on the theory of symmetric, alpha-stable processes [46, 39] and examine the design and the performance of signal detection algorithms developed within the new model.

Symmetric, alpha-stable processes form a class of random models which present several similarities to the Gaussian processes, such as the stability property and a generalized form of the Central Limit Theorem, and, in fact, contain the Gaussian processes as a subclass. However, several significant differences exist between the Gaussian and the non-Gaussian stable processes, as explained briefly in Section 2, which make the general stable processes very attractive statistical models for time series that contain outliers [46, 47, 39]. The first mention of stable processes seems to date back to 1919, a time when the Danish astronomer Holtzmark observed that the fluctuations in the gravitational fields of stars followed an alpha-stable law of characteristic exponent $\alpha = 1.5$. In 1925, a complete characterization of the class of alpha-stable distributions was given in [21] as the class of distributions that are invariant under linear transformations. Mandelbrot made use of stable processes to model the variation of economic indices [24, 25, 26], the clustering of errors in telephone circuits [1], and the variation of hydrological quantities [22]. At the same time, the Cauchy distribution, which is a stable distribution, was considered in [44] as a model for severe impulsive noise, while Stuck and Kleiner [49] experimentally observed that the noise over certain telephone lines was best described by almost Gaussian, stable processes. Very recently, it was theoretically shown that, under general assumptions, the first-order distributions of a broad class of impulsive noise can, indeed, be described via an analytically tractable and mathematically appealing model based on the theory of symmetric stable distributions [39]. On the other hand, a number of statisticians have, in the mean time, contributed tools for the description and characterization of stable processes and the design of proper signal processing algorithms [45]. These facts have led us to systematically address signal processing problems within the framework of symmetric, alpha-stable processes in an attempt to develop algorithms which will perform robustly when applied to data containing impulsive components and outliers.

More specifically, the performance of optimum and suboptimum receivers in the presence of SαS impulsive interference was examined by Tsihrintzis and Nikias [54], both theoretically and via Monte-Carlo simulation, and a method was presented for the real time implementation of the optimum nonlinearities. From this study, it was found that the corresponding optimum receivers perform in the presence of SαS impulsive interference quite well, while the performance of Gaussian and other suboptimum receivers is unacceptably low. It was also shown that a receiver designed on a Cauchy assumption for the first-order distribution of the impulsive interference performed only slightly below the corresponding optimum receiver, provided that a reasonable estimate of the noise dispersion was available.

The study in [54] was, however, limited to coherent reception only, in which the amplitude and phase of the signals is assumed to be exactly known. The optimum demodulation algorithm for reception of signals with random phase in impulsive intereference and its corresponding performance was derived in [55] and tested against the traditional incoherent Gaussian receiver [43, Ch. 4]. Finally, the performance of asymptotically optimum multichannel structures for incoherent detection of amplitude-fluctuating bandpass signals in impulsive noise modeled as a bivariate, isotropic, symmetric, alpha-stable (BISαS) process was evaluated in [56]. In particular, the attention in [56] was directed to detector structures in which the different observation channels corresponded to spatially diverse receiving elements. However, the findings hold for communication receivers of arbitrary diversity. Tsihrintzis and Nikias derived the proper test statistic, by generalizing the detector proposed by Izzo and Paura [17] to take into account the infinite variance in the noise model, and showed that exact knowledge of the noise distribution was *not* required for almost optimum performance. They also showed that receiver diversity did *not* improve the performance of the Gaussian receiver when operating in non-Gaussian impulsive noise and, therefore, a non-Gaussian detection algorithm needed to substitute for receiver diversity.

These findings clearly indicated a need for real-time estimators of the parameters of impulsive interference, a problem previously addressed in the literature mainly within the framework of Modern Statistics. However, major difficulties are encountered when the classical estimation methods of Statistics are applied to this particular problem, mainly due to the lack of closed-form expressions for the general SαS pdf. Mandelbrot [26] and, in more detail, Fama [10] proposed a graphical procedure for estimating the characteristic exponent α of the stable distribution. Mandelbrot [27] also proposed approximating the stable distribution with a mixture of a uniform and a Pareto distribution and then applying the method of maximum likelihood in the estimation of the characteristic exponent α. DuMouchel [7] obtained approximate expressions for the maximum likelihood estimates of the characteristic exponent α and the dispersion γ of the SαS pdf under the assumption of zero location parameter ($\delta = 0$) and gave a table of the asymptotic standard deviations and correlations of the maximum likelihood estimates. In [8], DuMouchel considered the estimation of all the parameters of a SαS pdf, including the location parameter δ, and showed that the corresponding likelihood function has no maximum for arbitrary observations if the true characteristic exponent is allowed to range in the entire interval $0 < \alpha \leq 2$. However, restriction of the characteristic exponent α to the range $0 < \epsilon < \alpha \leq 2$, where ϵ can be arbitrarily small, provides (restricted) maximum likelihood estimates which are consistent and asymptotically normal, provided that the true characteristic exponent lies within the specified range of values. The actual estimation algorithm, even when the characteristic exponent can be restricted, is not readily available due to the lack of closed-form expressions for the SαS pdf. Zolotarev [62, 3] proposed a numerical method which begins with an integral form for the SαS pdf and iterative minimization. This approach was investigated via Monte-Carlo simulation by Brorsen and Yang [3] with fairly good results. However, this approach is extremely computation-intensive, since an iterative algorithm needs to be implemented at each step of which a number of numerical integrations need to be performed. Additionally, no initialization or convergence analysis is available. As alternatives to the maximum likelihood method, the method of sample quantiles has been proposed by Fama and Roll [11] and later generalized by McCulloch [28]. These methods are not computational in that they require table look-up at a certain point and interpolation between table values. Press [42], Paulson, Holcomb, and Leitch [41], and Koutrouvelis [18, 19] proposed estimation methods based on the empirical characteristic function of the data. It has been shown that, in terms of consistency, bias, and efficiency, Koutrouvelis's regression method is better than the other two. However, neither the sample quantile- nor the empirical characteristic function-based methods are suitable for real-time signal processing. Alternative estimators for the parameters of SαS distributions were proposed by Tsihrintzis and Nikias [51], which relied on asymptotic extreme value theory, order statistics, and certain relations between fractional, lower-order moments and the parameters of the distribution. These estimators were shown to maintain acceptable performance, while at the same time they were simple enough to be computable in real time. These two properties of the proposed estimators render them very useful for the design of algorithms for statistical signal processing applications.

In this Chapter, we summarize some of the findings of our research to date in the area of signal processing with symmetric, alpha-stable processes, with particular emphasis put on signal detection applications. The Chapter is organized as follows: In Section 2, we briefly review the basic definitions and properties of symmetric, alpha-stable distributions and processes and, in particular, those aspects of the theory that will be needed in the understanding of later sections of the Chapter. In Section 3, we derive symmetric, alpha-stable models for impulsive interference by following a statistical-physical approach and making realistic generic assumptions about the distribution of the sources of the interference[2]. Section 4 is devoted to the presentation of new, robust algorithms for signal detection in alpha-stable interference. The proposed algorithms cover both maximum likelihood and moment-based approaches and are tested on simulated data. The Chapter is completed with Section 5, in which a summary of its key findings is given, conclusions are drawn, and suggestions for possible future research along similar directions are made.

[2]The material in Section 3 has been excerpted from: M. Shao, *Symmetric, Alpha-Stable Distributions: Signal Processing with Fractional, Lower-Order Statistics*, Ph.D. Dissertation, University of Southern California, Los Angeles, CA, 1993.

II. Univariate and multivariate alpha-stable random processes

A. Symmetric, alpha-stable distributions

A univariate symmetric, α-stable (SαS) pdf $f_\alpha(\gamma, \delta; \cdot)$ is best defined via the inverse Fourier transform integral [21, 46]

$$f_\alpha(\gamma, \delta; x) = \frac{1}{2\pi} \int_{-\infty}^{\infty} \exp(i\delta\omega - \gamma|\omega|^\alpha) e^{-i\omega x} \, d\omega \tag{2-1}$$

and is completely characterized by the three parameters α (*characteristic exponent*, $0 < \alpha \leq 2$), γ (*dispersion*, $\gamma > 0$), and δ (*location parameter*, $-\infty < \delta < \infty$).

The characteristic exponent α relates directly to the heaviness of the tails of the SαS pdf: the smaller its value, the heavier the tails. The value $\alpha = 2$ corresponds to a Gaussian pdf, while the value $\alpha = 1$ corresponds to a Cauchy pdf. For these two pdfs, closed-form expressions exist, namely

$$f_2(\gamma, \delta; \xi) = \frac{1}{\sqrt{4\pi\gamma}} \exp[-\frac{(\xi - \delta)^2}{4\gamma}] \tag{2-2}$$

$$f_1(\gamma, \delta; \xi) = \frac{1}{\pi} \frac{\gamma}{\gamma^2 + (\xi - \delta)^2}. \tag{2-3}$$

For other values of the characteristic exponent, no closed-form expressions are known. All the SαS pdfs can be computed, however, at arbitrary argument with the real time method developed in [54]. The dispersion γ is a measure of the spread of the SαS pdf, in many ways similar to the variance of a Gaussian pdf and, indeed, equal to half the variance of the pdf in the Gaussian case ($\alpha = 2$). Finally, the location parameter δ is the point of symmetry of the SαS pdf and equals its median. For $\alpha > 1$, the SαS pdf has finite mean, equal to its location parameter δ.

The non-Gaussian ($\alpha \neq 2$) SαS distributions maintain many similarities to the Gaussian distribution, but at the same time differ from it in some significant ways. For example, a non-Gaussian SαS pdf maintains the usual bell shape and, more importantly, non-Gaussian SαS random variables satisfy the linear stability property [21]. However, non-Gaussian SαS pdfs have much sharper peaks and much heavier tails than the Gaussian pdf. As a result, only their moments of order $p < \alpha$ are finite, in contrast with the Gaussian pdf which has finite moments of arbitrary order. These and other similarities and differences between Gaussian and non-Gaussian SαS pdfs and their implications on the design of signal processing algorithms are presented in the tutorial paper [46] or, in more detail, in the monograph [39] to which the interested reader is referred. For illustration purposes, we show in Fig. 1 plots of the SαS pdfs for location parameter $\delta = 0$, dispersion $\gamma = 1$, and for characteristic exponents $\alpha = 0.5, 1, 1.5, 1.99$, and 2. The curves in Fig. 1 have been produced by calculation of the inverse Fourier transform integral in Eq.(2-1).

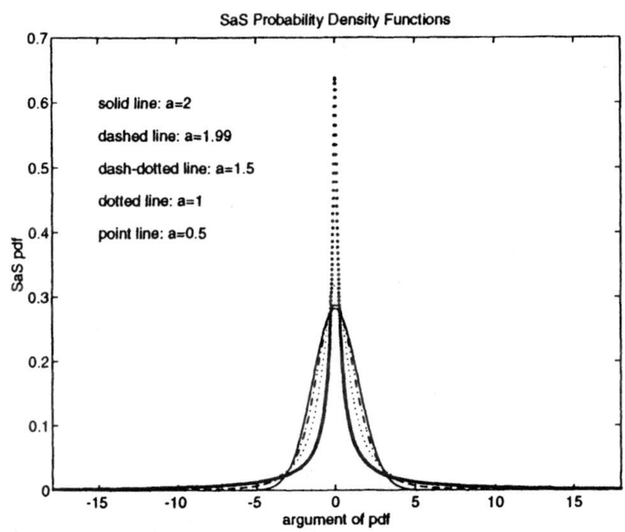

Figure 1: SαS distributions of zero location parameter and unit dispersion for various characteristic exponents

B. Bivariate, isotropic, symmetric, alpha-stable distributions

Multivariate stable distributions are defined as the class of distributions that satisfy the *linear stability property*. In particular, a n-dimensional distribution function $F(\mathbf{x})$, $\mathbf{x} \in \mathcal{R}^n$ is called *stable* if, for any independent, identically distributed random vectors $\mathbf{X}_1, \mathbf{X}_2$, each with distribution $F(\mathbf{x})$, and arbitrary constants a_1, a_2, there exist $a \in \mathcal{R}$, $\mathbf{b} \in \mathcal{R}^n$, and a random vector $\mathbf{X}$ with distribution $F(\mathbf{x})$, such that $a_1\mathbf{X}_1 + a_2\mathbf{X}_2$ has the same distribution as $a\mathbf{X} + \mathbf{b}$. Unfortunately, the class of multivariate stable distributions cannot be parameterized[3]. Fortunately, however, the subclass of multivariate stable distributions that arise in impulsive noise modeling fall within the family of *isotropic* multivariate stable distributions. More specifically, the bivariate isotropic symmetric alpha-stable (BISαS) probability density function (pdf) $f_{\alpha,\gamma,\delta_1,\delta_2}(x_1, x_2)$ is defined by the inverse Fourier transform

$$f_{\alpha,\gamma,\delta_1,\delta_2}(x_1, x_2) = \frac{1}{(2\pi)^2} \int_{-\infty}^{\infty} \int_{-\infty}^{\infty} \exp[i(\delta_1\omega_1 + \delta_2\omega_2) - \gamma(\omega_1^2 + \omega_2^2)^{\alpha/2}] \, e^{-i(x_1\omega_1 + x_2\omega_2)}, \tag{2-4}$$

where the parameters α and γ are termed its *characteristic exponent* and *dispersion*, respectively, and δ_1 and δ_2 are *location parameters*. The characteristic exponent generally ranges in the interval $0 < \alpha \leq 2$ and relates to the heaviness of the tails, with a smaller exponent indicating heavier tails. The dispersion γ is a positive constant relating to the spread of the pdf. The two marginal distributions obtained from the bivariate distribution in Eq.(2-4) are univariate SαS with characteristic exponent α, dispersion γ, and location parameters δ_1 and δ_2, respectively [46, 39]. We are going to

[3]The characteristic function of any multivariate stable distribution can be shown to attain a certain nonparametric form. The details can be found in [46, 39] and references therein.

assume $\delta_1 = \delta_2 = 0$, without loss of generality, and drop the corresponding subscripts from all our expressions.

Unfortunately, no closed-form expressions exist for the general BISαS, pdf except for the special cases of $\alpha = 1$ (Cauchy) and $\alpha = 2$ (Gaussian):

$$f_{\alpha,\gamma}(x_1, x_2) = \begin{cases} \frac{\gamma}{2\pi(\rho^2+\gamma^2)^{3/2}} & \text{for } \alpha = 1 \\ \frac{1}{4\pi\gamma} \exp(-\frac{\rho^2}{4\gamma}) & \text{for } \alpha = 2, \end{cases} \tag{2-5}$$

where $\rho^2 = x_1^2 + x_2^2$. For the remaining (non-Gaussian, non-Cauchy) BISαS distributions, power series exist [46, 39], but are not of interest to this Chapter and, therefore, are not given here.

C. Amplitude probability distribution

A commonly used statistical measure of noise impulsiveness is the amplitude probability distribution (APD), defined as the probability that the noise magnitude exceed a threshold. Hence, if X is the instantaneous amplitude of impulsive noise, then its APD is given by $P(|X| > x)$ as a function of x. The APD can easily be measured in practice by counting the percentage of time for which the given threshold is exceeded by the noise magnitude during the period of observation.

In the case of SαS distributed X with dispersion γ, its APD can be calculated as

$$P(|X| > x) = 1 - \frac{2}{\pi} \int_0^\infty \frac{\sin \omega x}{\omega} \exp(-\gamma \omega^\alpha) \, d\omega. \tag{2-6}$$

It can also be shown that

$$\lim_{x \to \infty} x^\alpha P(|X| > x) = (2/\alpha) D(\alpha, \gamma), \tag{2-7}$$

where $D(\alpha, \gamma)$ is independent of x. Hence, the APD of SαS noise decays asymptotically as $x^{-\alpha}$. As we will see later, this result is consistent with experimental observations.

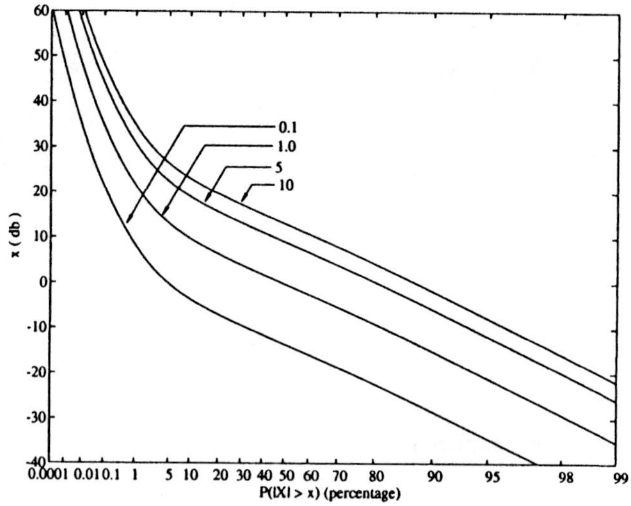

Figure 2: The APD of the instantaneous amplitude of $S\alpha S$ noise for $\alpha = 1.5$ and various values of γ

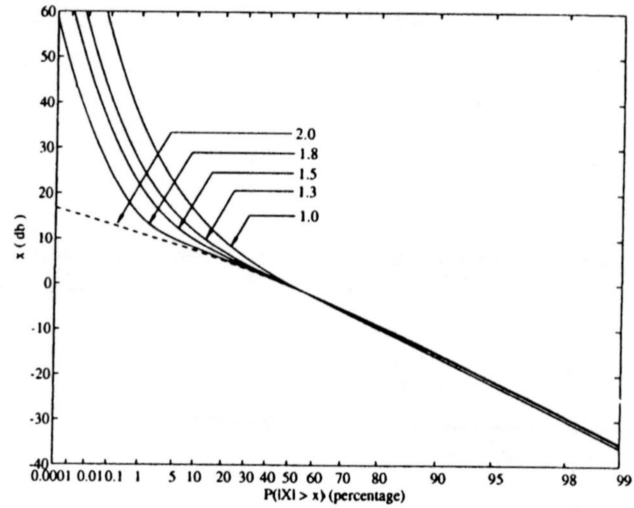

Figure 3: The APD of the instantaneous amplitude of $S\alpha S$ noise for $\gamma = 1$ and various values of α

Figs. 2 and 3 plot the APD of SαS noise for various values of α and γ. To fully represent the large range of the exceedance probability $P(|X| >$

x), the coordinate grid used in these two figures employs a highly folded abscissa. Specifically, the axis for $P(|X| > x)$ is scaled according to $-\log(-\log P(|X| > x))$. As clearly shown in Fig. 3, SαS distributions have a Gaussian behavior when the amplitude is below a certain threshold.

D. Symmetric, alpha-stable processes

A collection of r.v.'s $\{Z(t) : t \in T\}$, where T is an arbitrary index set, is said to constitute a real SαS stochastic process if all real linear combinations $\sum_{j=1}^{n} \lambda_j Z(t_j)$, $\lambda_j \in \mathcal{R}^1$, $n \geq 1$, are SαS r.v.'s of the same characteristic exponent α. A complex-valued r.v. $Z = Z' + iZ''$ is rotationally invariant SαS if Z', Z'' are jointly SαS and have a radially symmetric distribution. This is equivalent to requiring that for any $z \in \mathcal{C}^1$:

$$\mathcal{E}\{e^{i\Re(\overline{z}Z)}\} = \exp(-\gamma|z|^\alpha) \tag{2-8}$$

for some $\gamma > 0$. A complex-valued stochastic process $\{Z(t) : t \in T\}$ is SαS if all linear combinations $\sum_{j=1}^{n} \overline{z_j} Z(t_j)$, $z_j \in \mathcal{C}^1$, $n \geq 1$, are complex-valued SαS r.v.'s. Note that the overbar denotes the complex conjugate.

A concept playing a key role in the theory of SαS (with $1 < \alpha \leq 2$) r.v.'s and processes is that of the *covariation*. The covariation of two complex-valued SαS r.v.'s Z_1, Z_2 can be defined as the quantity

$$[Z_1, Z_2]_\alpha = \frac{\mathcal{E}\{Z_1 Z_2^{<p-1>}\}}{\mathcal{E}\{|Z_2|^p\}} \gamma_2, \quad 1 \leq p < \alpha, \quad 1 < \alpha \leq 2, \tag{2-9}$$

where γ_2 is the dispersion in the characteristic function of the r.v. Z_2 and for any $z \in \mathcal{C}^1$: $z^{<p>} \equiv |z|^{p-1}\overline{z}$, $\overline{z}$ being the complex conjugate of z. By letting $Z_1 = Z_2$, we observe that $[Z_2, Z_2]_\alpha = \gamma_2$, i.e., the covariation of a r.v. with itself is simply equal to the dispersion in its characteristic function.

The above definition of a covariation is mathematically equivalent [4] to the definition given in [5] and relates to a concept of orthogonality in a Banach space [48]. Since it can be shown [5] that there exists a constant

$C(p, \alpha)$, [4] depending solely on p and α ($1 \leq p < \alpha$ and $1 < \alpha \leq 2$), such that for any SαS r.v. Z_2: $\gamma_2 = C(p, \alpha)\mathcal{E}\{|Z_2|^p\}^{\alpha/p}$, we have

$$[Z_1, Z_2]_\alpha = C(p, \alpha)\mathcal{E}\{Z_1 Z_2^{<p-1>}\}^{\alpha/p}\mathcal{E}\{|Z_2|^p\}^{\alpha/p-1} \tag{2-10}$$

The covariation function of a SαS random process $\{Z(t) : t \in T\}$ is in turn defined as the covariation of the r.v.'s $Z(t)$ and $Z(s)$ for $t, s \in T$. The concept of covariation is a generalization of the usual concept of covariance of Gaussian random variables and processes and reduces to it when $\alpha = 2$. However, several of the properties of the covariance fail to hold in the non-Gaussian SαS case of $\alpha < 2$ [46, 39].

E. Fractional, lower-order statistics of alpha-stable processes

E.1 pth-order processes

We consider a random variable ζ such that its fractional, lower-order pth moment is finite, i.e.,

$$\mathcal{E}\{|\zeta|^p\} < \infty, \tag{2-11}$$

where $0 < p < \infty$. We will call ζ a pth-order random variable. We now consider two pth-order random variables, ζ and η. We define their pth-order fractional correlation as [5]

$$< \zeta, \eta >_p \equiv \mathcal{E}\{\zeta(\eta)^{(p-1)}\}, \tag{2-12}$$

where

$$(\cdot)^{(p-1)} \equiv |\cdot|^{(p-1)} \operatorname{sgn}(\cdot) \tag{2-13}$$

for real-valued random variables and

$$(\cdot)^{(p-1)} \equiv |\cdot|^{(p-2)}\overline{(\cdot)} \tag{2-14}$$

[4] In particular, $C(p, \alpha) = \frac{2^p\Gamma(1-p/\alpha)\Gamma(\frac{p+1}{2})}{\sqrt{\pi}\Gamma(1-p/2)}$, for real S$\alpha$S r.v.'s, and $C(p, \alpha) = \frac{2^p\Gamma(1-p/\alpha)\Gamma(1+p/2)}{\sqrt{\pi}\Gamma(1-\pi/2)}$, for isotropic complex SαS r.v.'s, with $\Gamma(\cdot)$ indicating the Gamma function.

for complex-valued random variables. In Eqs.(2-13) and (2-14), sgn($\cdot$) denotes the signum function, while the overbar denotes complex conjugation, respectively.

The above definitions are clearly seen to reduce to the usual SOS and HOS in the cases where those exist and can be easily extended to include random processes and their corresponding fractional correlation sequences. For example, if $\{x_k\}$, $k = 1, 2, 3, \ldots$, is a discrete-time random process, we can define its fractional, pth-order correlation sequence as

$$\rho_p(n, m) =< x_n, x_m >_p= \mathcal{E}\{x_n(x_m)^{(p-1)}\}, \tag{2-15}$$

which, for $p = 2$, gives the usual autocorrelation sequence.

The FLOS of a random process will be useful in designing algorithms that exhibit resistance to outliers and allow for robust processing of impulsive, as well as Gaussian, data.

A pth-order random process $\{x_k\}$, $k = 1, 2, 3, \ldots$, will be called *pth-order stationary* if its corresponding pth-order correlation sequence $\rho_p(n, m)$ in Eq.(2-15) depends only on the difference $l = m - n$ of its arguments. Sample averages can be used to define the FLOS of an *ergodic* stationary observed time series $\{x_k\}$, $k = 1, 2, 3, \ldots$, similarly to ensemble averages:

$$r_p(l) = \lim_{N\to\infty} \frac{1}{2N+1} \sum_{k=-N}^{N} x_k(x_{k+l})^{(p-1)}. \tag{2-16}$$

All the properties of the ensemble average definition carry over to the sample average case.

Proposition 1 *For a stationary pth-order random process $\{x_k\}$, $k = 1, 2, 3, \ldots$, its pth-order correlation and the corresponding sample average satisfy*

$$\rho_p(l) \leq \rho_p(0), \tag{2-17}$$

$$l = 0, \pm 1, \pm 2, \ldots.$$

$$r_p(l) \leq r_p(0), \tag{2-18}$$

Proof To prove Eq.(2-17), we set $l = m - n$ and start with

$$\begin{aligned} \rho_p(l) &= < y_n, y_m >_p = \mathcal{E}\{y_n(y_m)^{(p-1)}\} \\ &= \mathcal{E}\{y_n|y_m|^{p-1}\operatorname{sgn}(y_m)\} \le \mathcal{E}\{|y_n||y_m|^{p-1}\}. \end{aligned}$$

Applying the Hölder inequality [23, p. 29] to the rightmost part of the above expression gives

$$\rho_p(l) \le \mathcal{E}^{1/p}\{|y_n|^p\}\mathcal{E}^{1/q}\{|y_m|^{(p-1)q}\},$$

where $1/q = 1 - 1/p = (p-1)/p$. Therefore,

$$\begin{aligned} \rho_p(l) &\le \mathcal{E}^{1/p}\{|y_n|^p\}\mathcal{E}^{(p-1)/p}\{|y_m|^{(p-1)p/(p-1)}\} \\ &= \mathcal{E}^{1/p}\{|y_n|^p\}\mathcal{E}^{(p-1)/p}\{|y_m|^p\} = \mathcal{E}\{|y_n|^p\}, \end{aligned}$$

or, in final form

$$\rho_p(l) \le \rho_p(0).$$

Eq.(2-18) can be proved in a similar manner. •

E.2 Properties of FLOS of SαS processes

With the definitions of Section II.E.1, the FLOS of a random variable or process satisfy the following two properties, which are proved in Appendix A:

P.1 For any ζ_1 and ζ_2, we have

$$< a_1\zeta_1 + a_2\zeta_2, \eta >_p = a_1 < \zeta_1, \eta >_p + a_2 < \zeta_2, \eta >_p$$

P.2 If ζ and η are independent, then

$$< \zeta, \eta >_p = 0,$$

while the converse is not true.

The FLOS of a SαS random variable or process additionally satisfy the property [5]

P.3 If η_1 and η_2 are independent, then

$$< a_1\eta_1 + a_2\eta_2, a_1\eta_1 + a_2\eta_2 >_p = |a_1|^p < \eta_1, \eta_1 >_p + |a_2|^p < \eta_2, \eta_2 >_p$$

These properties of the FLOS of a random variable or process will be useful in designing algorithms that exhibit resistance to outliers and allow for robust processing of impulsive, as well as Gaussian, data.

F. SubGaussian symmetric, alpha-stable processes

One class of multivariate SαS processes is the class of subGaussian processes. A subGaussian random vector $\underline{X}$ can be defined as a random vector with characteristic function of the general form

$$\phi(\underline{\omega}) = \exp[-\frac{1}{2}(\underline{\omega}^T \underline{\underline{R}}\underline{\omega})^{\alpha/2}], \tag{2-19}$$

where $\underline{\underline{R}}$ is a positive-definite matrix. Unfortunately, closed-form expressions for the joint pdf of subGaussian random vectors are known only for the Gaussian ($\alpha = 2$) and Cauchy($\alpha = 1$) cases:

$$f_G(\underline{X}) = \frac{1}{\sqrt{(2\pi)^L ||\underline{\underline{R}}||}} \exp(-\underline{X}^T \underline{\underline{R}}^{-1} \underline{X}) \quad \text{(Gaussian)} \tag{2-20}$$

$$f_C(\underline{X}) = \frac{c||\underline{\underline{R}}||^{-1/2}}{[1 + \underline{X}^T \underline{\underline{R}}^{-1} \underline{X}]^{(L+1)/2}} \quad \text{(Cauchy),} \tag{2-21}$$

where L is the length of the random vector, $||\underline{\underline{R}}||$ is the determinant of $\underline{\underline{R}}$, and $c = \frac{1}{\pi^{(L+1)/2}}\Gamma(\frac{L+1}{2})$.

The following proposition relates Gaussian and subGaussian random vectors and can be used to generate subGaussian random deviates [45]:

Proposition 2 *Any subGaussian random vector is a SαS random vector. In addition, any subGaussian random vector can be expressed in the form*

$$\underline{X} = w^{\frac{1}{2}} \underline{G}, \tag{2-22}$$

where w *is a positive* $\frac{\alpha}{2}$*-stable random variable and* $\underline{G}$ *is a Gaussian random vector of mean zero and covariance matrix* $\underline{\underline{R}}$.

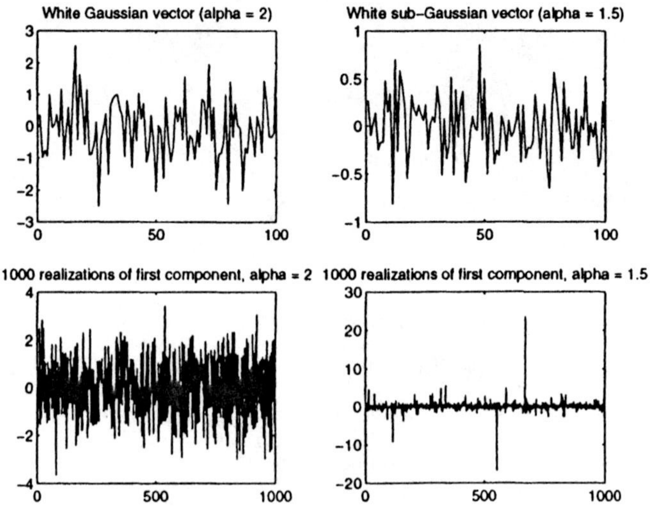

Figure 4: Typical realizations of subGaussian random vectors

SubGaussian SαS processes combine the capability to model statistical dependence with the capability to model the presence of outliers in observed time series of various degrees of severity. The example in Fig. 4 is indicative of the concept. Consider a subGaussian vector of length $L = 100$ and diagonal underlying covariance matrix $\underline{\underline{R}} =$ diag $\{1, 1, \ldots, 1\}$. Typical realizations of the vector are shown in Figs. 4(a) and 4(b) for characteristic exponents $\alpha = 2$ and $\alpha = 1.5$, respectively. Clearly, it is difficult to distinguish one vector from the other visually. However, if we look over 1000 independent realizations of the first component of the vector, we obtain Figs. 4(c) and 4(d), respectively, in which a clear difference is observed: In the strictly subGaussian case ($\alpha < 2$), each component in the vector can attain large values with probability significantly higher than the corresponding probability for the Gaussian case ($\alpha = 2$).

G. Estimation of the underlying matrix of a subGaussian vector

The following proposition expresses the underlying matrix of a subGaussian vector in terms of its covariation matrix and can, therefore, be used to obtain estimates of the underlying matrix of the vector from independent observations.

Proposition 3 *Let* $\underline{X} = [X_1, X_2, \ldots, X_L]^T$ *be a subGaussian random vector with underlying matrix* $\underline{\underline{R}}$. *Then, its covariation matrix* $\underline{\underline{C}}$ *will consist of the elements*

$$C_{ij} \equiv [X_i, X_j]_\alpha = 2^{-\frac{2}{\alpha}} R_{ij} R_{jj}^{\frac{\alpha-2}{2}} \tag{2-23}$$

Proof See [45]. ●

The usefulness of the proposition lies in finding consistent estimators of the underlying covariance matrix of a subGaussian vector from independent realizations $\underline{X}^1, \underline{X}^2, \ldots, \underline{X}^K$ of the vector. The elements C_{ij} of the covariation matrix $\underline{\underline{C}}$ can be estimated from Eq.(2-8) as [50]

$$\hat{C}_{ij} = C(p,\alpha)[\frac{1}{K}\sum_{k=1}^{K} X_i^k (X_j^k)^{<p-1>}]^{\alpha/p}[\frac{1}{K}\sum_{k=1}^{K} |X_j^k|^p]^{\alpha/p-1}, \tag{2-24}$$

where any $p < \frac{\alpha}{2}$ will result in a consistent estimate[5] [50]:

Proposition 4 *The estimator*

$$\hat{C}_{ij} = C(p,\alpha)[\frac{1}{K}\sum_{k=1}^{K} X_i^k (X_j^k)^{<p-1>}]^{\alpha/p}[\frac{1}{K}\sum_{k=1}^{K} |X_j^k|^p]^{\alpha/p-1}$$

of the covariation matrix elements, where $p < \alpha/2$, *is consistent and asymptotically normal with mean* C_{ij} *and covariance* $\mathcal{E}\{(\hat{C}_{ij} - C_{ij})(\hat{C}_{lm} - C_{lm})^*\}$.

Eq.(2-23) can then be used to compute an estimate of the underlying matrix $\underline{\underline{R}}$ from the estimate of the covariation matrix $\underline{\underline{C}}$.

---

[5] We have empirically found that a good choise is $p = \frac{\alpha}{3}$.

Proposition 5 *Let*

$$\hat{C}_{ij} = C(p,\alpha)[\frac{1}{K}\sum_{k=1}^{K} X_i^k (X_j^k)^{<p-1>}]^{\alpha/p}[\frac{1}{K}\sum_{k=1}^{K} |X_j^k|^p]^{\alpha/p-1}$$

be the estimator of the covariation matrix elements, where $p < \alpha/2$. The estimates

$$\begin{aligned} \hat{R}_{jj} &= [2^{\frac{\alpha}{2}}\hat{C}_{jj}]^{\frac{2}{\alpha}} \\ \hat{R}_{ij} &= 2^{\frac{\alpha}{2}}\hat{C}_{ij}/\hat{R}_{jj}^{\frac{\alpha-2}{2}} \end{aligned} \tag{2-25}$$

are consistent and asymptotically normal with means R_{jj} and R_{ij} and variances $\mathcal{E}\{|\hat{R}_{jj} - R_{jj}|^2\}$ and $\mathcal{E}\{|\hat{R}_{ij} - R_{ij}|^2\}$, respectively.

The procedure is illustrated with the following simulation study: Consider a subGaussian random vector of length $L = 32$ and underlying matrix $\underline{\underline{R}} =$ diag $\{1, 1, \ldots, 1\}$. We assume that $K = 100$ independent realizations of the vector are available and compute and plot the 16th row of the mean over 1000 Monte-Carlo simulations of the following two estimates:

$$\hat{\hat{\underline{\underline{R}}}} = \frac{1}{K}\sum_{k=1}^{K} \underline{X^k X^k}^T \tag{2-26}$$

$$\hat{\underline{\underline{R}}} = \text{as in (Eq.(2-25))}. \tag{2-27}$$

We examined the cases of $\alpha = 2$ and $\alpha = 1.5$ and used a FLOS estimator of order $p = 0.6$. Figs. 5(a) and 5(b) show the performance of the estimators $\hat{\hat{\underline{\underline{R}}}}$ and $\hat{\underline{\underline{R}}}$, respectively, for $\alpha = 2$, while Figs. 5(c) and 5(d) show the performance of the same estimators for $\alpha = 1.5$. Clearly, the FLOS-based estimator performs well in both cases and remains robust to the presence of outliers in the observations.

Proposition 6 *Consider the collection of K vectors $\underline{X}^k = A\underline{s} + \underline{N}^k$, $k = 1, 2, \ldots, K$, where $\underline{s}^T\underline{s} = 1$. Form the least-squares estimates $\hat{A}_k = \underline{s}^T\underline{X}^k =$*

$\underline{s}^T A \underline{s} + \underline{s}^T \underline{N}^k = A + \underline{s}^T \underline{N}^k$, $k = 1, 2, \ldots, K$. *Define* $\hat{A} =$ *sm* $\{\hat{A}_1, \hat{A}_2, \ldots, \hat{A}_K\}$, *where* $sm\{\cdots\}$ *indicates the sample median of its arguments. The estimate* $\hat{A}$ *is consistent and asymptotically normal with mean equal to the true signal amplitude* A *and variance* $\frac{1}{K}[\frac{\pi\alpha\gamma^{1/\alpha}}{2\Gamma(1/\alpha)}]^2$, *where* $\gamma = 2^{-2/\alpha} \sum_{i=1}^{L} \sum_{j=1}^{L} s_i s_j^* R_{ij}$.

Proof See Appendix B. ●

For an illustration of the performance of the estimator in Proposition 6 for $L = 1$, $K = 100$, and various values of α, see [51].

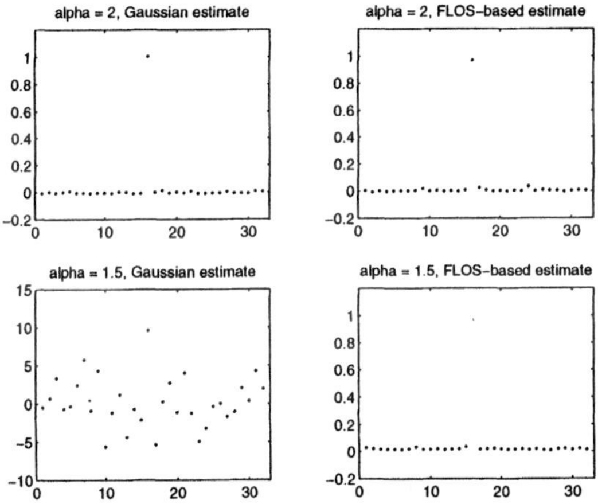

Figure 5: Illustration of the performance of estimators of the underlying matrix of a subGaussian vector

III. Alpha-stable models for impulsive interference

This Section has been excerpted from: M. Shao, *Symmetric, Alpha-Stable Distributions: Signal Processing with Fractional, Lower-Order Statistics*, Ph.D. Dissertation, University of Southern California, Los Angeles, CA, 1993.

A. Classification of statistical models

Over the last forty years, there have been considerable efforts to develop accurate statistical models for non-Gaussian, impulsive noise. The models that have been proposed so far may be roughly categorized into two groups: empirical models and statistical-physical models. Empirical models are the results of attempts to fit the experimental data to familiar mathematical expressions without considering the physical foundations of the noise process. Commonly used empirical models include the hyperbolic distribution and Gaussian mixtures [29, 58]. Empirical models are usually simple and, thus, lead to analytically tractable signal processing algorithms. However, they may not be motivated by the physical mechanism that generates the noise process. Hence, their parameters are often physically meaningless. In addition, applications of the empirical models are usually limited to specific situations.

Statistical-physical models, on the other hand, are derived from the underlying physical process giving rise to the noise and, in particular, take into account the distribution of noise sources in space and time and the propagation characteristics from the sources to the receiver. The stable model for impulsive noise, that we present in this Section, is of this nature. In particular, we show how the stable model can be derived from the familiar filtered-impulse mechanism of the noise process under appropriate assumptions on the spatial and temporal distributions of noise sources and the propagation conditions.

B. Filtered-impulse mechanism of noise processes

Let us assume, without loss of generality, that the origin of the spatial coordinate system is at the point of observation. The time axis is taken in the direction of past with its origin at the time of observation, i.e., t is the time length from the time of pulse emission to the time of observation. Consider a region Ω in $\mathcal{R}^n$, where $\mathcal{R}^n$ may be a plane ($n = 2$) or the entire three-dimensional space ($n = 3$). For simplicity, we assume that

Ω is a semi-cone with vertex at the point of observation. Inside this region, there is a collection of noise sources (e.g., lightning discharges) which randomly generate transient pulses. It is assumed that all sources share a common random mechanism so that these elementary pulses have the same type of waveform, $aD(t;\underline{\theta})$, where the symbol $\underline{\theta}$ represents a collection of time-invariant random parameters that determine the scale, duration, and other characteristics of the noise, and a is a random amplitude. We shall further assume that only a countable number of such sources exist inside the region Ω, distributed at random positions $\mathbf{x}_1, \mathbf{x}_2, \cdots$. These sources independently emit pulses $a_i D(t;\underline{\theta}_i), i = 1, 2, \cdots$, at random times $t_1, t_2, \cdots$, respectively. This implies that the random amplitudes $\{a_1, a_2, \cdots\}$ and the random parameters $\{\underline{\theta}_1, \underline{\theta}_2, \cdots\}$ are both i.i.d. sequences, with the prespecified probability densities $p_a(a)$ and $p_{\underline{\theta}}(\underline{\theta})$, respectively. The location $\mathbf{x}_i$ and emission time t_i of the ith source, its random parameter $\underline{\theta}_i$ and amplitude a_i are assumed to be independent for $i = 1, 2, \cdots$. The distribution $p_a(a)$ of the random amplitude a is assumed to be symmetric, implying that the location parameter of the noise is zero.

When an elementary transient pulse $aD(t;\underline{\theta})$ passes through the medium and the receiver, it is distorted and attenuated. The exact nature of the distortion and the attenuation can be determined from knowledge of the beam patterns of the source and the antenna, the source locations, the impulse response of the receiver, and other related parameters [32]. For simplicity, we will assume that the effect of the transmission medium and the receiver on the transient pulses may be separated into two multiplicative factors, namely filtering and attenuation. Without attenuation, the medium and the receiver together may be treated as a deterministic linear, time-invariant filter. In this case, the received transient pulse is the convolution of the impulse response of the equivalent filter and the original pulse waveform $aD(t;\underline{\theta})$. The result is designated by $aE(t;\underline{\theta})$. The attenuation factor is generally a function of the source location relative to the receiver. For simplicity, we shall assume that the sources within the region of consideration have the same isotropic radiation pattern and the receiver has an omnidirectional antenna. Then the attenuation factor is simply a decreasing function of the distance from the source to the receiver. A good approximation is that the attenuation factor varies inversely with a power of the distance [15, 32], i.e.,

$$g(\mathbf{x}) = c_1/r^p, \tag{3-1}$$

where $c_1, p > 0$ are constants and $r = |\mathbf{x}|$. Typically, the attenuation rate exponent p lies between $\frac{1}{2}$ and 2.

Combining the filtering and attenuation factors, one finds that the waveform of a pulse originating from a source located at $\mathbf{x}$ is $aU(t;\mathbf{x},\underline{\theta})$, where

$$U(t;\mathbf{x},\underline{\theta}) = \frac{c_1}{r^p}E(t;\underline{\theta}). \tag{3-2}$$

Further assuming that the receiver linearly superimposes the noise pulses, the observed instantaneous noise amplitude at the output of the receiver and at the time of observation is

$$X = \sum_{i=1}^{N} a_i U(t_i;\mathbf{x}_i,\underline{\theta}_i), \tag{3-3}$$

where N is the total number of noise pulses arriving at the receiver at the time of observation.

In our model, we maintain the usual basic assumption for the noise generating processes that the number N of arriving pulses is a Poisson point process in both space and time, the intensity function of which is denoted by $\rho(\mathbf{x},t)$ [13, 15, 32]. The intensity function $\rho(\mathbf{x},t)$ represents approximately the probability that a noise pulse originating from a unit area or volume and emitted during a unit time interval will arrive at the receiver at the time of observation. Thus, it may be considered as the spatial and temporal density of the noise sources. In this Chapter, we shall restrict our consideration to the common case of time-invariant source distribution, i.e., we set $\rho(\mathbf{x}, t) = \rho(\mathbf{x})$. In most applications, $\rho(\mathbf{x})$ is a non-increasing function of the range $r = |\mathbf{x}|$, implying that the number of sources that occur close to the receiver is usually larger than the number of sources that occur farther away. This is certainly the case, for example, for the tropical atmospheric noise where most lightning discharges occur locally, and relatively few discharges occur at great distances [15]. If the source distribution is isotropic about the point of observation, i.e., if there is no preferred direction from which the pulses arrive, then it is reasonable to assume that $\rho(\mathbf{x})$ varies inverse-proportionately with a certain power of the distance r [32, 15]:

$$\rho(\mathbf{x}, t) = \frac{\rho_0}{r^{\mu}}, \tag{3-4}$$

where μ and $\rho_0 > 0$ are constants.

C. Characteristic function of the noise amplitude

Our method for calculating the characteristic function $\varphi(\omega)$ of the noise amplitude X is similar to the one used in [62] for the model of point sources of influence. We first restrict our attention to noise pulses emitted from sources inside the region $\Omega(R_1, R_2)$ and within the time interval $[0, T)$, where $\Omega(R_1, R_2) = \Omega \cap \{\mathbf{x} : R_1 < |\mathbf{x}| < R_2\}$. The amplitude of the truncated noise is then given by

$$X_{T,R_1,R_2} = \sum_{i=1}^{N_{T,R_1,R_2}} a_i U(t_i; \mathbf{x}_i, \underline{\theta}_i), \tag{3-5}$$

where $N_{T,R1,R2}$ is the number of pulses emitted from the space-time region $\Omega(R_1, R_2) \times [0, T)$. The observed noise amplitude X is understood to be the limit of X_{T,R_1,R_2} as $T, R_2 \to \infty$ and $R_1 \to 0$ in some suitable sense.

Note that N_{T,R_1,R_2} is a Poisson random variable with parameter

$$\lambda_{T,R_1,R_2} = \int_{\Omega(R_1,R_2)} \int_0^T \rho(\mathbf{x},t) d\mathbf{x} dt \tag{3-6}$$

and its *factorial* moment-generating function is given by

$$\mathbf{E}(t^{N_{T,R_1,R_2}}) = \exp[\lambda_{T,R_1,R_2}(t-1)]. \tag{3-7}$$

Let the actual source locations and their emission times be $(\mathbf{x}_i, t_i), i = 1, \cdots, N_{T,R_1,R_2}$. Then, the random pairs $(\mathbf{x}_i, t_i)$, $i = 1, \cdots, N_{T,R_1,R_2}$, are i.i.d., with a common joint density function given by

$$f_{T,R_1,R_2}(\mathbf{x},t) = \frac{\rho(\mathbf{x},t)}{\lambda_{T,R_1,R_2}}, \qquad \mathbf{x} \in \Omega(R_1,R_2), \quad t \in [0,T). \tag{3-8}$$

In addition, N_{T,R_1,R_2} is independent of the locations and emission times of all the sources. All of the above results are the consequences of the basic Poisson assumption [40].

By our previous assumptions, $\{(a_i, t_i, \mathbf{x}_i, \underline{\theta}_i)\}_{i=1}^{\infty}$ is an i.i.d. sequence. Hence, X_{T,R_1,R_2} is a sum of i.i.d. random variables with a random number of terms. Its characteristic function can be calculated as follows:

$$\begin{aligned} \varphi_{T,R_1,R_2}(\omega) &= \mathbf{E}\{\exp(i\omega X_{T,R_1,R_2})\} \\ &= \mathbf{E}\{[\psi_{T,R_1,R_2}(\omega)]^{N_{T,R_1,R_2}}\}, \end{aligned} \tag{3-9}$$

where

$$\psi_{T,R_1,R_2}(\omega) = \mathbf{E}\{\exp(i\omega a_1 U(t_1; \mathbf{x}_1, \underline{\theta}_1)) \mid T, \Omega(R_1,R_2)\}. \tag{3-10}$$

By Eq.(3-7),

$$\varphi_{T,R_1,R_2}(\omega) = \exp(\lambda_{T,R_1,R_2}(\psi_{T,R_1,R_2}(\omega) - 1)). \tag{3-11}$$

Since $a_1, \underline{\theta}_1$ and $(\mathbf{x}_1, t_1)$ are independent, with pdfs $p_a(a), p_{\underline{\theta}}(\underline{\theta})$ and $f_{T,R_1,R_2}(\mathbf{x},t)$, respectively, one obtains

$$\begin{aligned} \psi_{T,R_1,R_2}(\omega) &= \int_{-\infty}^{\infty} p_a(a) da \int_{\Theta} p_{\underline{\theta}}(\underline{\theta}) d\underline{\theta} \int_0^T dt \\ & \int_{\Omega(R_1,R_2)} \frac{\rho(\mathbf{x},t)}{\lambda_{T,R_1,R_2}} \exp(i\omega a U(t; \mathbf{x}, \underline{\theta})) d\mathbf{x}. \end{aligned} \tag{3-12}$$

Combining (3-29), (3-31), (3-39) and (3-38), one can easily show that the logarithm of the characteristic function of X_{T,R_1,R_2} is

$$\log \varphi_{T,R_1,R_2}(\omega) = \rho_0 \int_{-\infty}^{\infty} p_a(a)da \int_{\Theta} p_{\underline{\theta}}(\underline{\theta})d\underline{\theta} \int_0^T dt \int_{\Omega(R_1,R_2)} \frac{\exp(i\omega a c_1 r^{-p} E(t;\underline{\theta})) - 1}{r^{-\mu}} d\mathbf{x}, \quad (3\text{-}13)$$

where $r = |\mathbf{x}|$.

After some tedious algebraic manipulations [39], one can finally show that the characteristic function of the instantaneous noise amplitude attains the form

$$\varphi(\omega) = \exp(-\gamma|\omega|^{\alpha}), \quad (3\text{-}14)$$

where

$$0 < \alpha = \frac{n-\mu}{p} \leq 2 \quad (3\text{-}15)$$

is an effective measure of an average source density with range [32] and determines the degree of impulsiveness of the noise. Hence, we have shown that *under a set of very mild and reasonable conditions, impulsive noise follows, indeed, a stable law.*

Similarly, in the case of narrowband reception, one can show [39] that the joint characteristic function of the quadrature components of the noise attains the form

$$\varphi(\omega_1, \omega_2)) = \exp(-\gamma|\omega_1^2 + \omega_2^2|^{\frac{\alpha}{2}}), \quad (3\text{-}16)$$

where

$$\gamma > 0, \quad 0 < \alpha \leq 2.$$

Hence, *the joint distribution of the quadrature components is bivariate isotropically stable.* From this result, one can derive the first-order statistics of the envelope and phase [39] of impulsive noise. Specifically, it can be shown that the random phase is *uniformly distributed* in $[0, 2\pi]$ and is *independent* of the envelope. The density of the envelope, on other hand, is given by

$$f(a) = a \int_0^\infty s \exp(-\gamma s^\alpha) J_0(as) ds, \quad a \geq 0. \tag{3-17}$$

By integrating Eq.(3-17), one obtains the envelope distribution function

$$F(a) = a \int_0^\infty \exp(-\gamma t^\alpha) J_1(at) dt, \quad a \geq 0. \tag{3-18}$$

We note that when $\alpha = 2$, the well-known Rayleigh distribution is obtained for the envelope.

The APD of the envelope can now be computed as

$$P(A > a) = 1 - a \int_0^\infty \exp(-\gamma t^\alpha) J_1(at) dt, \quad a \geq 0. \tag{3-19}$$

From [63], it follows that *the envelope distribution and density functions are again heavy-tailed.*

Figs. 4 and 5 plot the APD of SαS noise for various values of α and γ. Note that when $\alpha = 2$, i.e., when the envelope distribution is Rayleigh, one obtains a straight line with slope equal to $-\frac{1}{2}$. Fig. 5 shows that at low amplitudes SαS noise is basically Gaussian (Rayleigh).

D. Application of the stable model on real data

The stable model has been found consistent with empirical models. Many of the empirical models are based on experimental observations that most of the impulsive noises, such as atmospheric noise, ice-cracking noise and automotive ignition noise, are approximately Gaussian at low amplitudes and impulsive at high amplitudes. A typical empirical model then approximates the probability distribution of the noise envelope by a Rayleigh distribution at low levels and a heavy-tailed distribution at high levels. In many cases, it has been observed that the heavy-tailed distribution can be assumed to follow some algebraic law x^{-n}, where n is typically between 1 and 3 [20, 29, 16].

The behavior of the SαS model coincides with these empirical observations, i.e., SαS distributions exhibit Gaussian behavior at low amplitudes and decay algebraically at the tails. Unlike the empirical models, however, the SαS model provides physical insight into the noise generation process and is not limited to particular situations. It is certainly possible that other probability distributions could be formulated exhibiting these behaviors, but the SαS model is preferred because of its appealing analytical properties. In addition, it agrees very well with the measured data of a variety of man-made and natural noises, as demonstrated in the following example.

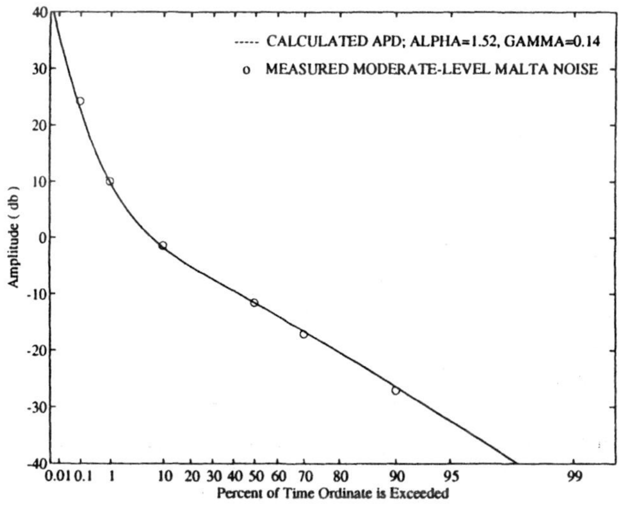

Figure 6: Comparison of a measured APD of ELF atmospheric noise with the SαS model (experimental data taken from [9].)

The example refers to atmospheric noise, which is the predominant noise source at ELF and VLF. Fig. 6 compares the SαS model with experimental data for typical ELF noise. The measured points for moderate-level Malta ELF noise in the bandwidth from 5 to 320 Hz have been taken from [9]. The characteristic exponent α and the dispersion γ are selected to best fit the data. Fig. 7 is analogous to Fig. 6 and compares the SαS model with experimental data for typical VLF noise. The experimental APD is replotted from [32] and the theoretic APD is calculated by selecting best values for α and γ. These two figures show that the two-parameter representation of the APD by SαS distributions provides an excellent fit to measurements of atmospheric noise. The stable model has been extensively tested on a variety of real, impulsive noise data with success [39], including a recent application on real sea-clutter [53].

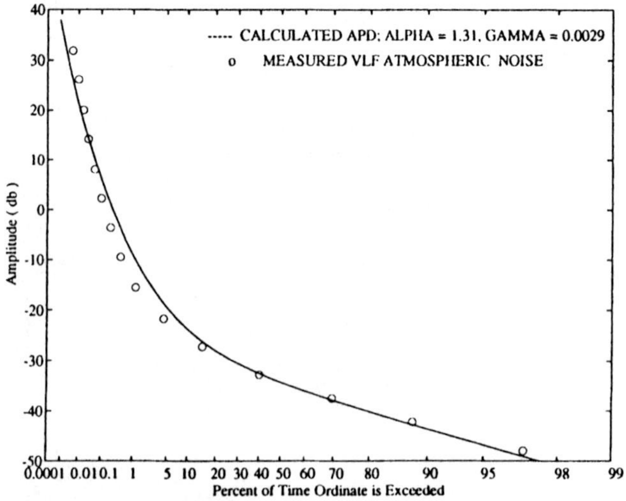

Figure 7: Comparison of a measured envelope APD of VLF atmospheric noise with the $S\alpha S$ model (experimental data taken from [32].)

IV. Algorithms for signal detection in impulsive interference

As an illustration of the concepts of the previous Sections, we look into problems of detection of signals in alpha-stable impulsive interference and develop both maximum likelihood and FLOS-based algorithms.

A. Generalized likelihood ratio tests

We consider the hypothesis testing problem

$$\begin{aligned} H_0 &: \underline{X}^k = \underline{N}^k \\ & \qquad\qquad\qquad\qquad k = 1, 2, \ldots, K \\ H_1 &: \underline{X}^k = \underline{S} + \underline{N}^k, \end{aligned} \tag{4-1}$$

where all the vectors have dimension (length) L and $k = 1, 2, \ldots, K$ indexes independent, identically distributed realizations.

We make the following assumptions:

1. The noise vectors $\underline{N}^k$ have a subGaussian distribution, i.e.,

$$\underline{N}^k = w_k^{\frac{1}{2}} \underline{G}^k,$$

where w_k are positive $(\alpha/2)$-stable random variables of unit dispersion, $\underline{G}^k$ are Gaussian random vectors of mean zero and covariance matrix $\underline{\underline{R}}$, and w_k and $\underline{G}^k$ are independent.

2. The signal vector $\underline{S} = A\underline{s}$ consists of a known shape $\underline{s}$ (for which $\underline{s}^T \underline{s} = 1$) and an unknown amplitude A.

The proposed test statistic is a generalized likelihood ratio test that makes use of the multidimensional Cauchy pdf defined in Eq.(2-21):

$$t_C = \sum_{k=1}^{K} \log\left[\frac{1 + \underline{X}^T \hat{\underline{\underline{R}}}^{-1} \underline{X}}{1 + (\underline{X} - \hat{A}\underline{s})^T \hat{\underline{\underline{R}}}^{-1} (\underline{X} - \hat{A}\underline{s})}\right] \tag{4-2}$$

For the estimates $\hat{A}$ and $\hat{\underline{\underline{R}}}$, we choose the procedures outlined in Propositions 6 and 5, respectively.

Assuming Gaussian noise of unknown covariance matrix $\underline{\underline{R}}$ and unknown signal amplitude, the optimum detector attains the form of an adaptive matched-filter [57], i.e., it computes the test statistic

$$t_G = (2/K)\sum_{k=1}^{K} \hat{A}\underline{s}^T \hat{\underline{\underline{R}}}^{-1} \underline{X}^k - |\hat{A}|^2 \underline{s}^T \hat{\underline{\underline{R}}}^{-1} \underline{s}, \tag{4-3}$$

where $\hat{A} = (1/K)\sum_{k=1}^{K} \frac{\underline{s}^T \underline{\underline{R}}^{-1} \underline{X}^k}{\underline{s}^T \underline{\underline{R}}^{-1} \underline{s}}$ and $\hat{\underline{\underline{R}}} = \frac{1}{K}\sum_{k=1}^{K}(\underline{X}^k - \hat{A}\underline{s})(\underline{X}^k - \hat{A}\underline{s})^T$.

The small sample performance of both the Gaussian and the proposed Cauchy detectors can be accurately assessed only via Monte-Carlo simulation. To this end, we chose an observation vector of length $L = 8$ and $K = 10$ independent copies of it, while for the signal we chose a shape of a square pulse of unit height and an amplitude of $A = 1$. The sub-Gaussian interference was assumed to be of characteristic exponent $\alpha = 2, 1.75, 1.5, 1.25, 1$, and 0.75 and underlying matrix $\underline{\underline{R}} =$ diag $\{1, 1, \ldots, 1\}$. The performance of the Gaussian and the Cauchy detectors was assessed via 10,000 Monte-Carlo runs.

In Fig. 8, we compare the performance of the Gaussian and the Cauchy detectors for different values of the characteristic exponent α. We see that, for $\alpha = 2$, the Gaussian detector, as expected, outperforms the Cauchy detector; however, for all other values of α, the Cauchy detector maintains a high performance level, while the performance of the Gaussian detector deteriorates down to unacceptably low levels. In Fig. 9, we show the performance of the Gaussian and the Cauchy detectors for different values of the characteristic exponent α.

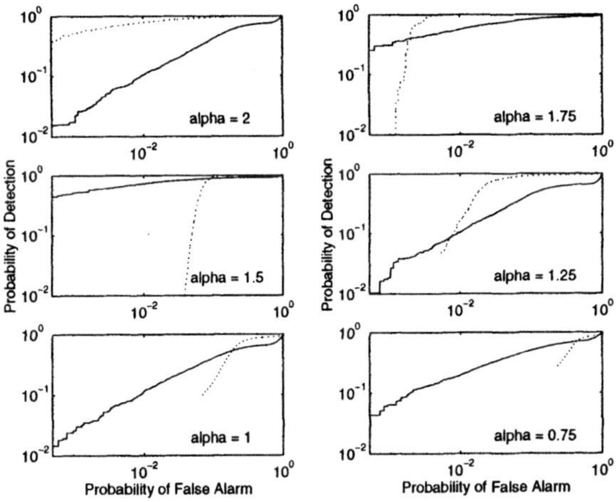

Figure 8: Comparison of the small sample performance of the Gaussian (dotted line) and the Cauchy (solid line) detector.

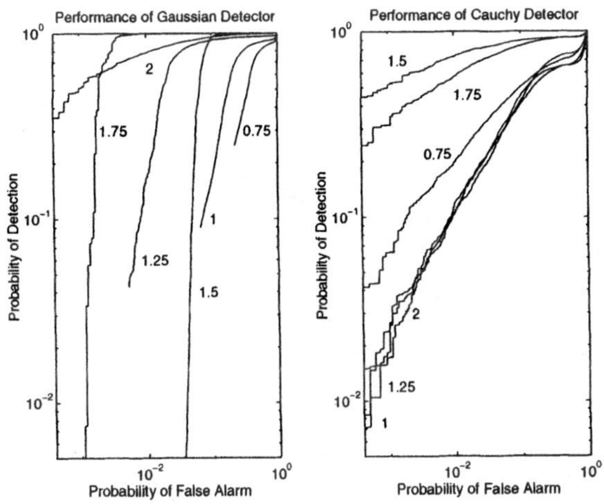

Figure 9: Performance of the Gaussian (left column) and the Cauchy (right column) detector as a function of the characteristic exponent α.

B. Fractional, lower-order statistics-based tests

The concepts of FLOS and their properties can be used to address a number of statistical signal processing problems. In particular, the FLOS-based algorithms are expected to perform very robustly in the presence of severe outliers in the observed time series. FLOS-based algorithms are also expected to converge to their asymptotic performance much faster than SOS- or HOS-based algorithms and, thus, be applicable even in the case of short data. In this section, we examine nonparametric, FLOS-based algorithms for detection of FIR random signals in noise of arbitrary pth-order correlation structure. The assumption we make is that the signal and the noise are statistically independent SαS processes, each of arbitrary pth-order correlation structure.

More specifically, we derive a decision rule for the hypothesis testing problem

$$\begin{aligned} H_0 &: x_l = w_l \\ & \qquad\qquad\qquad\qquad l = 0, 1, 2, \ldots, N, \qquad (4\text{-}4) \\ H_1 &: x_l = \sum_{k=0}^{q} s_k u_{l-k} + w_l, \end{aligned}$$

where $\{u_k\}$ is a sequence of iid SαS random variables, $\{s_k\}$, $k = 0, 1, 2, \ldots, q$, is a known signal sequence, and $\{w_k\}$ is a sequence of SαS random noise variables independent of the FIR signal. Finally, we are going to assume that $N > q$. For the dependence structure of the signal and the noise, we are not making any assumptions beyond those stated above.

To proceed with the derivation of a nonparametric decision rule, we follow a methodology similar to the one described in [14]. In particular, we consider the FIR filter with impulse respone $\{h_k = s_{q-k}\}$, $k = 0, 1, 2, \ldots, q$, i.e., the filter matched to the sequence $\{s_k\}$, $k = 0, 1, 2, \ldots, q$. Let $\{c_k\}$, $k = 0, \pm 1, \pm 2, \ldots, \pm q$, be the convolution of the sequences $\{s_k\}$ and $\{s_{q-k}\}$. Alternatively, $\{c_k\}$ is the autocorrelation sequence of the sequence $\{s_k\}$. With these definitions in mind, the output of the matched filter for input

$\{x_l\}$ under the two hypotheses will be

$$
\begin{aligned}
H_0 &: \quad y_n = \sum_{l=0}^{q} s_{q-l} w_{n-l} = v_n \\
& \qquad\qquad\qquad\qquad\qquad\qquad n = 0, 1, 2, \ldots, N, \qquad \text{(4-5)} \\
H_1 &: \quad y_n = \sum_{l=-q}^{q} c_l u_{n-l} + \sum_{l=0}^{q} s_{q-l} w_{n-l} = \sum_{l=-q}^{q} c_l u_{n-l} + v_n,
\end{aligned}
$$

The detection statistic that we propose to use relies on the properties of FLOS that were summarized in Section 2. In particular, we propose the use of the zeroth lag of the pth-order correlation sequence of the matched filter output $< y_n, y_n >_p = \mathcal{E}\{|y_n|^p\}$ as the basis for developing a test statistic. The power of the procedure to discriminate between the two hypotheses lies in the following theorem:

Proposition 7 *The statistic $t \equiv < y_n, y_n >_p$ is independent of n and, under the two hypotheses H_0 and H_1, equals*

$$
\begin{aligned}
H_0 &: \quad < y_n, y_n >_p = < v_n, v_n >_p \equiv \gamma_v \\
& \qquad\qquad\qquad\qquad\qquad\qquad\qquad\qquad \text{(4-6)} \\
H_1 &: \quad < y_n, y_n >_p = \gamma_u \sum_{l=-q}^{q} |c_l|^p + \gamma_v,
\end{aligned}
$$

where $\gamma_u = \mathcal{E}\{|u_k|^p\}$.

Proof The fact that the statistic t is independent of n under either hypothesis arises from the stationarity assumption for the signal and noise processes. Therefore, under hypothesis H_0 (noise only), the statistic t will have some value γ_v, as indicated above. We, thus, need to derive the expression given above under hypothesis H_1.

From properties **P.1** and **P.3** of FLOS and the independence assumption for the signal and noise processes, it is clear that

$$t \equiv < y_n, y_n >_p = < \sum_{l=-q}^{q} c_l u_{n-l}, \sum_{l=-q}^{q} c_l u_{n-l} >_p + < v_n, v_n >_p .$$

Since the sequence $\{u_k\}$ is iid and $\mathcal{E}\{|u_k|^p\} = \gamma_u$, properties **P.1** and **P.3** give

$$t = \gamma_u \sum_{l=-q}^{q} |c_l|^p + \gamma_v .$$

●

Therefore, we propose a detection rule that consists of computing the test statistic

$$r_p = \frac{1}{N} \sum_{n=0}^{N} |y_n|^p \tag{4-7}$$

and comparing it to a threshold. If the threshold is exceeded, hypothesis H_1 is declared, otherwise hypothesis H_0 is declared. The success of the test statistic r_p is based on the following fact:

Proposition 8 *Under either hypothesis H_0 or H_1 and assuming that $\mathcal{E}\{|x_l|^p\}$, $\mathcal{E}\{|x_l|^{2p}\} < \infty$, the test statistic r_p in Eq.(4-7) is a consistent and asymptotically normal estimator of the pth-order correlation $< y_n, y_n >_p$ with mean $< y_n, y_n >_p$ and variance $\frac{1}{N}(m_{2p} - m_p^2)$, where $m_p = \mathcal{E}\{|y_n|^p\}$ and $m_{2p} = \mathcal{E}\{|y_n|^{2p}\}$.*

Proof The assumptions $\mathcal{E}\{|x_l|^p\}$, $\mathcal{E}\{|x_l|^{2p}\} < \infty$ imply that m_p, $m_{2p} < \infty$ and, therefore, $\text{var}\{|y_n|^p\} = m_{2p} - m_p^2 < \infty$. Since the test statistic r_p consists of the sum of finite-variance random variables, the Central Limit Theorem can be invoked to guarantee that the asymptotic distribution of r_p will be Gaussian.

We can immediately compute the mean and variance of the test statistic r_p as

$$\mathcal{E}\{r_p\} = \mathcal{E}\{|y_n|^p\} = < y_n, y_n >_p$$

$$\mathrm{var}\{r_p\} = \frac{1}{N}[\mathcal{E}\{|y_n|^{2p}\} - \mathcal{E}^2\{|y_n|^p\}] = \frac{1}{N}(m_{2p} - m_p^2).$$

As a Corollary of Proposition 8, we can deduce the asymptotic performance of the proposed new detector as follows:

Proposition 9 *The asymptotic (for $N \to \infty$) Receiver Operating Characteristic of the test statistic r_p in Eq.(4-7) is*

$$P_d = \frac{1}{2}\,\mathrm{erfc}[\frac{\eta - (\gamma_u \sum_{l=-q}^{q} |c_l|^p + \gamma_v)}{\sqrt{2\sigma_{H_1}^2}}] \tag{4-8}$$

$$P_{fa} = \frac{1}{2}\,\mathrm{erfc}[\frac{\eta - \gamma_v}{\sqrt{2\sigma_{H_0}^2}}], \tag{4-9}$$

where P_d and P_{fa} are the probabilities of detection and of false alarm, respectively, $\mathrm{erfc}(x) = \frac{2}{\sqrt{\pi}}\int_x^\infty e^{-\xi^2}\, d\xi$ *is the complementary error function, and*

$$\sigma_{H_0}^2 = \mathrm{var}\{r_p|H_0\} = \frac{1}{N}(m_{2p,H_0} - m_{p,H_0}^2) \tag{4-10}$$

$$\sigma_{H_1}^2 = \mathrm{var}\{r_p|H_1\} = \frac{1}{N}(m_{2p,H_1} - m_{p,H_1}^2). \tag{4-11}$$

Eqs.(4-8) and (4-9) can be combined into

$$P_d = \frac{1}{2}\,\mathrm{erfc}[\frac{\sqrt{2\sigma_{H_0}^2}\,\mathrm{erfc}^{-1}(2P_{fa}) - \gamma_u \sum_{l=-q}^{q} |c_l|^p}{\sqrt{2\sigma_{H_1}^2}}]. \tag{4-12}$$

Proof Proposition 8 guarantees that the test statistic r_p is asymptotically Gaussian with mean γ_v and variance $\sigma_{H_0}^2$ under H_0 and mean $\gamma_u \sum_{l=-q}^{q} |c_l|^p + \gamma_v$ and variance $\sigma_{H_1}^2$ under H_1. Therefore,

$$P_d = \Pr\{r_p > \eta|H_1\} = \frac{1}{2}\,\mathrm{erfc}[\frac{\eta - (\gamma_u \sum_{l=-q}^{q} |c_l|^p + \gamma_v)}{\sqrt{2\sigma_{H_1}^2}}]$$

$$P_{fa} = \Pr\{r_p > \eta|H_0\} = \frac{1}{2}\,\mathrm{erfc}[\frac{\eta - \gamma_v}{\sqrt{2\sigma_{H_0}^2}}],$$

where η is the detector threshold. •

The performance of our proposed FLOS-based detector relative to its SOS- and HOS-based counterparts is illustrated with the following example. The test signal is the stochastic FIR signal $x_l = 0.3u_l + 0.2u_{l-1} - 0.1u_{l-2} + 0.1u_{l-3}$, where the variables $\{u_l\}$ are i.i.d., Laplace-distributed random variables of variance 0.5 and the sequence $\{w_l\}$ are i.i.d. samples from a S($\alpha = 1.5$) distribution of dispersion 0.15. We chose $N = 50$ samples per block, a FLOS of order $p = 1$, and a HOS statistic based on fourth-order cumulants [14]. The ROC of the three detectors were evaluated from 10,000 Monte-Carlo runs and are shown in Fig. 10. Clearly, the performance of the fourth-order cumulant-based detector is the lowest of the three. The proposed FLOS-based detector gives the highest performance.

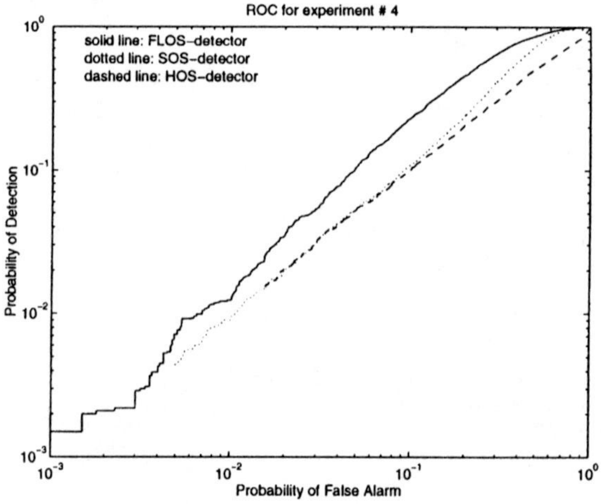

Figure 10: ROC of FLOS- (solid line), SOS- (dotted line), and HOS- (dashed line) based detector.

V. Summary, conclusions, and future research

In this chapter, we introduced symmetric, alpha-stable distributions and processes in a mathematically rigorous manner and proceeded to highlight their usefulness as statistical models for time series that contain outliers of various degrees of severity. The modeling capabilities of the symmetric, alpha-stable processes was illustrated on real data sets and tools, namely fractional, lower-order statistics, were presented, with which signal processing algorithms can be designed. Finally, we applied these concepts to signal detection problems and we illustrated the use of both maximum likelihood and moment-based methods.

From our findings, we can conclude that, unlike second- or higher-order statistics-based signal processing, the proposed algorithms are resistant to the presence of outliers in the observations and maintain a high performance level in a wide range of interferences. Additionally, the proposed algorithms perform close to optimum in the case of Gaussian interference. Given the above observations, we can, in summary, state that fractional, lower-order statistics-based signal processing is robust in a wide range of interferences.

Future research in the area seems to be leading towards problems in system identification, interference mitigation, adaptive beamforming, time-frequency analysis, and time-scale analysis within the framework of this chapter. This and related research is currently underway and its findings are expected to be announced soon.

Appendix A: Properties of Fractional, Lower-Order moments

Proof of P.1 From the definition in Eq.(2-2), we get

$$\begin{aligned} < a_1\zeta_1 + a_2\zeta_2, \eta >_p &= \mathcal{E}\{(a_1\zeta_1 + a_2\zeta_2)(\eta)^{(p-1)}\} \\ &= a_1\mathcal{E}\{\zeta_1(\eta)^{(p-1)}\} + a_2\mathcal{E}\{\zeta_2(\eta)^{(p-1)}\} \\ &= a_1 < \zeta_1, \eta >_p + a_2 < \zeta_2 >_p . \end{aligned}$$

Proof of P.2 From the definition in Eq.(2-2), we get

$$\begin{aligned} < \zeta, \eta >_p &= \mathcal{E}\{\zeta(\eta)^{(p-1)}\} \\ &= \mathcal{E}\{\zeta\}\mathcal{E}\{(\eta)^{(p-1)}\} \\ &= 0. \end{aligned}$$

Proof of P.3 The proof of this property is complicated and requires several advanced concepts from the theory of SαS processes. The proof can be found in [5, pp. 45-46].

Appendix B: Proof of Proposition 6

The random variables $\hat{A}_k$, $k = 1, 2, \ldots, K$, will be independent, each of pdf $f(x)$, which can be computed as follows. From Eq.(2-22)

$$\hat{A}_k = A + \underline{s}^T \underline{N}^k = A + w^{\frac{1}{2}} \underline{s}^T \underline{G}^k,$$

where $\underline{G}^k$, $k = 1, 2, \ldots, K$, are independent Gaussian random vectors, each of mean zero and covariance matrix $\underline{\underline{R}}$. Therefore, $\underline{s}^T \underline{G}^k$, $k = 1, 2, \ldots, K$, are independent Gaussian random variables of mean zero and variance

$\underline{s}^T \underline{\underline{R}}\underline{s}$, which implies that $\underline{s}^T \underline{N}^k$, $k = 1, 2, \ldots, K$, are independent sub-Gaussian random variables of length $L = 1$ and dispersion $\gamma = 2\underline{s}^T \underline{\underline{R}}\underline{s}$. Thus, $f(x) = f_\alpha(\gamma, A; x)$.

From [6, p. 369], it follows that the sample median of $\hat{A}_k$, $k = 1, 2, \ldots, K$, is asymptotically (for $K \to \infty$) normal with mean equal to the true median (A) and variance $(\frac{1}{K}[\frac{1}{2f_\alpha(\gamma,A;A)}]^2$. But, $f_\alpha(\gamma, \delta; x) = \gamma^{-1/\alpha} f_\alpha[1, 0; (x - \delta)\gamma^{-1/\alpha}]$ and $f_\alpha(1, 0; 0) = \frac{1}{\pi\alpha}\Gamma(\frac{1}{\alpha})$ [62]. Combining the last two relations, we get

$$\frac{1}{K}[\frac{1}{2f_\alpha(\gamma, A; A)}]^2 = \frac{1}{K}[\frac{\pi\alpha\gamma^{1/\alpha}}{2\Gamma(1/\alpha)}]^2$$

as the asymptotic variance of the estimator $\hat{A}$.

References

[1] J M Berger and B B Mandelbrot. A new model of error clustering on telephone circuits. *IBM J. Res. and Dev.*, 7:224–236, 1963.

[2] L A Berry. Understanding Middleton's canonical formula for class A noise. *IEEE Transactions on Electromagnetic Compatibility*, EMC-23:337–344, 1981.

[3] B W Bronsen and S R Yang. Maximum likelihood estimates of symmetric stable distributions. *Comm. Stat. Sim.*, 19:1459–1464, 1990.

[4] S Cambanis, C.D. Hardin Jr., and A Weron. Innovations and Wold decompositions of stable sequences. *Probab. Th. Rel. Fields*, 79:1–27, 1988.

[5] S Cambanis and G Miller. Linear problems in pth order and stable processes. *SIAM J. Appl. Math.*, 41:43–69, 1981.

[6] H. Cramér. *Mathematical Methods of Statistics.* Princeton University Press, Princeton, NJ, 1946.

[7] W H DuMouchel. *Stable Distributions in Statistical Inference.* Ph.D. Dissertation, Department of Statistics, Yale University, 1971.

[8] W H DuMouchel. On the asymptotic normality of the maximum likelihood estimate when sampling from a stable distribution. *Ann. Statistics*, 1:948–957, 1973.

[9] J.E. Evans and A. S. Griffiths. Design of a Sanguine noise processor based upon world-wide extremely low frequency (ELF) recordings. *IEEE Trans. Commun.*, 22:528–539, 1974.

[10] E F Fama. The behavior of stock market prices. *J. Bus. Univ. Chicago*, 38:34–105, 1965.

[11] E F Fama and R Roll. Parameter estimates for symmetric stable distributions. *J. Amer. Stat. Assoc.*, 66:331–338, 1971.

[12] J A Fawcett and B H Maranda. The optimal power law for the detection of a Gaussian burst in a background of Gaussian noise. *IEEE Trans. Inform. Theory*, IT-37:209–214, 1991.

[13] K. Furutsu and T. Ishida. On the theory of amplitude distribution of impulsive random noise. *J. of Applied Physics*, 32(7), 1961.

[14] G B Giannakis and M K Tsatsanis. Signal detection and classification using matched filtering and higher-order statistics. *IEEE Trans. Acoust. Speech, Sign. Proc.*, ASSP-38:1284–1296, 1990.

[15] A.A. Giordano and F. Haber. Modeling of atmospheric noise. *Radio Science*, 7:1101–1123, 1972.

[16] O. Ibukun. Structural aspects of atmospheric radio noise in the tropics. *Proc. IRE*, 54:361–367, 1966.

[17] L Izzo and L Paura. Asymptotically optimum space-diversity detection in non-gaussian noise. *IEEE Trans. Comm.*, COM-34:97–103, 1986.

[18] I A Koutrouvelis. Regression-type estimation of the parameters of stable laws. *J. Amer. Stat. Assoc.*, 75:918–928, 1980.

[19] I A Koutrouvelis. An iterative procedure for the estimation of the parameters of stable laws. *Comm. Stat. Sim.*, 10:17–28, 1981.

[20] R. M. Lerner. Design of signals. In E. J. Baghdady, editor, *Lectures on Communication System Theory*, pages 243–277. McGraw-Hill, New York, 1961.

[21] P Lévy. *Calcul des Probabilités*, volume II. Gauthier-Villards, Paris, 1925. chapter 6.

[22] S Lovejoy and B B Mandelbrot. Fractal properties of rain and a fractal model. *Tellus*, 37A:209–232, 1985.

[23] D G Luenberger. *Optimization by vector space methods.* J Wiley & Sons, New York, NY, 1969.

[24] B B Mandelbrot. The Pareto-Lévy law and the distribution of income. *International Economic Review*, 1:79–106, 1960.

[25] B B Mandelbrot. Stable Paretian random variables and the multiplicative variation of income. *Econometrica*, 29:517–543, 1961.

[26] B B Mandelbrot. The variation of certain speculative prices. *Journal of Business*, 36:394–419, 1963.

[27] B B Mandelbrot. The variation of some other speculative prices. *J. Bus. Univ. Chicago*, 40:393–413, 1967.

[28] J H McCulloch. Simple consistent estimates of stable distribution parameters. *Comm. Stat. Sim.*, 15:1109–1136, 1986.

[29] P Mertz. Model of impulsive noise for data transmission. *IRE Trans. Comm. Systems*, CS-9:130–137, 1961.

[30] D. Middleton. First-order probability models of the instantaneous amplitude, Part I. Report OT 74-36, Office of Telecommunications, 1974.

[31] D. Middleton. Statistical-physical models of man-made and natural radio noise, Part II: First-order probability models of the envelope and phase. Report OT 76-86, Office of Telecommunications, 1976.

[32] D. Middleton. Statistical-physical models of electromagnetic interference. *IEEE Trans. Electromagnetic Compatibility*, EMC-19(3):106–127, 1977.

[33] D. Middleton. Statistical-physical models of man-made and natural radio noise, Part III: First-order probability models of the instantaneous amplitude of Class B interference. Report NTIA-CR-78-1, Office of Telecommunications, 1978.

[34] D. Middleton. Canonical non-Gaussian noise models: Their implications for measurement and for prediction of receiver performance. *IEEE Transactions on Electromagnetic Compatibility*, EMC-21(3), 1979.

[35] D. Middleton. Procedures for determining the parameters of the first-order canonical models of class A and class B electromagnetic interference. *IEEE Trans. Electromagnetic Compatibility*, EMC-21(3):190–208, 1979.

[36] D. Middleton. Threshold detection in non-Gaussian interference environments: Exposition and interpretation of new results for EMC applications. *IEEE Transactions on Electromagnetic Compatibility*, EMC-26(1), 1984.

[37] J H Miller and J B Thomas. Detectors for discrete-time signals in non-Gaussian noise. *IEEE Trans. Inform. Theory*, IT-18:241–250, 1972.

[38] C L Nikias and A Petropulu. *Higher-Order Spectra Analysis: A Nonlinear Signal Processing Framework.* Prentice-Hall, Englewood Cliffs, NJ, 1993.

[39] C L Nikias and M Shao. *Signal Processing with Alpha-Stable Distributions and Applications.* John Wiley & Sons, Inc., New York, NY, 1995.

[40] E. Parzen. *Stochastic Process.* Holden-Day, San Francisco, CA, 1962.

[41] A S Paulson, E W Holcomb, and R Leitch. The estimation of the parameters of the stable laws. *Biometrika*, 62:163–170, 1975.

[42] S J Press. Estimation in univariate and multivariate stable distributions. *J. Amer. Stat. Assoc.*, 67:842–846, 1972.

[43] J.G. Proakis. *Digital Communications*. McGraw–Hill, New York, 1983.

[44] S S Rappaport and L Kurz. An optimal nonlinear detector for digital data transmission through non-Gaussian channels. *IEEE Trans. Comm. Techn.*, COM-14:266–274, 1966.

[45] G Samorodnitsky and M S Taqqu. *Stable, Non-Gaussian Random Processes: Stochastic Models with Infinite Variance*. Chapman & Hall, New York, NY, 1994.

[46] M Shao and C L Nikias. Signal processing with fractional lower-order moments: Stable processes and their applications. *Proc. IEEE*, 81:986–1010, 1993.

[47] M Shao and C L Nikias. Detection and adaptive estimation of stable processes with fractional lower order moments. In *Proceedings of Sixth IEEE Workshop on Statistical Signal and Array Processing*, pages 94–97, Victoria, BC, Canada, October, 1992.

[48] I Singer. *Bases in Banach Spaces*, volume I. Springer-Verlag, New York, 1970.

[49] B W Stuck and B Kleiner. A statistical analysis of telephone noise. *The Bell System Technical Journal*, 53:1262–1320, 1974.

[50] G A Tsihrintzis and C L Nikias. Data-adaptive algorithms for signal detection in sub-gaussian impulsive interference. *IEEE Trans. Signal Processing*. (submitted on Jan. 20, 1996, pp. 25).

[51] G A Tsihrintzis and C L Nikias. Fast estimation of the parameters of alpha-stable impulsive interference. *IEEE Trans. Signal Processing*. accepted for publication.

[52] G A Tsihrintzis and C L Nikias. On the detection of stochastic impulsive transients over background noise. *Signal Processing*, 41:175–190, January 1995.

[53] G A Tsihrintzis and C L Nikias. Modeling, parameter estimation, and signal detection in radar clutter with alpha-stable distributions. In *1995 IEEE Workshop on Nonlinear Signal and Image Processing*, Neos Marmaras, Halkidiki, Greece, June 1995.

[54] G A Tsihrintzis and C L Nikias. Performance of optimum and sub-optimum receivers in the presence of impulsive noise modeled as an α–stable process. *IEEE Trans. Comm.*, COM-43:904–914, March 1995.

[55] G A Tsihrintzis and C L Nikias. Incoherent receivers in alpha-stable impulsive noise. *IEEE Trans. Signal Processing*, SP-43:2225–2229, September 1995.

[56] G A Tsihrintzis and C L Nikias. Asymptotically optimum multichannel detection of fluctuating targets in alpha-stable impulsive interference. *Signal Processing*, (submitted on July 19, 1995, pp. 20).

[57] H L Van Trees. *Detection, Estimation, and Modulation Theory, Part I.* Wiley, New York, 1968.

[58] E J Wegman, S G Schwartz, and J B Thomas, editors. *Topics in Non-Gaussian Signal Processing.* Academic Press, New York, 1989.

[59] P Zakarauskas. Detection and localization of non-deterministic transients in time series and application to ice-cracking sound. *Digital Signal Processing*, 3:36–45, 1993.

[60] P Zakarauskas, C J Parfitt, and J M Thorleifson. Automatic extraction of spring–time arctic ambient noise transients. *J. Acoust. Soc. Am.*, 90:470–474, 1991.

[61] P Zakarauskas and R I Verall. Extraction of sea–bed reflectivity using ice–cracking noise as a signal source. *J. Acoust. Soc. Am.*, 94:3352–3357, 1993.

[62] V Zolotarev. *One-Dimensional Stable Distributions.* American Mathematical Society, Providence, RI, 1986.

[63] V. M. Zolotarev. Integral transformations of distributions and estimates of parameters of multidimensional spherically symmetric stable laws. In J. Gani and V. K. Rohatgi, editors, *Contribution to Probability: A Collection of Papers Dedicated to Eugene Lukacs*, pages 283–305. Academic Press, 1981.

INDEX

B

C

D

E

F

Q

R

U

V

W

Z

ISBN 0-12-012778-4

90065

9 780120 127788